MICROQUASARS

Proceedings of the Third Microquasar Workshop
Granada Workshop on Galactic Relativistic Jet Sources
Granada, Spain, 11–13 September 2000

Edited by

A.J. CASTRO-TIRADO, J. GREINER and J.M. PAREDES
Instituto de Astrofísica de Andalucía (IAA-CSIC), Granada, Spain
Astrophysikalisches Institut, Potsdam, Germany
Universidad de Barcelona, Spain

Reprinted from *Astrophysics and Space Science*
Supplement to Volume 276, 2001

KLUWER ACADEMIC PUBLISHERS
DORDRECHT / BOSTON / LONDON

Library of Congress Cataloging-in-Publication Data

Microquasar Workshop (3rd: 2000: Granada, Spain)
Microquasars: proceedings of the third Microquasar Workshop: Granada workshop on galactic relativistic jet sources, Granada, Spain, 11–13 Sep 2000 / edited by Alberto J. Castro-Tirado, Jochen Greiner, Josep M. Paredes.
p. cm.
Includes index.
ISBN-0-7923-6923-8 (alk. paper)
1. Quasars-Congresses. 2. Astrophysical jets-Congresses. I. Castro-Tirado, Alberto J.
II. Greiner, Jochen, 1959- III. Paredes, J.M. (Josep M.) IV. Title

QB860. M57 2000
523.1′15–dc21 2001029973

Published by Kluwer Academic Publishers,
P.O. Box 17, 3300 AA Dordrecht, The Netherlands.

Sold and distributed in the U.S.A. and Canada
by Kluwer Academic Publishers,
101 Philip Drive, Norwell, MA 02061, U.S.A.

In all other countries, sold and distributed
by Kluwer Academic Publishers,
P.O. Box 322, 3300 AH Dordrecht, The Netherlands.

Printed on acid-free paper

Printed in the Netherlands

TABLE OF CONTENTS

Chapter III. INNER DISK OSCILLATIONS

Chapter V. JET FORMATION AND DYNAMICS

Chapter VI. PROSPECTS FOR THE FUTURE

PREFACE

The Third Microquasar Workshop (or the 'Fifth' Workshop on Galactic Relativistic Jet Sources), was held in Granada, Andalucia (Spain) on 11–13 September 2000. The aim of this workshop in Granada, following the previous *Microquasar Workshops* in Greenbelt (1997) and Paris (1998) and the *Workshops on galactic sources with relativistic jets* in Jodrell Bank (1996) and Milton Keynes (1998), was to focus on the theoretical and observational aspects of microquasars.

The study of microquasars, the sources in our Galaxy displaying powerful relativistic jets, is a rapidly advancing field in astrophysics. The new instrumentation on ground (MERLIN, SCUBA, VLA, VLT) and aboard satellites (ASCA, BSAX, ISO, IXAE and RXTE) has provided important results, and much more is expected to come from Chandra and XMM-Newton. In the further future, powerful instrumentation will come online in the sub-mm (ALMA) and gamma-rays (INTEGRAL), extending our coverage to important regions for the study of microquasars. Energy transport via relativistic jets is one of the most important physical mechanisms taking place in compact objects. Large efforts have been devoted to properly understand the disk-jet connection, and even the effects of rotation or magnetic fields. Several new important advances have been made recently, both from the point of view of the theoretical treatment of jets and the different new observational tests.

Granada was the perfect place for hosting the workshop. Astronomy fluorished there since the 10th century, thanks to the moorish heritage: al-Qalusi (study of the lunar and solar movements), ibn al-Sahm (the instrument dubbed 'ecuatorium'), ibn al-Banna Al-garnati (treatises on astronomical instrumentation, planets and stars), Avempace (comments on Ptolomeus' work), Abentofail (a treatise on Astronomy) and ibn Abi Ali. Nowadays, the town hosts the superb Arabian Palaces of the Alhambra and the borough of the Albayzin which has been designed by UNESCO as Heritage of Humanity.

The workshop was organized by the Instituto de Astrofísica de Andalucía (IAA, CSIC) with support from the Spanish Consejo Superior de Investigaciones Científicas (CSIC), the Universidad de Barcelona (UB), and the Laboratorio de Astrofísica Espacial y Física Fundamental (LAEFF) with support from the Instituto Nacional de Técnica Aeroespacial (INTA). We want to thank the members of the Scientific Advisory Committee (SAC): T. Belloni, A.J. Castro-Tirado (chair), Ph. Durouchoux, R. Fender, J. Greiner, R.M. Hjellming (†), F. Mirabel, J.M. Paredes, A.R. Rao and J. Swank for arranging an excellent scientific programme, and ex-

Astrophysics and Space Science is the original source of publication of this article. It is recommended that this article is cited as: *Astrophysics and Space Science* **276** (suppl.): ix–x, 2001.

press our gratitude to the members of the Local Organizing Committe (LOC): A. Alberdi, A.J. Castro-Tirado, C. García-Miró, J.L. Gómez, L. Lara, J. Martí, B. Montesinos and C. Sánchez-Fernández for making of the staying of the 90 participants in Granada (four! social events included) an unforgettable experience.

We thank our colleagues T. Belloni, S. Brandt, A.J. Castro-Tirado, R. Fender, J.L. Gómez, J. Greiner, D. Hannikainen, R. Hynes, D. Meier, J.M. Paredes, A.R. Rao, J. Swank and M. Tagger, who served, on short notice, as 'referees' for all the contributions printed in this book.

We are indebted to Rafael Rodrigo for the facilities given at the IAA in order to organize the Workshop and to Asunción Jódar for the excellent hospitality offered by the Granada Town Hall. We appreciate the financial support of the CSIC, the Spanish Ministry of Science and Technology and the Junta de Andalucía. Thanks to them and to the voluntary payment of some of the SAC and LOC members, the LOC was able to devote about one third of the incomings for funding in order to help many scientists to attend the Workshop.

We thank Rosa García Román for designing the nice Workshop announcement poster that also serves as front cover of this book, Iván Agudo for acting as 'official photographer', and Genevieve Thiéry for her assistant during the Workshop. And we are much indebted to María Eva Alcoholado-Feltström, who worked for many months as the unpayed Workshop secretary in order to achieve a perfect organization, becoming 'again' the beloved wife of one of us (AJCT) after the Workshop's end.

Finally, we should mention that all of us were shocked when we knew that Bob Hjellming left us suddenly on 5 Aug 2000. He was a very active member of the SAC and the scheme we followed during the six scientific sessions was proposed by him. The Astrophysics community will deeply miss him. Our gratitude to Carole Hjellming (his widow) and Ralph Spencer for the notes written about him and to M. Ruppen and V. Dhawan for further help. The Granada Microquasar Workshop was celebrated on his honour and this book is dedicated to his memory.

Alberto J. Castro-Tirado *15 Nov 2000*
Jochen Greiner
Josep M. Paredes

Robert M. Hjellming (1938–2000) with his wife Carole.

R.M. HJELLMING (1938–2000): IN MEMORIAM

Bob Hjellming and I met and married when we were both undergraduates at University of Chicago. Six years and three children later he finished his doctorate and working together had become our way of life. It was in 1962, at Yerkes Observatory, that Bob met his first computer – an interest that would continue throughout his lifetime and account for half the hours in his 80-hour work weeks. It also freed us from reducing data using log tables!

His first professional work was a teaching appointment at Case Western Reserve in Cleveland, Ohio. He didn't believe he should close his office door to his students, so many evenings on campus were spent enthusiastically discussing everything from course work to Star Trek.

Moving to NRAO and Charlottesville meant research and mentoring graduate students from other institutions. These were the dim, dark, exciting days before email when astronomy was international but communication was by telephone. Our older daughter, who was still in elementary school at that time, remembers answering the phone but never knowing where or from which continent the call originated. (I met Rashid Sunyaev years later who said I was famous in Moscow – for being able to keep track of Hjellming during coordinated runs involving international observers.) And Bob really traveled during these years. However, no one could convince me that traveling with five children separated by only eight years was called a vacation!

Moving to Socorro and the VLA brought interesting changes to our family. I switched from being an English major who gardened and did arts and crafts to returning to college to get a degree in Chemistry. Without having to hire babysitters and after learning that some of the same editing programs were used on computers on campus and at the VLA, Bob and I had some interesting days observing at the site when there were only six telescopes on line. As the children grew and became more independent, we all began taking vacations camping in Southwest parks and monuments and driving to meetings in the western United States and Canada. After they left for college, I was able to go to more of Bob's meetings and meet many of the wonderful people working in astronomy.

An interest common to Bob and me was scientific publishing. We were still in Socorro when I found work in geology, editing a quarterly research journal. Bob eventually joined my professional organization and went with me to meetings; he really enjoyed our field trips and the sessions on electronic publishing! And I particularly enjoyed equivalent sessions and conversations at his meetings. Some other

Astrophysics and Space Science is the original source of publication of this article. It is recommended that this article is cited as: *Astrophysics and Space Science* **276** (suppl.): xiii–xv, 2001.

things are common to both astronomy and geology: amateurs and professionals can be equally knowledgeable and respected, and both fields deal with time. This latter observation accounts for much of the excitement in astronomy. That, although some events happened long ago, we observe them in our own real-time sequence. That, if some activity is missed, it might be missed for all time.

Bob saw no conflict in being deeply religious and being an active scientist. He read widely and had boundless enthusiasm for learning new things. Witness our archaeological trips to Egypt and Guatemala and the snorkeling at a 1996 meeting in Australia. We took scuba diving lessons this past summer preparing for optional activities after a November meeting in Hawaii. We were together doing open-water certification dives when Bob died of natural causes not related to diving. We had been together, through good times and hard times, for longer than the 41 years of our marriage, longer than two-thirds of our lives. He was my best friend.

Carole Hjellming *30 Nov 2000*

After completing his PhD at the University of Chicago, Bob became assistant professor at Case Western Reserve. He then joined the National Radio Astronomy Observatory, where he remained until his death. In his career he produced 154 papers and a number of major reports resulting from his leading work on the analysis of interferometer data and computer systems.

His first paper while he was a student presented the optical light curves of two Wolf-Rayet stars. Perhaps this was a foretaste of his later work, where light curves feature strongly, though in the radio. Bob's following papers were on the physics of HII regions, and indeed this work was referred to as the definitive answer in lectures on the interstellar medium given by the late Franz Kahn at the University of Manchester in 1967 while I was a student. He then went on to develop the use of radio recombination lines as an important diagnostic for HII regions.

However in spite of this early emphasis on theoretical work Bob was quick to realise the potential of radio techniques for stellar astronomy, with the first detections of Antares , Scorpius X-1 and Cygnus X-1 using the Green Bank interferometer in 1971. This opened up a whole new area of research, and stellar radio astronomy has dominated Bob's work since then. His research included definitive work on the evolution of radio emitting novae, as well as the pioneering work on the properties of radio emitting binary stars and the discovery of superluminal motion in GRO J1655-40, the second superluminal microquasar to be discovered.

Bob produced the book describing how to use the Green Bank interferometer in 1973; this document taught many astronomers the esoteric art of radio astronomy, including myself. He later led the group which developed software for the analysis of data from the fledgling VLA, which later became the AIPS system which we all love (and occasionally hate!). He has played a major part in the development of radio interferometry, and his expert opinion has often been sought and freely given.

Perhaps the most endearing character of Bob has been his infectious enthusiasm for his subject – from the delight in explaining a new model to explain the radio emission, to discussions of how to analyse the data. He will be sorely missed.

R.E. Spencer *31 Oct 2000*

LIST OF PARTICIPANTS

Iván Agudo (*IAA-CSIC, Spain*)	ivan@iaa.es
Antxón Alberdi (*IAA-CSIC, Spain*)	antxon@iaa.es
Reba Bandyopadhyay (*NRL, USA*)	reba@xeus.nrl.navy.mil
Didier Barret (*CESR, France*)	Didier.Barret@cesr.fr
Bianca Maria Belli (*IAS, Italy*)	bianca@ias.rm.cnr.it
Tomaso Belloni (*Brera Obs., Italy*)	belloni@merate.mi.astro.it
Gennadi Bisnovatyi-Kogan (*IKI, Russia*)	gkogan@mx.iki.rssi.ru
Soren Brandt (*DSRI, Denmark*)	sb@dsri.dk
Catherine Brocksopp (*Open Univ., UK*)	cb@astro-livjm.ac.uk
Michelle Buxton (*Mt. Stromlo Obs., Australia*)	buxton@mso.anu.edu.au
Fabien Casse (*Lab. d'Astrophys., France*)	Fabien.Casse@obs.ujf-grenoble.fr
Alberto J. Castro-Tirado (*IAA-CSIC, Spain*)	ajct@iaa.es
Stuart Caunt (*Univ. Oulu, Finland*)	scaunt@sun3.oulu.fi
Sandip K. Chakrabarti (*S. N. Bose Centre, India*)	chakraba@boson.bose.res.in
Phil A. Charles (*Univ. of Southampton, UK*)	pac@astro.soton.ac.uk
Silvayn Chaty (*Open Univ., UK*)	S.Chaty@open.ac.uk
Tapas K. Das (*IUCAA, India*)	tapas@iucaa.ernet.in
Evgeny Derishev (*Inst. of Applied Physics, Russia*)	derishev@appl.sci-nnov.ru
Vivek Dhawan (*NRAO, USA*)	vdhawan@aoc.nrao.edu
Philippe Durouchoux (*CEA/SAp, France*)	pdurouchoux@cea.fr
Ken Ebisawa (*NASA/GSFC, USA*)	ebisawa@olegacy.gsfc.nasa.gov
Stephen Eikenberry (*Cornell Univ., USA*)	sse2@cornell.edu
Rob Fender (*Univ. of Amsterdam, The Netherlands*)	rpf@astro.uva.nl
Marco Feroci (*IAS-CNR, Italy*)	feroci@ias.rm.cnr.it
Chris de la Force (*Jodrell Bank, UK*)	cjd@jb.man.ac.uk
Yael Fuchs (*CEA/SAp, France*)	yfuchs@cea.fr
Marek Gierlinski (*Copernicus Astr. Center, Poland*)	gier@camk.edu.pl
Gabriele Giovannini (*Ist. di Radioastronomia, Italy*)	ggiovann@ira.bo.cnr.it
Jose Luis Gómez (*IAA-CSIC, Spain*)	jlgomez@iaa.es
Javier Gorosabel (*DSRI, Denmark*)	jgu@dsri.dk
Jochen Greiner (*AIP, Germany*)	jgreiner@aip.de
Sergei Guziy (*Nikolaev Astr. Obs., Ukraine*)	gss@comcent.nikolaev.ua
Diana Hannikainen (*Univ. of Helsinki, Finland*)	dch@astro.soton.ac.uk
Carole Haswell (*Open Univ., UK*)	C.A.Haswell@open.ac.uk
Robert Hynes (*Univ. of Southampton, UK*)	rih@astro.soton.ac.uk
Anatoli Iyudin (*MPIE, Germany*)	ani@mpe.mpg.de
Christian Kaiser (*MPIA, Germany*)	ckaiser@mpa-garching.mpg.de
Andrew King (*Univ. of Leicester, UK*)	ark@star.le.ac.uk
Marc Klein-Wolt (*Univ. of Amsterdam, The Netherlands*)	klein@astro.uva.nl
Wlodzimierz Kluzniak (*Copernicus Astr. Center, Poland*)	wlodek@camk.edu.pl
Wolfgang Kundt (*IfA/Bonn Univ., Germany*)	wkundt@astro.uni-bonn.de

Astrophysics and Space Science is the original source of publication of this article. It is recommended that this article is cited as: *Astrophysics and Space Science* **276** (suppl.): xvii–xviii, 2001.

Lucas Lara (*IAA-CSIC, Spain*) lucas@iaa.es
William H. Lee (*UNAM, Mexico*) wlee@astroscu.unam.mx
Galina Lipunova (*Moscow State Univ., Russia*) galja@sai.msu.ru
Fangjun Lu (*Inst. of High Energy Physics, China*) lufj@astrosv1.ihep.ac.cn
Philip Maloney (*CASA, USA*) maloney@casa.colorado.edu
Craig Markwardt (*NASA/GSFC, USA*) craigm@pcasun3.gsfc.nasa.gov
Herman L. Marshall (*MIT, USA*) hermanm@space.mit.edu
Josep Martí (*Univ. de Jaén, Spain*) jmarti@ujaen.es
Enrico Massaro (*Univ. 'La Sapienza', Italia*) adnvma@tin.it
Maria Massi (*MPIR, Germany*) mmassi@mpifr-bonn.mpg.de
Giorgio Matt (*Univ. di Roma III, Italy*) matt@haendel.fis.uniroma3.it
Shane McKie (*Univ. of Manchester, UK*) shane@reynolds.ph.man.ac.uk
David Meier (*JPL, USA*) David.L.Meier@jpl.nasa.gov
Andrea Merloni (*Institute of Astronomy, UK*) am@ast.cam.ac.uk
Felix Mirabel (*CEA/SAp, France*) mirabel@discovery.saclay.cea.fr
Sergej Moiseenko (*IKI, Russia*) moiseenko@mx.iki.rssi.ru
Michael Muno (*MIT, USA*) muno@mit.edu
Lorenzo Natalucci (*IAS, Italy*) lorenzo@ias.rm.cnr.it
Luciano Nobili (*Univ. di Padova, Italia*) nobili@pd.infn.it
Michael Nowak (*JILA, USA*) mnowak@rocinante.colorado.edu
Michal Ostrowski (*Jagellonian Univ., Poland*) mio@oa.uj.edu.pl
Zsolt Paragi (*FÖMI Satel. Geodetic Observatory, Hungary*) paragi@sgo.fomi.hu
Josep M. Paredes (*Univ. de Barcelona, Spain*) josepmp@mizar.am.ub.es
Elena Pavlenko (*Crimean Astrophys. Obs., Ukraine*) pavlenko@crao.crimea.ua
Marta Peracaula (*Univ. de Barcelona, Spain*) mperacau@am.ub.es
Guy Pooley (*Cavendish Lab., UK*) ggp1@cam.ac.uk
Raghu Rao (*TIFR, India*) arrao@tifr.res.in
Arne Rau (*AIP, Germany*) arau@aip.de
Marc Ribó (*Univ. de Barcelona, Spain*) mribo@am.ub.es
Jerome Rodriguez (*CEA/SAp, France*) rodrigue@discovery.saclay.cea.fr
Samar Safi-Harb (*Univ. of Manitoba*) samar@physics.umanitoba.edu
Celia Sánchez-Fernández (*LAEFF-INTA, Spain*) celia@laeff.esa.es
Tariq Shahbaz (*IAC, Spain*) tsh@ll.iac.es
Aleksei Shlyapnikov (*Nikolaev State Univ., Ukraine*) alex@aox.mk.ua
Ravi Sood (*Australian Defence Force Academy, Australia*) r-sood@adfa.edu.au
Ralph Spencer (*Jodrell Bank, UK*) res@jb.man.ac.uk
Jean Swank (*NASA/GSFC, USA*) swank@pcasun1.gsfc.nasa.gov
Ewa Szuszkiewicz (*N. Copernicus Univ., Poland*) szusz@sus.univ.szczecin.pl
Michael Tagger (*CEA/SAp, France*) tagger@cea.fr
Marco Thiel (*AIP, Germany*) thiel@Mycroft.agnld.uni-potsdam.de
Lev Titarchuk (*NASA/GSFC, USA*) titarchuk@lheavx.gsfc.nasa.gov
Sergei Trushkin (*SAO, Russia*) satr@cats.sao.ru
Yoshihiro Ueda (*ISAS, Japan*) ueda@astro.isas.ac.jp
Peggy Varniere (*CEA/SAp, France*) pvarni@discovery.saclay.cea.fr
Osmi Vilhu (*Univ. of Helsinki, Finland*) osmi.vilhu@helsinki.fi
Xuebing Wu (*MPIA, Germany*) wuxb@mpa-garching.mpg.de
Jagdish S. Yadav (*TIFR, India*) jsyadav@tifr.res.in
Kazutaka Yamaoka (*ISAS, Japan*) yamaoka@astro.isas.ac.jp
Janus Ziolkowski (*Copernicus Astr. Center, Poland*) jz@camk.edu.pl
Piotr Zycki (*Copernicus Astr. Center, Poland*) ptz@camk.edu.pl

1. Dhawan
2. Sood
3. Lu
4. Ebisawa
5. Durouchoux
6. Ueda
7. Parago
8. Nowak
9. Pooley
10. Kaiser
11. Maloney
12. Mirabel
13. Lara
14. Marshall
15. Barret
16. Charles
17. Derishev
18. Varniere
19. Chaty
20. Titarchuk
21. Hynes
22. Rau
23. Massaro
24. Das
25 Moiseenko
26. Nobili
27. Giovannini
28. Thiel
29. Bandyopadhyay
30. Trushkin
31. Bisnovatyi-Kogan
32. Fender
33. Hannikaine
34. Spencer
35. Lipunova
36. Belloni
37. Buxton
38. Bello
39. Feroci
40. Paredes
41. McKie
42. Ribó
43. Gómez
44. Klein-Wolt
45. Lee
46 Martí
47. Markwardt
48. Swank
49. Ziolkowski
50. Vilhu
51. Natalucci
52. Meier
53. Caunt
54. Ostrowski
55. Kluzniak
56. Castro-Tirado
57. Greiner
58. Gierlinski
59. Brocksopp
60. Wu
61. de la Force
62. Rao
63. Zycki
64. Sánchez-Fernández
65. Rodriguez
66. Yadav
67. Haswell
68. King
69. Alcoholado-Feltström
70. Muno
71. Thiéry
72. Guziy
73. Shlyapnikov
74. Yamaoka
75. Iyudin
76. Pavlenko
77. Casse
78. Eikenberry
79. Shahbaz
80. Peracaula
81. Brandt
82. Agudo
83. Tagger
84. Massi
85. Kundt

Chapter I
MICROQUASARS AND X-RAY NOVAE

RECENT RESULTS ON X-RAY NOVAE AND MICROQUASARS AT RADIO AND INFRARED WAVELENGTHS

JOSEP MARTÍ

Universidad de Jaén, Departamento de Física, Escuela Politécnica Superior, Virgen de la Cabeza 2, E-23071 Jaén, Spain; E-mail: jmarti@ujaen.es

Abstract. Observations at radio and infrared wavelengths have provided a significant contribution to our knowledge of X-ray novae and microquasar systems. For instance, radio interferometers are routinely the first ones to provide accurate subarcsec positions allowing further detailed studies. Similarly, infrared spectroscopy of highly obscured systems is the best way to unveil the often elusive nature of their companion star. In this talk I will review some recent findings in this field obtained by both the Very Large Array (VLA) and the Very Large Telescope (VLT). The VLA has a long record of X-ray novae and microquasar observations and R.M. Hjellming did a very remarkable contribution to it. On the other hand, the VLT results have been available only from recent times, but they look no less promising for the future.

Keywords: XTE J1748−288, V4641 Sgr, Cyg X-3, 1E 1740.7−2942, GRS 1915+105

1. Introduction

X-ray novae are binary systems that eventually undergo a significant increase in their X-ray flux due to an episodic enhancement of mass transfer between the normal star and its compact companion (see e.g. Tanaka and Shibazaki, 1996). Their discovery normally occurs thanks to space-based observatories in the X-ray domain, where a significant fraction of their total luminosity is released. In many cases, optical-infrared and radio counterparts are also identified inside the error circles provided by X-ray telescopes. The term microquasar is reserved for these cases where the flaring radio counterpart is resolved into relativistic jets. It is conceivable that all X-ray novae or transient events would fall into this category if they were observed adequately before the radio ejecta fades away.

On the other hand, the microquasar group also includes X-ray binary stars with continuous X-ray activity and production of radio jets. We can think of them as persistent microquasar systems, in contrast to their transient relatives associated to X-ray novae events. Nevertheless, this distinction may not be so clear in some objects where repeated transient activity has been observed over the years.

In any case, the most relevant property of microquasars in the Galaxy is their astonishing scaled down similarity with extragalactic quasars and active galactic nuclei, both morphological and physical. The reader is referred to Mirabel and Rodríguez (1999) for a general review of the microquasar field. Here, I will con-

Astrophysics and Space Science is the original source of publication of this article. It is recommended that this article is cited as: *Astrophysics and Space Science* **276** (suppl.): 3–10, 2001.

centrate on a few individual sources where recent observational progress has been obtained at radio and infrared wavelengths with both the VLA and the VLT, respectively.

2. New Superluminal Microquasars in the Galaxy

The analogy quasar-microquasar was strongly reinforced when the first superluminal microquasar, GRS 1915+105, was discovered (Mirabel and Rodríguez, 1994). Very soon another microquasar, GRO J1655−40, displayed the same kind of behavior (Tingay *et al.*, 1995; Hjellming and Rupen, 1995). Based on preliminary results made public by Hjellming and collaborators, the census of superluminal microquasars appears to have doubled in the latest two years. The new members in the superluminal group are two transient X-ray sources, namely XTE J1748−288 and V4641 Sgr.

2.1. XTE J1748−288

This object is one of the 1998 X-ray transients detected at radio wavelengths. The VLA images by Hjellming *et al.* (1998) indicated one-sided collimated ejection of relativistic plasma, with proper motions of ≥ 20 mas d^{-1} and a high degree of linear polarization. For a distance of ≥ 8 kpc, based on H I absorption measurements, the ejection velocities inferred are $\geq 0.93c$ and possibly superluminal. The most remarkable fact concerning XTE J1748−288 is the evidence of interaction between the jets and the surrounding medium. This statement is based on the observation of deceleration and brightening at the leading edge of the jet, as illustrated in Figure 1.

2.2. V4641 Sgr

This star underwent a very fast X-ray transient event reaching up to 12 Crab units in 1999 September. The associated radio counterpart was extensively monitored by different radio observatories, and a detailed account of radio results is given by Hjellming *et al.* (2000). These authors report evidence for fast expansion of an elongated radio source, with proper motion estimates in the range 0.4-1.1 arcsecond d^{-1}. This is a very impressive value for an object outside the Solar System and translates into ejection velocities of $1.0\text{-}3.2c[d/500\text{pc}]$. Although it was not possible to image the detailed movie of the ejection events with the VLA, there is indication of the reliability of the jet structure and the high proper motion. The evidence comes mainly from direct model fitting to the VLA visibilities, together with a good agreement between the multi-frequency light curves and theoretical models for ejection of truncated conical jets.

The distance to V 4641 Sgr is currently a matter of debate. Hjellming *et al.* (2000) provide some arguments supporting a 500 pc value. However, the recent

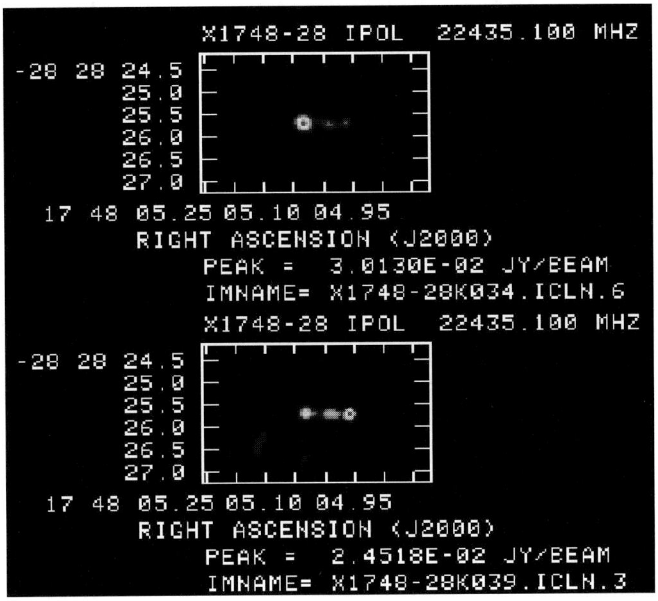

Figure 1. Sequence of VLA images by Hjellming *et al.* (1998) revealing the ejection and deceleration of a relativistic jet in XTE J1748−288 during 1998 August.

photometric and spectroscopic analysis by Chaty *et al.* in this volume strongly indicate a B3-A2 V spectral type and a distance value between 4 and 8 kpc.

2.3. Extended Radio Jets and Double Lobe Structures

The range of relativistic plasma ejections from microquasars is likely to extend up to very large linear scales, i.e., a fraction of a parsec and possibly beyond. The W50 radio nebula, associated with SS 433, is the only case where such effects are clearly seen up to $\sim$ 50 pc distances (Dubner *et al.* 1998). Actually, this is not very surprising. Consider for instance the low power jets from young stellar objects, which are able to produce collimated Herbig Haro flows up to several parsec from the exciting source (e.g. Bally and Devine, 1994). In this context, even more dramatic effects should be expected from relativistic jet sources.

Another example could be the case of Cygnus X-3, where only milli-arcsecond radio jets had been reported until not long ago. This microquasar has been recently observed with the VLA by Martí *et al.* (2000a) in the months following one of its major radio outbursts. After a process of removing the map artifacts introduced by the highly variable compact core, it has been possible to reveal the presence of previously unknown arcsecond radio jets (see Figure 2). Their inferred linear size and ejection velocities are estimated as $\sim$ 0.2 pc[d/10 kpc] and $\sim$ 0.6c [d/10 kpc], respectively.

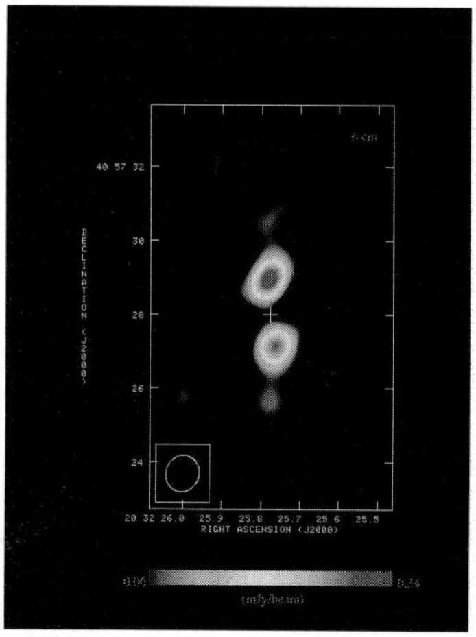

Figure 2. The Cygnus X-3 arcsecond radio jets evidenced after subtracting the variable compact core. The position of the core is indicated by the central cross.

On the other hand, both GRS 1915+105 and Cygnus X-3 have a pair of H II regions that may be extended lobes produced by the relativistic jets (Chaty *et al.*, 2001; Martí *et al.*, 2000a) at huge distances of $\sim$ 100 pc. Unfortunately, the main evidence for a physical connection remains to be the remarkable symmetrical location and position angle coincidence with the jets, thus a chance alignment cannot be strictly ruled out.

3. Microquasars and High Energy γ-Ray Sources

It has been proposed that the luminous star LS 5039 is a new microquasar systems related to one of the unidentified $\geq$ 100 MeV sources in the 3rd EGRET catalog (Paredes *et al.*, 2000). No other microquasar had been reported before to emit such energetic photons. Further details are given by Paredes *et al.* in this same volume.

4. Infrared Studies of 1E 1740.7−2942 and GRS 1915+105

After many years, there is yet no consensus on the nature of the donor star in these two prototypical microquasars. The new generation of 8 to 10 m telescopes, operating at infrared wavelengths, is possibly the best approach to finally solve the problem. Some interesting results have been already obtained with the 8.2 m unit (Antu) of the VLT and its ISAAC spectro-imager.

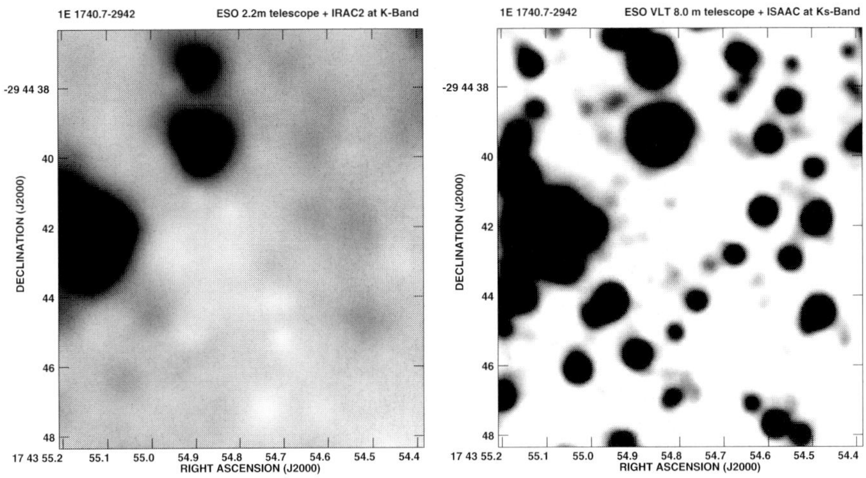

Figure 3. The field of the Galactic Center microquasar 1E 1740.7−2942 as seen in the infrared (K-band) with the ESO 2.2 m telescope (left) and with the VLT (right). The images are centered at the microquasar radio position.

4.1. The infrared counterpart of 1E 1740.7−2942

The remarkable improvement in sensitivity obtained by the VLT is clearly evident when comparing the two panels in Figure 3 In spite of this progress, the VLT has not revealed until now a firm candidate to be the 1E 1740.7−2942 counterpart. The VLT does detect, however, a few sources whose position may be consistent with the radio and X-ray counterpart. These sources are better seen in the expanded view of Figure 4. All the VLT images in this paper, as well as the information plotted on them, are taken from Martí *et al.* (2000b) and references therein.

None of the possible candidates, those labelled as 4, 6, 7 and 8 in Figure 4, has experienced any significant variability that could betray one of them as the counterpart. Under such circumstances, the main obstacle for the identification is the astrometric uncertainty, at the ~ 0.6 arcsecond level, when plotting the accurate VLA radio and Chandra X-ray positions on the infrared frames.

The infrared magnitude of the possible candidate sources is in the range $K =$ 18.5-19.0 mag. These values would agree with a Cygnus X-1 type companion (O9.7 Iab, $M_K = -5.8$) provided that an infrared absorption $A_K \simeq 10$ mag exists. This is equivalent to a hydrogen column density of $N_H \simeq 1.5 \times 10^{23}$ cm^{-2}, i.e., consistent with constrains derived from X-ray spectral fits. Alternatively, if there is no detected counterpart, the $K \geq 19.4$ limiting magnitude (3σ) implies a companion star later than B8 V or earlier than G5 III.

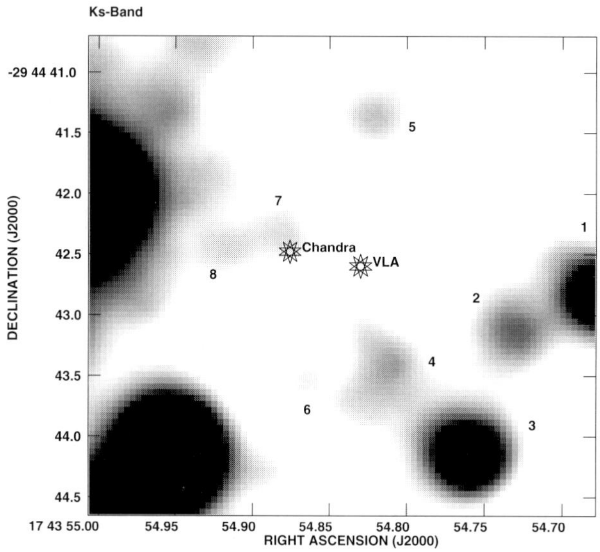

Figure 4. Close up view of the 1E 1740.7−2942 field from the infrared VLT images by Martí *et al.* (2000b). The VLA radio and Chandra X-ray positions are plotted as star symbols. Only a few sources discussed in the text may be consistent with being the microquasar counterpart.

4.2. GRS 1915+105: A HIGH MASS X-RAY BINARY?

The infrared counterpart of the first superluminal microquasar has been intensively observed by different authors, but a reliable spectral class is not yet available. The main problems to achieve this goal are mainly: a) the spectral morphology overlap of some star classes in the infrared; b) the difficulties to separate the photospheric and accretion disk contribution; c) the difficulties of correcting for telluric absorptions in the infrared windows. It is also necessary to obtain good spectra, with high signal-to-noise ratio (SNR), in order to facilitate the process of comparison with spectral standards. Therefore, the large collecting area of the VLT is really of great assistance for this task.

In Figure 5 the K-band spectrum of GRS 1915+105, as observed with the VLT+ISAAC in low resolution mode, is compared against those of some massive luminous stars. The spectrum in this figure is the result of combining two epochs of observations for better sensitivity (see Martí *et al.*, 2000c for details). At this resolution GRS 1915+105 strongly resembles a high mass star, specially taking into account the new Na I detection. The tentative classification proposed by Martí *et al.* (2000c) is that of a Of/WN. The possibility of a massive companion is also supported by some spectral features detected in the individual epochs, i.e., a P Cygni profile in quiescence and a blue emission wing in outburst. These features require the existence of an expanding envelope, being blown out during outbursts, and such of circumstellar material is more naturally understood in a high mass system.

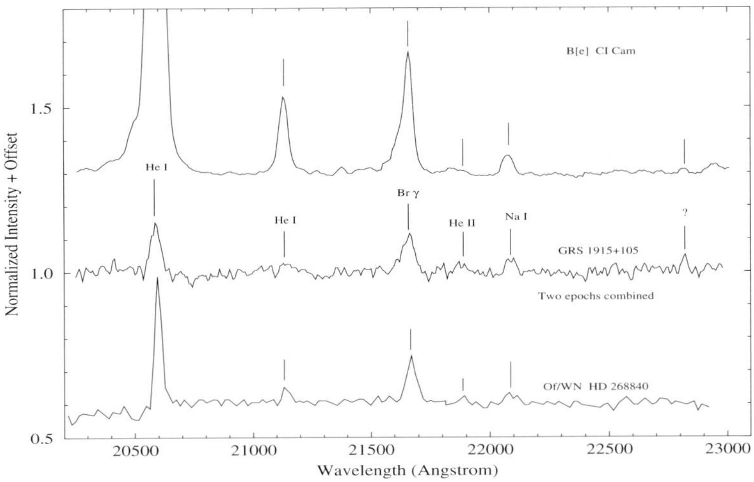

Figure 5. Comparison of the VLT spectrum of GRS 1915+105 in the K-band with those of selected luminous stars (Martí *et al.* 2000c).

Finally, Greiner *et al.* in this proceedings also report ISAAC observations of GRS 1915+105. Their observed in a higher resolution mode, where high SNR is not always easy to achieve, and conclude that a late type giant companion is the most likely scenario. Further sensitive spectroscopy at all resolutions is therefore required to solve this issue.

Acknowledgements

JM acknowledges partial support by DGICYT (PB97-0903) (Spain).

References

Bally, J. and Devine, D.: 1994, *ApJ*, **428**, L65.
Chaty, S. *et al.*: 2001, *AandA*, **366**, 1035.
Dubner, G. M. *et al.*: 1998, *AJ*, **116**, 1842.
Hjellming, R. M. and Rupen, M. P.: 1995, *Nature*, **375**, 464.
Hjellming, R. M. *et al.*: 1998, *AAS*, 19310308H.
Hjellming, R. M. *et al.*: 2000, *ApJ*, **544**, 977.
Martí, J., Paredes, J. M. and Peracaula, M.: 2000a, *ApJ*, **545**, 939.
Martí, J., Mirabel, I. F., Chaty, S. and Rodríguez, L. F.: 2000b, *AandA*, **363**, 184.
Martí, J., Mirabel, I. F., Chaty, S. and Rodríguez, L. F.: 2000c, *AandA*, **356**, 943.
Mirabel, I. F. and Rodríguez, L. F.: 1994, *Nature*, **371**, 46.
Mirabel, I. F. and Rodríguez, L. F.: 1999, *ARAandA*, **37**, 409.

Paredes, J. M., Martí, J., Ribó, M. and Massi, M.: 2000, *Science*, **288**, 2340.
Tingay, S. J. *et al.*: 1995, *Nature*, **374**, 141.
Tanaka, Y. and Shibazaki, N.: 1996, *ARAandA*, **34**, 607.

ACCRETION DISK SPECTRA OF SUPER-LUMINAL JET SOURCES AND ULTRA-LUMINOUS COMPACT X-RAY SOURCES IN NEARBY SPIRAL GALAXIES

KEN EBISAWA*
NASA/GSFC, Code 662, Greenbelt, MD 20771, USA; E-mail: ebisawa@subaru.gsfc.nasa.gov

AYA KUBOTA, TSUNEFUMI MIZUNO
Department of Physics, The University of Tokyo, 7-3-1 Hongo Bunkyo-ku, Tokyo 113, Japan

PIOTR ŻYCKI
N. Copernicus Astronomical Center, Bartycka 18, 00-716 Warsaw, Poland

Abstract. The Ultra-luminous Compact X-ray Sources (ULXs) in nearby spiral galaxies and the Galactic super-luminal jet sources share the common spectral characteristic that they have extremely high disk temperatures which cannot be explained in the framework of the standard accretion disk model in the Schwarzschild metric. We have calculated an extreme Kerr disk model to examine if the Kerr disk model can instead explain the observed 'too hot' accretion disk spectra. We found that the Kerr disk spectrum becomes significantly harder compared to the Schwarzschild disk only when the disk is highly inclined. For super-luminal jet sources, which are known to be inclined systems, the Kerr disk model may thus work if we choose proper values for the black hole angular momentum. For the ULXs, however, the Kerr disk interpretation will be problematic, as is is highly unlikely that their accretion disks are preferentially inclined.

Keywords: accretion disk, black hole, relativistic jet, X-ray sources

1. The 'Too Hot' Accretion Disk Problem

Although X-ray energy spectra of GRS 1915+105 and GRO J1655–40 may be fitted with an optically thick accretion disk plus power-law, which is a canonical model for the high-state black hole candidates, their disk color temperatures can be as high as $\sim$1.3 – 2.0 keV (e.g., Belloni *et al.*, 1997; Zhang *et al.*, 1997; Tomsick *et al.*, 1999); this is significantly higher than other non-jet, well-studied black hole candidates such as Cyg X-1 and LMC X-3, whose disk temperatures are almost always less than $\sim$ 1 keV.

Interestingly, the unusually high disk color temperature, $\sim$1.5–1.8 keV, is also observed from several ultra-luminous compact X-ray sources (ULXs) in IC 342 (Okada *et al.*, 1998), NGC 4565 (Mizuno *et al.*, 1999) and other nearby spiral galaxies; in fact the high accretion disk temperature seems to be a common spectral characteristic of the ULXs (Makishima *et al.*, 2000).

* also Universities Space Research Association

Astrophysics and Space Science is the original source of publication of this article. It is recommended that this article is cited as: *Astrophysics and Space Science* **276** (suppl.): 11–14, 2001.

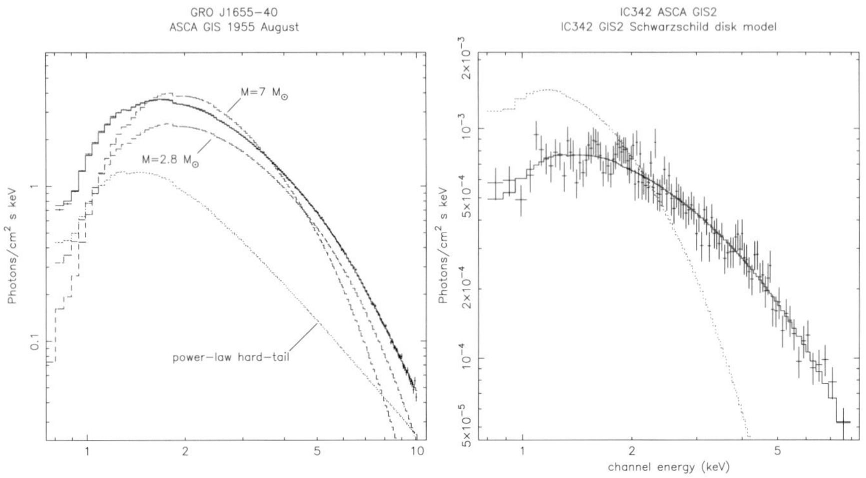

Figure 1. ASCA GIS energy spectra of GRO J1655–40 (left) and the Source 1 in IC 342 (right panel).

The maximum disk color temperature in the Schwarzschild metric may be written as

$$T_{col}^{(max)} \approx 1.3\ \text{keV} \left(\frac{T_{col}/T_{eff}}{1.7}\right)\left(\frac{\dot{M}}{\dot{M}_{Edd}}\right)^{1/4}\left(\frac{M}{7\,M_\odot}\right)^{-1/4},$$

where we define the Eddington mass accretion rate as $\dot{M}_{Edd} = 3 \times 10^{18}(M/M_\odot)$ g s^{-1}, so that $\dot{M}/\dot{M}_{Edd} = 1$ gives the Eddington luminosity. Precise radiative transfer calculation indicates the spectral hardening factor, T_{col}/T_{eff}, is in the range of 1.5 to 1.9, as long as $\sim 1 - 10$ keV band is concerned (Shimura and Takahara 1995; Ross and Fabian 1996). From this formula, we can see that an optically thick accretion disk around a 7 $M_\odot$ Schwarzschild black hole may not be hotter than the color temperature $\sim$ 1.3 keV.

We show examples of the observed 'too-hot' accretion disks in GRO J1655–40 (left panel, Figure 1). We applied a Schwarzschild disk model by Hanawa (1989) with the distance and inclination angle being fixed at 3.2 kpc and 70° respectively (Orosz and Bailyn 1997), and $T_{col}/T_{eff} = 1.7$. The best-fit model disk model was obtained by allowing both mass and mass accretion rate to be free; we have obtained a too small mass $M = 2.8M_\odot$, with $\dot{M} = 2.1 \times 10^{18}$ g s^{-1} ($\dot{M}/\dot{M}_{Edd}$=0.25). If we fix at the most probable mass $M = 7M_\odot$ (Orosz and Bailyn, 1997), we may not achieve the acceptable fit, as the model spectrum is too soft as indicated in the figure.

The right panel in Figure 1 indicates an ASCA spectrum of the 'Source 1' in IC342 observed in September 1993 when the source is in the high state (Okada *et al.*, 1998). We fit the observed spectrum with the Schwarzschild disk model by Hanawa (1989), allowing the mass and mass accretion rate to be free parameters, and assuming the face-on geometry. The best-fit model gives $M = 12.6M_\odot$ and $\dot{M} = 3 \times 10^{20}$ g s^{-1} ($\dot{M}/\dot{M}_{Edd} = 7.9$). Obviously, such a super-Eddington lu-

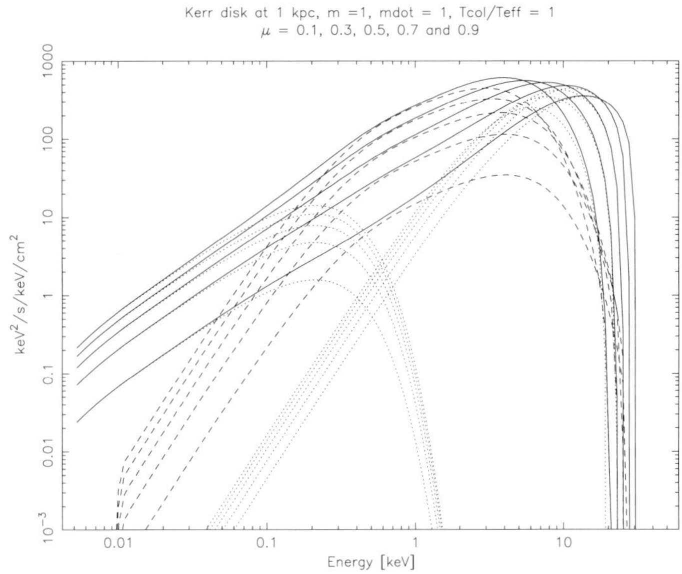

Figure 2. The Kerr accretion disk model with an extreme angular momentum ($a = 0.998$) for the inclination angles $\mu \equiv \cos i = 0.1, 0.3, 0.5, 0.7$ and 0.9. The distance and mass are assumed to be 1 kpc and 1 $M_{\odot}$ respectively. The Eddington luminosity is assumed, and $T_{col}/T_{eff} = 1$. Solid lines indicate the total disk spectra, and contribution from inner ($1.26\,r_g < r < 7\,r_g$), middle ($7\,r_g < r < 400\,r_g$), and outer parts ($400\,r_g < r$) are also plotted seperately either by dotted or broken lines.

minosity is very unlikely. For a comparison, we have shown a Schwarzschild disk model spectrum with $M = 100M_{\odot}$ at the Eddington limit; it is obvious that the observed spectrum is much harder than the model spectrum.

2. Does Kerr Disk Model Explain the Too-Hot Disks?

We have calculated the Kerr disk spectrum using the transfer function computed by Laor, Netzer and Piran (1990) for $a = 0.998$. The calculated energy spectra are shown in Figure 2 for several inclination angles.

We have applied the Kerr disk model to GRO J1655–40 and the Source 1 in IC342. For GRO J1655–40, the inclination angle is fixed to $i = 70°$ and $T_{col}/T_{eff} = 1.7$ is assumed. If we allow the mass and mass accretion rate to be free, we obtain $M = 16.4M_{\odot}$ and $\dot{M} = 3.5 \times 10^{17}$ g s^{-1} ($\dot{M}/\dot{M}_{Edd} = 0.046$). Compared to the fit with Schwarzschild disk model ($M = 2.8M_{\odot}$), significant increase of the mass does indicate the spectral hardening of the Kerr disk model with a large inclination angle. In fact, the derived mass is too large, and further tuning of the angular momentum a will be required to achieve the realistic mass (7 $M_{\odot}$). Gieliński *et al.*, (2000) calculated the Kerr disk model for several different a values, and found that $a = 0.75$ gives $6.4M_{\odot}$ for GRO J1655–40.

IC 342 fit with the face-on Kerr disk model gives (assuming d= 4 Mpc) $M = 27.6 M_\odot$ and $\dot{M} = 2.2 \times 10^{20}$ g s^{-1} ($L = 17\ L_{Edd}$). Only a factor of ~ 2 increase of the mass is due to slight hardening of the face-on Kerr disk model compared to the Schwarzschild one. Note that the super-Eddington problem is not solved. If we assume $i = 80°$, we obtain $M = 340 M_\odot$ and $\dot{M} = 1.7 \times 10^{20}$ g s^{-1} ($L = 1.1\ L_{Edd}$). Then the super-Eddington problem will be almost solved, but such a large mass seems to be unlikely. Also, it is very unlikely that most of the accretion disks in ULXs are inclined when seen from the earth. Therefore, Kerr disk interpretation of the ULXs will be problematic, unless considering some other mechanisms to make the disk spectra apparently harder.

References

Belloni, T. *et al.*: 1997, *ApJ* **488**, L109.
Gierlinski, M. *et al.*: 2000, *MNRAS*, submitted.
Hanawa, T.: 1989, *ApJ* **341**, 948.
Laor, A., Netzer, H. and Piran, T.: 1990, *MNRAS* **242**, L560.
Makishima K. *et al.*: 2000, *ApJ* **535**, 632.
Mizuno, T. *et al.*: 1999, *PASJ* **51**, 663.
Okada, K. *et al.*: 1998, *PASJ* **50**, 25.
Orosz, J.A. and Bailyn, C.D.: 1997, *ApJ* **477**, 876.
Ross, R.R. and Fabian, A.C.: 1996, *MNRAS* **281**, 637.
Shimura, T. and Takahara, F.: 1985, *ApJ* **445**, 780.
Tomsick J. *et al.*: 1999, *ApJ* **512**, 892.
Zhang, S.N. *et al.*: 1997, *ApJ* **479**, 381.

BEPPOSAX OBSERVATIONS OF GRS 1915+105

M. FEROCI, P. CASELLA, E. COSTA, E. MASSARO and P. SOFFITTA
Istituto di Astrofisica Spaziale, CNR, Roma, Italy

G. MATT
Dipartimento di Fisica, Univ. Roma Tre, Roma, Italy

T. BELLONI
Osservatorio Astronomico di Brera, Merate, Italy

A.J. CASTRO-TIRADO
Istituto de Astrofísica de Andalucía, Granada, Spain

V. DHAWAN
NRAO, Socorro, New Mexico, USA

F. FRONTERA
Istituto TESRE, CNR, Bologna, Italy

A. HARMON
Marshall Space Flight Center, NASA, Huntsville, Alabama, USA

F. MIRABEL
Service d' Astrophysique, CEA, Saclay, France & IAFE/Conicet, Argentina

G. POOLEY
Mullard Radio Astronomy Observatory, Cambridge, UK

M. TAVANI
Istituto di Fisica Cosmica, CNR, Milano, Italy

Abstract. The galactic superluminal source GRS 1915+105 was observed by BeppoSAX at several occasions from October 1996 to April 2000. In April 1999 the 2–10 keV light curve showed a long series of quasi regular pulses with a recurrence time of about 50 s. Some preliminary results of a spectral analysis are presented.

Keywords: X-ray sources, accretion disk, microquasar, GRS 1915+105

1. Introduction

The transient source GRS 1915+105 was discovered in 1992 (Castro-Tirado; Brandt and Lund, 1992). VLA images of the radio counterpart showed two opposite radio jets moving at an apparent superluminal velocity (Mirabel and Rodriguez, 1994) and from their proper motions a distance of about 12 kpc and an inclination angle of about 70 ° were inferred. The X-ray behaviour of GRS 1915+105 is characterised by alternating bright, very active phases with quiescent periods. Several different

Astrophysics and Space Science is the original source of publication of this article. It is recommended that this article is cited as: *Astrophysics and Space Science* **276** (suppl.): 15–18, 2001.

modes of variations have been classified by Belloni *et al.* (2000). This extreme variability (e.g. Greiner *et al.* 1996), both in flux and spectrum, has been interpreted by Belloni *et al.* (1997a) in terms of abrupt emptying of the inner portion of an accretion disk around a black hole, followed by a refilling phase. In this contribution we present the results of a recent observation of GRS 1915+105 performed with three of the Narrow Field Instruments (LECS, MECS and PDS) on board the Italian-Dutch X-ray satellite BeppoSAX.

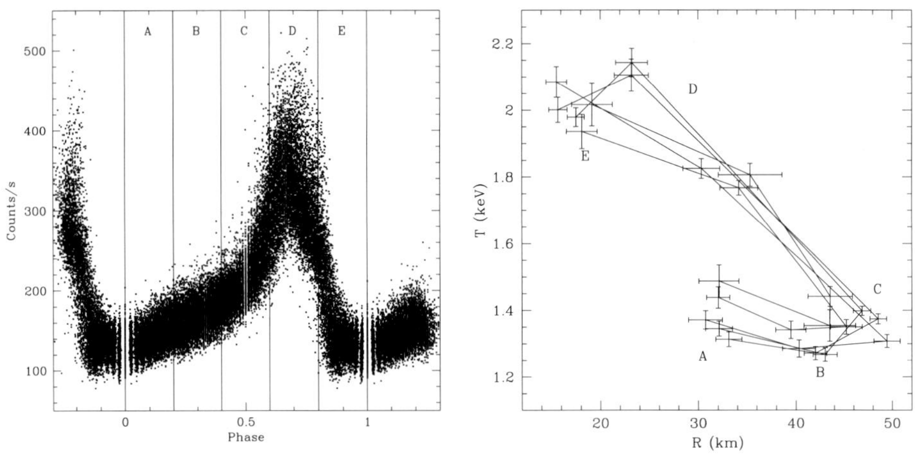

Figure 1. Five phase intervals dividing the folded light curve (left) and the corresponding best-fit spectral parameters for the disk component (right).

2. Observations and Data Analysis

BeppoSAX pointed at GRS 1915+105 several times from October 1996 up to April 2000. In April 1999, during a pointing about 100,000 s long, the source showed a quasi regular train of pulses with a typical recurrence time in the interval from 40 to 60 s. A similar behaviour was observed by Taam *et al.* (1997) with RXTE and by Paul *et al.* (1998) with IRS-P3. Individual pulses were evident in the MECS data and not in those of the LECS (< 2 keV) and PDS (> 13 keV), although peaks at the pulse and first harmonic frequencies are prominent in the power spectra. Simultaneous radio observations were carried out at 15 GHz with the Ryle Telescope (UK) and at 8.46 GHz with the VLA: at 15 GHz the source was at the detection limit and at the lower frequency the radio flux was about 1 mJy.

The X-ray spectrum of GRS 1915+105 is complex and strongly variable. Flares are dominated by a soft emission, as apparent from the hardness ratio time series (Belloni *et al.*, 1997a), while a more stable hard component is present. The spectral distribution adopted for these components is a power law extending up to a few hundreds keV superposed to a multitemperature blackbody originating from an accretion disk around a black hole. Spectral fits were carried out using the *diskbb* model included in the *XSPEC* package, based on an equilibrium disk having a radial temperature distribution like $r^{-3/4}$ (Mitsuda *et al.* 1984). A fixed interstellar column density of 5.5×10^{22} cm^{-2} was also included in the fit.

Once assumed the source distance (12 kpc) and the inclination of the disk to the line of sight (70 °) from radio measurements, the only two free parameters of this component are the innermost disk radius R_{in} and the corresponding temperature kT_{in}. These parameters were estimated in five "phase" intervals defined from the folded pulse shapes. Although the pulse duration and shape are both variable, the folded profiles, integrated over a duration of the order of one satellite orbit (100 minutes) remained rather stable and allow to define five intervals of equal length (approximately 10 seconds) starting from the minimum flux time: A, B, C in the rising portion, D centred around the burst maximum, and E during the decay (Figure 1, left).

The hard component, dominant above 10 keV, can be detected in the PDS data up to about 200 keV. A simple power law fit gives a typical photon index of 3, practically constant over the entire pulse duration. Better fits were obtained using either a power law with an exponential high energy cutoff or a mildly curved spectrum but these improved models did not affect significantly the parameters' values of the thermal component. The estimates of R_{in} and kT_{in} are variable during the pulse. In the first three intervals, when the source luminosity increases slowly, the temperature remains practically constant at about 1.3–1.4 keV while R_{in} changes from 32 to 45–50 km. During the burst kT_{in} grows to 1.9–2.1 keV and, correspondingly, R_{in} becames smaller than 20 km (Figure 1, right). Comparable values were also found for the 1996 November BeppoSAX observation (Feroci *et al.*, 1999). Paul *et al.* (1998), despite the similar variability mode, obtained different values: R_{in}, approximately constant during the rising phase, is less than 15 km and rapidly decreasing during the decay, kT_{in} varies between 1 and 2 keV and increases above 3 keV after the maximum.

3. Discussion

The results presented in this contribution, although preliminary, confirm the high complexity of the phenomena observed in GRS 1915+105. Spectral fitting has shown that different at least two components, but possibly more, contribute to its X-ray emission. The generally adopted spectral model is a multitemperature blackbody plus a power law component. Altough the fits are statistically acceptable, the physical meaning of the parameters' best estimates might be questioned. For example, our results imply that in the phase intervals from A to C the R_{in} increases whereas the temperature remains practically constant. According to the equilibrium model (Mitsuda *et al.*, 1984), this finding implies a nearly simultaneous increase of the temperature in the whole disk, or at least the portion emitting the X rays. Furthermore, if taken at face value, a R_{in} value smaller than about 20 km corresponds to a mass for a central Schwarzschild black hole (assuming $R_{in} = 6\ R_g$, i.e. the last stable orbit) less than 2.4 solar masses. This could be due to the crude approximation of the *diskbb* model when the inner disk edge radius and temperature are concerned. A more detailed modelling of the temperature profile at the inner edge of the disk should be used for estimating these parameters.

A crucial question is if the model of a steady, optically thick, multitemperature disk can be applied to our data because the light curve indicates that the source is describing a limit-cycle during a unstable phase. Taam *et al.* (1997) computed light curves similar to ours in the framework of thermal-viscous instabilities and Szuszkiewicz and Miller (1998) found that during this limit-cycle the disk does not remain optically thick. A future step will be then to perform a spectral analysis by taking into account a time-dependent (non-equilibrium) temperature distribution, changing in the course of the instability cycle.

Acknowledgments: Italian authors are grateful to Italian Space Agency (ASI) for the financial support. NRAO/VLA is a facility of the U.S. National Science Foundation, operated under Cooperative Agreement by Associated Universities Incorporated.

References

Belloni T. and Klein-Wolt M. *et al.*: 2000, *AandA* **355**, 271.
Belloni T. and Mendez M. *et al.*: 1997a, *ApJ* **479**, L145.
Belloni T. and Mendez M. *et al.*: 1997b, *ApJ* **488**, L109.
Castro-Tirado A.J., Brandt S. and Lund N. 1992, *IAU* Circ N.5590.
Feroci M. and Matt G. *et al.*: 1999, *AandA* **351**, 985.
Greiner J., Morgan E. and Remillard R.A.: 1996, *ApJ* **473**, L107.
Mirabel I.F. and Rodriguez L.F.: 1994, *Nature* **371**, 46.
Mitsuda K. and Inoue H. *et al.*: 1984 *PASJ* **36**, 741.
Paul B. and Agrawal P.C. *et al.*: 1998, *ApJ* **492**, L63.
Szuszkiewicz E. and Miller J.C.: 1998, *MNRAS* **298**, 888.
Taam R.E., Chen X. and Swank J.H.: 1997, *ApJ* **485**, L83.

ASCA OBSERVATIONS OF THE GALACTIC SUPERLUMINAL SOURCES GRO J1655−40 AND GRS 1915+105

KAZUTAKA YAMAOKA[1], YOSHIHIRO UEDA[1], HAJIME INOUE[1], FUMIAKI NAGASE[1], KEN EBISAWA[2], TARO KOTANI[2] and YASUO TANAKA[3]
1 *Institute of Space and Astronautical Science, Yoshinodai 3-1-1, Sagamihara, Kanagawa 229-8501, Japan*
[2]*NASA Goddard Space Flight Center, Greenbelt, MD 20771, U.S.A.*
[3]*Max-Plank-Institut für extraterrestrische Physik, 85740 Garching, Germany*

Abstract. We summarize the *ASCA* observations of the two Galactic superluminal sources GRO J1655−40 and GRS 1915+105, focusing on the absorption line features. The high spectroscopic capability of *ASCA* enabled us to detect iron-K absorption line features from both objects. This is direct evidence for the presence of highly ionized plasma in a non-spherical configuration around a black hole, and is considered to be a unique signature of superluminal jet systems. The candidate of the line-absorber would be a part of an accretion disk at a distance of 10^{9-11} cm from the central X-ray source.

Keywords: GRO J1655−40, GRS 1915+105, X-ray:stars

1. Introduction

The *Advanced Satellite for Cosmology and Astrophysics (ASCA)* (Tanaka, Inoue and Holt 1994) observed the two microquasars GRO J1655−40 and GRS 1915+105 at five and seven occasions, respectively, up to Sept. 2000. Table I shows a summary of *ASCA* observations of the two sources. In particular, a continuous multi-wavelength campaign over a week was performed in April 2000. The preliminary results are reported by Ueda *et al.*, in this proceeding. The high spectroscopic capability of *ASCA* ($\Delta E/E \sim 2\%$ at 6keV for the SIS (Burke *et al.*, 1994)) enabled us to detect the iron-K absorption line features for the first time from accretion powered X-ray sources. In Table I we indicate for each epoch whether the absorption line features were clearly detected or not from the *ASCA* observations. The results of the detailed absorption-line-analyses were reported by Ueda *et al.* (1998) and Yamaoka *et al.* (2000) for GRO J1655−40, and Kotani *et al.* (2000) for GRS 1915+105.

2. Detection of Iron K-absorption Lines

The left panel in Figure 1 shows the ratio of the SIS data to a continuum model of GRS 1915+105 in the 1994 and 1995 observations when we fit the SIS spectra

Astrophysics and Space Science is the original source of publication of this article. It is recommended that this article is cited as: *Astrophysics and Space Science* **276** (suppl.): 19–22, 2001.

TABLE I

ASCA observation summary of GRO J1655−40 and GRS 1915+105

Objects	Epoch	Exposure* [ksec]	Ave. flux†	Iron Kα abs.
GRO J1655−40	1994 Aug.	9	5.6	He-like
	1994 Sep.	19	12	He-like
	1995 Aug.	20	46	H-like
	1996 Mar.	51	0.002	...
	1997 Feb.	57	25	He-like/H-like
GRS 1915+105	1994 Sep.	10	9.8	He-like/H-like
	1995 Apr.	19	21	He-like/H-like
	1996 Oct.	13	24	?‖
	1997 Apr.	16	8.3	?
	1998 Apr.	1	18	?
	1999 Apr.	13	19	?
	2000 Apr.	157	10	?

∗ High bit-rate only.

† In unit of 10^{-9} ergs cm^{-2} s^{-1} in the 2 − 10 keV band. The flux of the Crab Nebula corresponds to 2.16×10^{-8} erg cm^{-2} s^{-1} in the same energy band.

‖ Not clearly detected, although weaker Kα absorption lines due to H-like iron ions may be present.

with a Multi-Color Disk (Mitsuda *et al.,* 1986) plus a power law model. As may be seen from this figure, we detected clear absorption line features due to highly ionized iron ions. The central energies of the absorption line features correspond to the Kα line of He-like iron (6.70 keV), and that of H-like iron (6.97 keV).

The right panel in Figure 1 shows the same plot for GRO J1655−40 in the 1994 August, 1994 September, 1995 and 1997 observations. In addition, we found that the central energy of the absorption lines varied with the continuum X-ray flux from GRO J1655−40. Namely, absorption lines were detected at the energy of the Kα line of He-like iron, of a blend of those of He-like and H-like iron, and that of H-like iron, as the continuum flux increases. This fact strongly suggests that the line-absorber was photoionized by an X-ray irradiation.

In February 1997, we observed GRO J1655−40 with *ASCA* covering one orbital period (2.62 d) of the binary. We divided the whole observation period into four phases and made spectral analysis of each spectrum. Figure 2 shows the equivalent widths of the absorption lines of H-like iron and He-like iron, plotted against the orbital phase. We found no significant changes of the equivalent width associated with the orbital phase, suggesting that the absorber is centered on the black hole and is not affected by the orbital motion of the companion star.

In case of GRS 1915+105, clear absorption line features due to He-like iron ion were not detected except for the 1994 and 1995 observation, although the presence

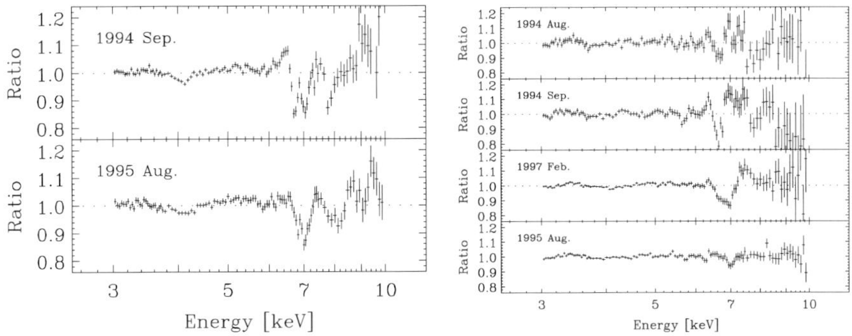

Figure 1. Left panel: SIS spectra of GRS 1915+105 in the 1994 and 1995 observations. Right panel: SIS spectra of GRO J1655−40 in the 1994 Aug.,1994 Sep., 1995 and 1997 observations. The corresponding X-ray intensity increases from upper to lower.

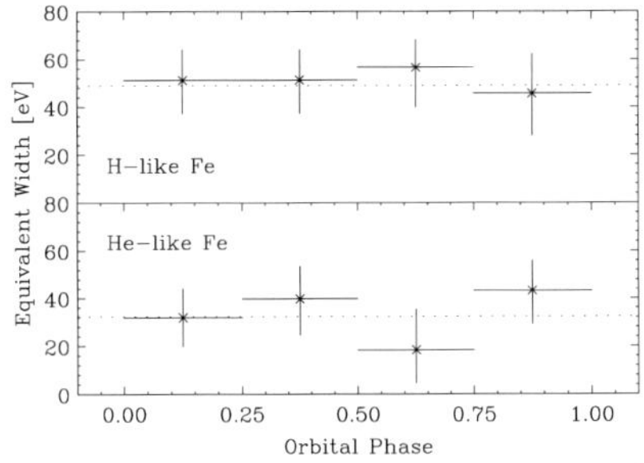

Figure 2. Orbital phase dependence of the equivalent width of H-like and He-like iron Kα absorption lines observed in the 1997 observation of GRO J1655−40.

of a H-like iron absorption lines could be possible. In these epochs, hard X-rays in the 20 − 100 keV band were high at a level of 0.05 photons cm^{-2} s^{-1}, as detected by the BATSE onboard *CGRO*. This suggests that the hard X-ray emissions might photoionize the plasma to a high ionization stage, which is consistent with the photoionization scenario proposed by Ueda *et al.* (1998).

3. Origin of the Absorption Lines

We can derive two facts from the detection of the absorption lines. One is the presence of highly ionized plasma between an X-ray source and the observer. The other is that the plasma has a non-spherical configuration around a black hole. If the plasma had a spherically symmetric configuration, then the emission lines from outside of our line of sight would fill the absorption lines completely.

Next, we discuss the distance of the line-absorber to the central X-ray source. Let n and r be the number density and the distance of the absorber, the column density nr can be obtained from the curve of growth analysis. Such analyses were already done by Ueda *et al.* (1998), for GRO J1655−40 and Kotani *et al.* (2000) for

GRS 1915+105, and nr was estimated to be 10^{23-24} cm^{-2}. In addition, so-called ionization parameter ξ ($\equiv L/nr^2$: Tarter, Tucker and Salpeter, 1969) should be around 10^{3-4} ergs cm s^{-1} in order for iron ion to be H-like or He-like according to Kallman and McCray (1982). Using the observed X-ray luminosity L of $\sim 10^{36}$ ergs s^{-1} for GRO J1655$-$40 and $\sim 10^{37}$ ergs s^{-1} for GRS 1915+105 in difinition of Kallman and McCray (1982), we estimate r to be 10^{9-10} cm and 10^{11} cm, respectively. The higher luminosity of GRS 1915+105 than that of GRO J1655$-$40 would cause the larger distance of the line-absorber of GRS 1915+105.

The 10^{9-11} cm distance and the constant equivalent width independent of the binary motion would indicate that the line-absorber could be a part of an accretion disk at that distance.

4. Summary

We detect absorption line features due to highly ionized iron from the two Galactic superluminal sources GRO J1655$-$40 and GRS 1915+105. These may have a connection with superluminal jets and further investigation of the absorption lines will be needed. We summarize the characteristics of the line-absorber as follows. The plasma

- has a non-spherical configuration around a black hole.
- was photoionized by X-ray irradiation.
- has a size of 10^{9-10} cm for GRO J1655$-$40 and 10^{11} cm for GRS 1915+105.
- shows no orbital phase dependence in GRO J1655$-$40.

From these observed facts, we conclude that the candidate of the line-absorber would be a part of an accretion disk at a distance of 10^{9-11} cm from the central X-ray source.

References

Burke, B. *et al.*: 1994, *IEEE Trans. Nucl. Sci.* **41**, 375.
Kallman, T.R. and McCray, R.: 1982, *ApJ S* **50**, 263.
Kotani, T. *et al.*: 2000, *ApJ* **539**, 413.
Mitsuda, K. *et al.*: 1986, *PASJ* **36**, 741.
Tanaka, Y., Inoue, H. and Holt, S.S.: 1994, *PASJ*, 430.
Tarter, C.B., Tucker, W. and Salpeter, E.E.: 1969, *ApJ* **156**, 943.
Ueda, Y. *et al.*: 1998, *ApJ* **492**, 782.
Yamaoka, K. *et al.*: 2000, *PASJ*, submitted.

OBSERVATIONS OF GRS 1915+105 FROM THE USA EXPERIMENT ON ARGOS

REBA M. BANDYOPADHYAY
Naval Research Laboratory/NRC, Washington, DC 20375

K.S. WOOD, P. HERTZ, M.N. LOVELLETTE, P.S. RAY and M.T. WOLFF
Naval Research Laboratory, Washington, DC 20375

B. GIEBELS and E.D. BLOOM
Stanford Linear Accelerator Center, Stanford, CA 94309

Abstract. We present first results from a monitoring campaign of GRS 1915+105 undertaken with the USA X-ray timing experiment on the ARGOS satellite. A variety of behaviour has been observed, ranging from low, steady X-ray emission to rapid quasi-periodic flaring on timescales of approximately 10–120 seconds.

1. The USA Experiment

The Unconventional Stellar Aspect (USA) Experiment is one of nine experiments aboard the Advanced Research and Global Observation Satellite (ARGOS) which was launched on 23 February 1999. USA is a collimated proportional counter X-ray telescope with 1000 cm^2 of effective area, sensitive to photons in the energy range 1-15 keV. A unique feature of USA is that all photon events are time tagged by reference to an onboard GPS receiver allowing precise absolute time and location determination.

2. Observations

Here we present first results from a monitoring campaign on the microquasar GRS 1915+105 undertaken with USA in 1999 and 2000. We have obtained a total of 255 ks of data, with individual observations ranging in length from 6-23 minutes. One USA count is approximately equal to 0.3 mCrab. A variety of behaviour is observed, including steady X-ray emission at several flux levels, rapid irregular $\sim$10 second flaring, quasi-periodic flaring (generally on a timescale of $\sim$60 seconds), and short intervals of 'quenching' of the X-ray emission. We have also combined some of our data with simultaneous X-ray data from the RXTE ASM and radio data from the GBI to examine the relationship between the X-ray and radio emission during a radio flaring episode which occurred in November 1999.

Astrophysics and Space Science is the original source of publication of this article. It is recommended that this article is cited as: *Astrophysics and Space Science* **276** (suppl.): 23–24, 2001.

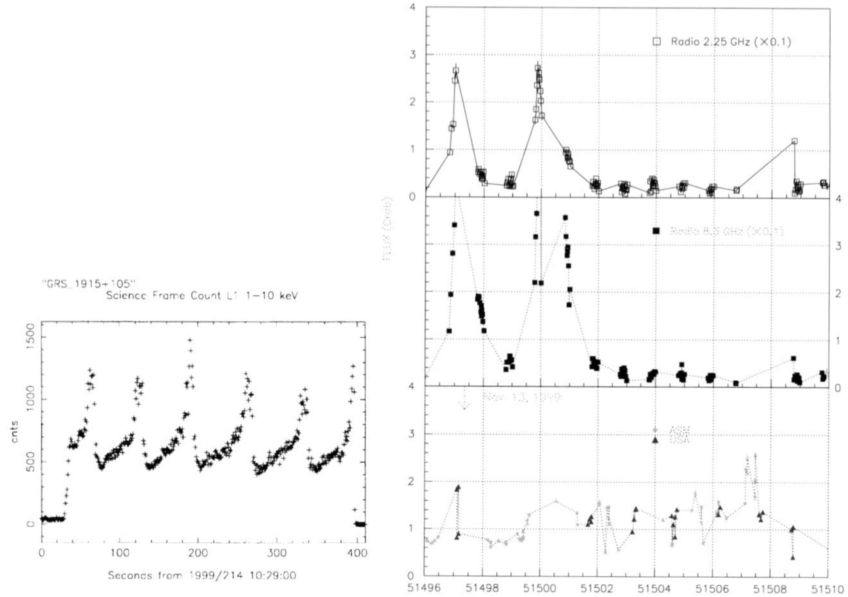

Figure 1. Left: USA X-ray light curve showing quasi-periodic flaring. Right: Radio (GBI, upper panels) and X-ray (combined USA and RXTE ASM, bottom panel) light curves from November 1999. All fluxes have been normalized to 1 Crab.

During July–August 1999 and June 2000 USA observed extended intervals of regular quasi-periodic flaring in GRS 1915+105 of the type shown in the left panel of Figure 1. The recurrence time for these flares varies between 30–150 seconds. Bursts of this type have previously been observed by RXTE (Vilhu and Nevalainen, 1998) and IAXE (Yadav *et al.* 1999). Our data indicate that in each of these two epochs this bursting behaviour persisted for several *weeks*.

The right panel of Figure 1 shows simultaneous multi-wavelength data from November 1999, with 2.25 and 8.3 GHz radio data from the GBI and X-ray monitoring with USA and the RXTE ASM. The radio flare on MJD 51495 has a clear counterpart in the X-ray light curve, and it appears that the X-ray flare on MJD 51507 is present in the radio emission. Therefore during this two week interval the X-ray and radio flares appear to be correlated, although some previous observations have shown similar radio flares to be anti-correlated with the X-ray emission (e.g., Bandyopadhyay *et al.*, 1998).

References

Bandyopadhyay R. *et al.*: 1998, *MNRAS* **295**, 623.
Vilhu O. and Nevalainen J.: 1998, *ApJ* **508**, L85.
Yadav J.S. *et al.*: 1999, *ApJ* **517**, 935.

THE 2000 APRIL MULTIWAVELENGTH CAMPAIGN OF GRS 1915+105

Y. UEDA[1], K. YAMAOKA[1], J.E. GROVE[2], M. McCOLLOUGH[3], P. DUROUCHOUX[4], J. RODRIGUEZ[4], F. MIRABEL[4,5], J. SWANK[6], M. FEROCI[7], P. CASELLA[7], A.J. CASTRO-TIRADO[8,9], C. SÁNCHEZ-FERNÁNDEZ[9], S. CHATY[10], H. CASTANEDA[11], K. KOHNO[12], V. DHAWAN[13], S.A. TRUSHKIN[14], K. EBISAWA[6], T. KOTANI[6], H. INOUE[1] and THE MULTIWAVELENGTH CAMPAIGN TEAM

[1] *Institute of Space and Astronautical Science, Japan;* [2] *Naval Research Lab, USA;* [3] *MSFC/NASA, USA;* [4] *Service d'Astrophysique, Saclay, France;* [5] *IAFE/CONICET, Argentina;* [6] *GSFC/NASA, USA;* [7] *Istituto di Astrofisica Spaziale, C.N.R., Italy;* [8] *IAA-CSIC, Granada, Spain;* [9] *LAEFF-INTA, Madrid, Spain;* [10] *Department of Physics and Astronomy, The Open University, UK;* [11] *Observatorio Astronomico Nacional de Mexico, Mexico;* [12] *Nobeyama Radio Observatory, NAO, Japan;* [13] *NRAO, USA;* [14] *SAO, Russia*

Abstract. We report results from multiwavelength observations of the microquasar GRS 1915+105 performed during the 2000 April campaign. This is one of the biggest campaigns ever made for this source covering the broad band from radio to γ-rays. Multiwavelength light curves compiled from all the data reduced up to date and broad band spectra obtained with ASCA and RXTE are presented.

Keywords: GRS 1915+105, multiwavelength observations

1. Observations

The ASCA satellite observed GRS 1915+105 in the 0.5–10 keV band from 2000 April 17 12:00 to April 25 12:00 for a net exposure of 271 ksec. Coordinated with the ASCA observations, simultaneous observations were made with RXTE on April 17, 22 and 23, with BeppoSAX from April 21 to 24 (PI: Feroci), and with OSSE/CGRO from April 18 to 25. From ground, VLA observations were made (PI: Mirabel) on April 17, 22, and 23, IR (K band) photometric monitoring at Calar Alto from April 18 to 25 (PI: Castro-Tirado), at the ESO from April 17 to 23 (PI: Chaty) , and at Mexico (PI: Castaneda) from April 22 to 25, and radio observations (at 3.2 mm) at the Nobeyama Millimeter Array from April 17 to 19 (PI: Kohno). In addition, monitoring data of the GBI(2.25 GHz and 8.3 GHz), ASM/RXTE (1.5–12 keV), and BATSE/CGRO (20–100 keV) are available. Results of timing analysis using the RXTE data are reported by Rodriguez *et al.* (2001). A detailed analysis of the IR K' data is reported in Sánchez-Fernández (2001).

Astrophysics and Space Science is the original source of publication of this article. It is recommended that this article is cited as: *Astrophysics and Space Science* **276** (suppl.): 25–28, 2001.

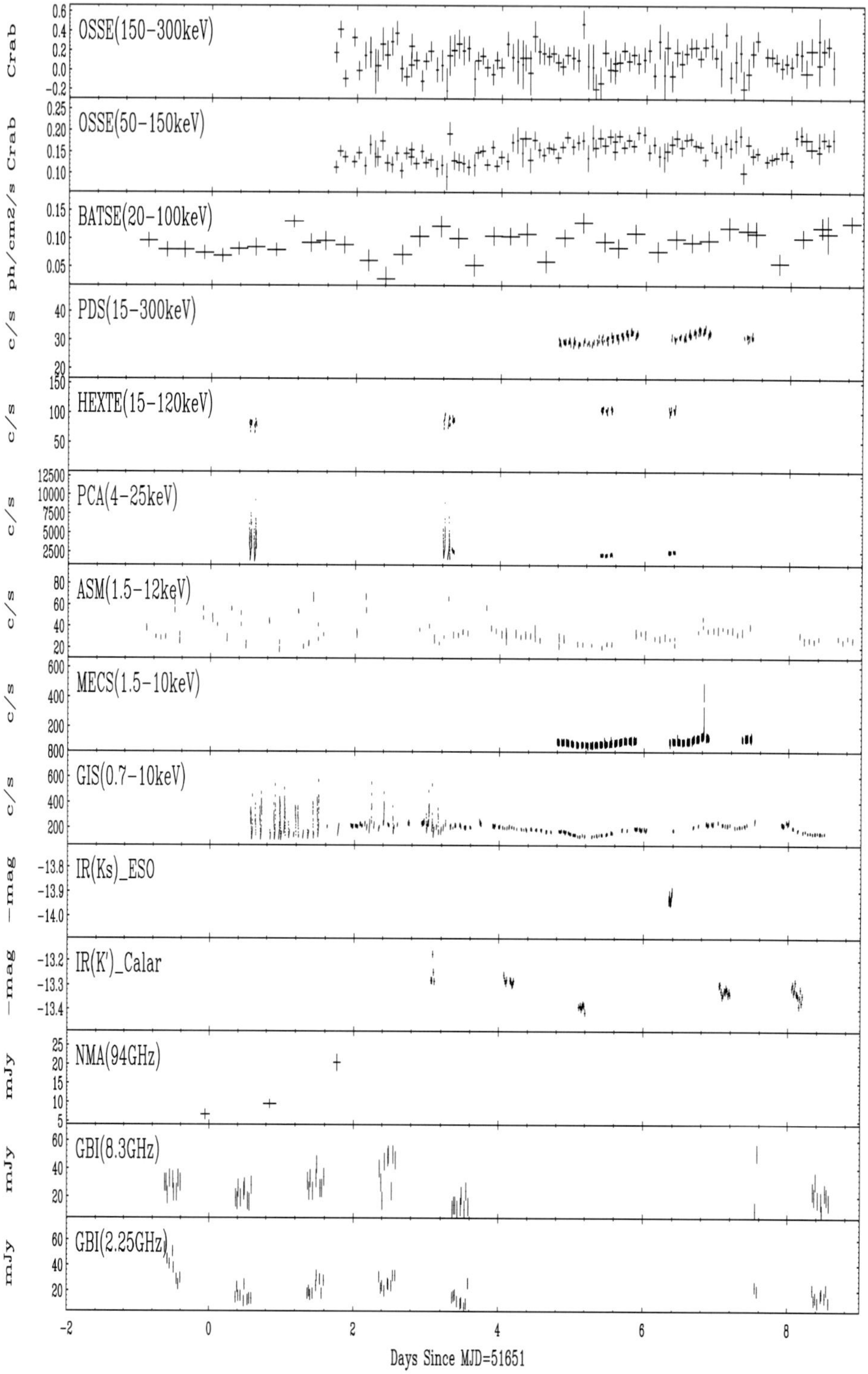

Figure 1. Multiwavelength light curves of GRS 1915+105 during the 2000 April campaign, covering between April 16 and 25. They are sorted in energy band (higher energy in upper). The reference day (MJD 51651) corresponds to 2000 April 17.0 (UT).

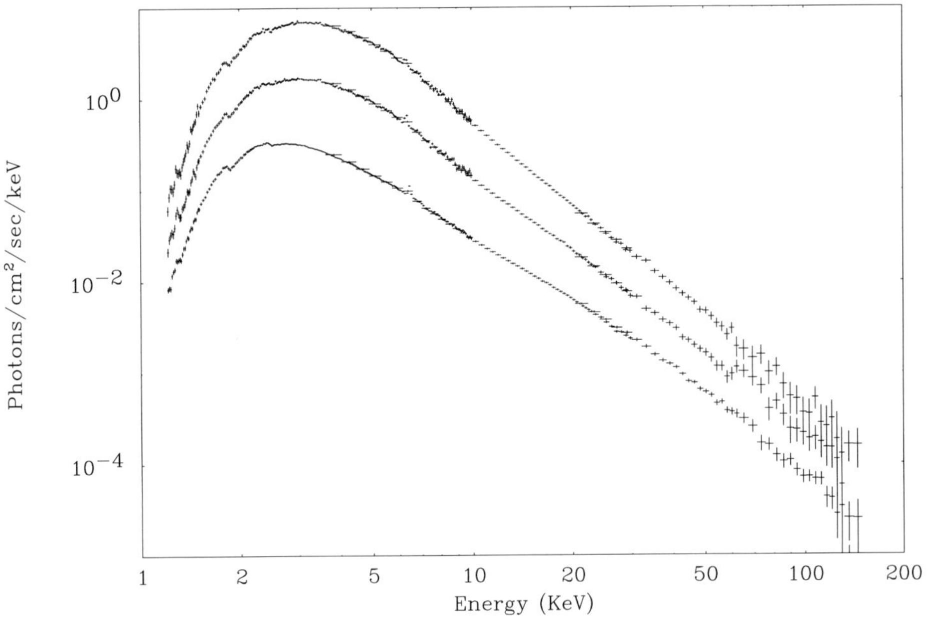

Figure 2. The unfolded spectra of GRS 1915+105 in different states, obtained by joint analysis of the GIS, PCA, and HEXTE spectra. Upper: oscillation high (normalization $\times$ 9), middle: oscillation low ($\times$ 3), and bottom: rise ($\times$ 1). The continuum is modeled by the multicolor disk model plus a Comptonized blackbody.

2. Results

Multiwavelength light curves obtained from this campaign are shown in Figure 1. We list several results noticed from the light curves.

- The soft X-rays showed rapid oscillations in the first half of the campaign (see below). Daily averaged 2–10 keV fluxes are $(0.9 - 1.1) \times 10^{-8}$ erg s^{-1} cm^{-2}.
- The hard X-ray flux (20–100 keV) and the radio fluxes (2.25 GHz and 8.3 GHz) were around the levels of the 'plateau' state defined by Foster *et al.* (1996).
- The OSSE flux (50–300 keV) showed a significant variability on hours to days. The photon index is about 3.
- The IR K-band fluxes in these data were mostly at a quiescence level (e.g., Fender *et al.* 1997) and seemed to correlate with the soft X-ray intensity on timescale of hours. A weak flare was detected at MJD = 51654.08.

Before MJD$\simeq$51655, the source showed a peculiar pattern of variability (class ν in Belloni *et al.* 2000), characterized by a rapid oscillation lasting for typically 500–1000 sec, followed by a decline of $\sim$ 200 sec and a slow rise of $\sim$ 1000 sec. The soft X-ray intensity and the hardness have correlations: when the intensity is high, the spectrum becomes hard in the 0.7–10 keV range but soft in the 3–31 keV range. We here define three characteristic states: (1) the 'oscillation high' state, when the soft X-ray flux is high during the oscillation (> 1800 c s^{-1} per PCU in the 16 sec resolution light curve) , (2) the 'oscillation low' state (< 1800 c s^{-1} per PCU), and (3) the 'rise' state, when the intensity increases linearly with time between oscillations.

We make a joint spectral analysis of the ASCA/GIS, RXTE/PCA, and RXTE/ HEXTE data, covering the 1–150 keV band. The spectra are created separately for the 3 states defined above, although the ASCA and RXTE data are not exactly simultaneous. Figure 2 shows unfolded spectra for each state. The normalization of the oscillation-high and -low state spectra is multiplied by 9 and 3, respectively, for plotting purpose. Preliminary analysis indicates that the contribution of a disk component is much weaker in the rise state than in the oscillation state although the difference of the hard band spectrum above 20 keV is not conspicuous between these states. Detailed spectral analysis and the implications of the results will be presented in future papers.

3. Acknowledgements

This research has made use of the data of the Green Bank Interferometer (GBI), which is a facility of the National Science Foundation operated by the NRAO in support of NASA High Energy Astrophysics programs.

References

Belloni, T. *et al.*: 2000, *A&A* **355**, 271.
Fender, R.P. *et al.*: 1997, *MNRAS* **290**, L65.
Foster, R.S. *et al.*: 1996, *ApJ* **467**, L81.
Mirabel, I.F. *et al.*: 1998, *A&A* **330**, L9.
Rodriguez, J. *et al.*: 2001, in preparation.
Sánchez-Fernández, C.: 2001, PhD. Thesis, Madrid

IDENTIFICATION OF THE DONOR IN GRS 1915+105

J. GREINER and M.J. McCAUGHREAN
Astrophysical Institute Potsdam, 14482 Potsdam, Germany

J.G. CUBY
ESO, Alonso de Córdova 3107, Santiago 19, Chile

A.J. CASTRO-TIRADO
Instituto de Astrofísica de Andalucía, E-18080 Granada, Spain
Laboratorio de Astrofísica Espacial y Física Fundamental, E-28080 Madrid, Spain

R.E. MENNICKENT
Universidad de Concepción, Casilla 160-C Concepcion, Chile

Abstract. We report on medium-resolution H and K band spectroscopy of GRS 1915+ 105 using the 8m VLT at ESO. We clearly identify absorption bandheads from $^{12}CO/^{13}CO$. Together with other features we classify the donor as a K-M III star.

Keywords: X-ray binary, GRS 1915+105, infrared observations

1. Introduction

GRS 1915+ 105 (Castro-Tirado *et al.*, 1992, 1994) is the prototypical microquasar, a galactic X-ray binary ejecting plasma clouds at v≈0.92 c (Mirabel and Rodriguez 1994). It exhibits unique X-ray variability patterns (Greiner *et al.*, 1996) which have been interpreted as instabilities leading to an infall of parts of the inner accretion disk (Belloni *et al.*, 1997). It is therefore of great importance to know some details about the system components in order to understand the conditions which lead to these unique characteristics.

2. Observations and Results

We obtained H and K band infrared spectroscopy of GRS 1915+ 105 with the aim of searching for absorption signatures due to the donor. Since GRS 1915+ 105 is a strongly variable infrared source, believed to predominantly caused by synchrotron emission of ejected material (Eikenberry *et al.*, 2000; Greiner *et al.*, 2001), this required high signal-to-noise as well as good spectral resolution in order to beat the strong veiling. We therefore used the infrared spectrometer ISAAC on the 8m VLT Antu telescope on Paranal (ESO, Chile).

The short wavelength (0.9–2.5 μm) arm of ISAAC is equipped with a 1024×1024 pixel Rockwell HgCdTe array with an image scale of 0$''$.147/pixel. Using the grating

Astrophysics and Space Science is the original source of publication of this article. It is recommended that this article is cited as: *Astrophysics and Space Science* **276** (suppl.): 31–34, 2001.

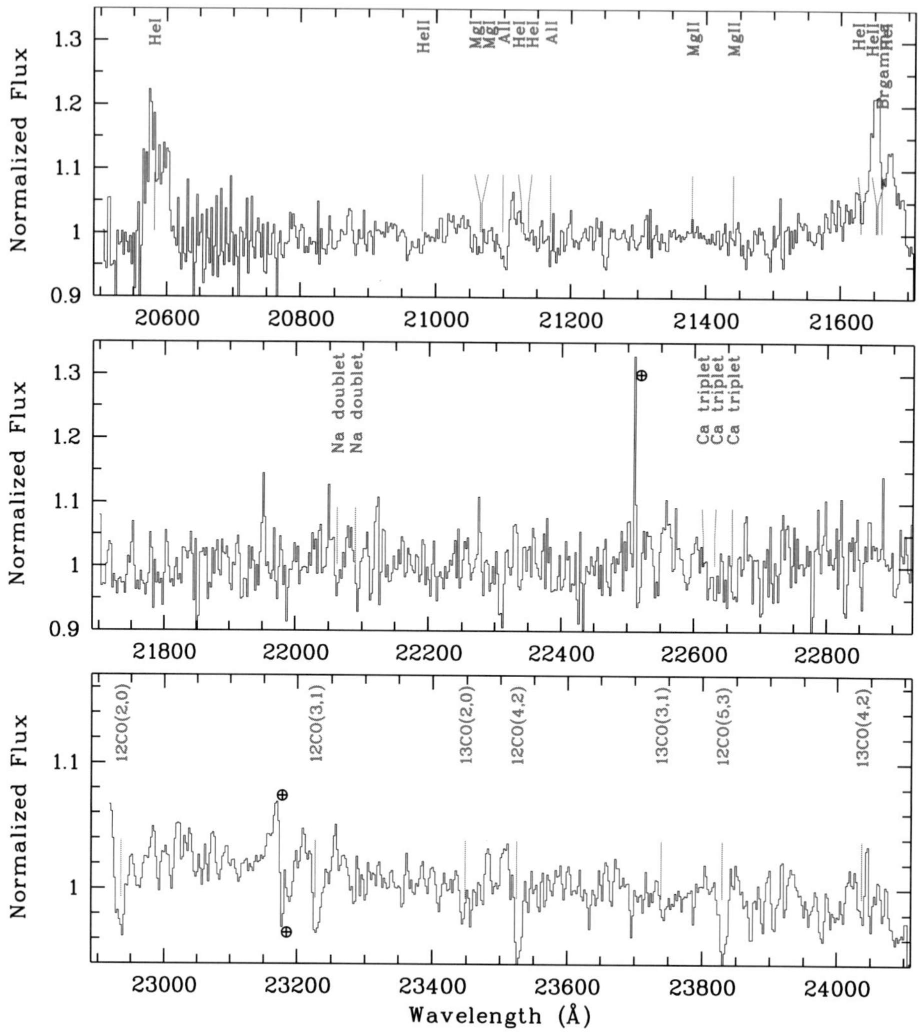

Figure 1. Spectrum of GRS 1915+105 in the K band. The top two panels correspond to 3000 sec exposure, while the bottom panel shows a sum of 4 images with 2000 sec exposure each. All spectra were rebinned by a factor of 2.

with the highest possible spectral resolution (~3000 for a 1″ slit) yields 0.8Å/pixel in the H band and 1.2Å/pixel in the K band. Observations were performed in 1 (2) adjacent H bands, and 2 (3) adjacent K bands on 20/21 July 1999 (24/25 July 2000).

Science exposures consisted of several 250–300 sec individual exposures which were dithered along the slit by ±30″. In order to correct for atmospheric absorption, the nearby star HD 179913 (A0 V) was observed either before or after each science exposure. The initial data reduction steps like debiasing, flatfielding and co-adding were performed within the *Eclipse* package (Devillard, 2000). The extraction and wavelength calibration was done using an optimal extraction routine within the MIDAS package.

The spectra of the 3 grating settings covering the K band are shown in Figure 1. The top panel shows the strong emission lines of Br γ and He I known already from previous low-resolution spectroscopy (Castro-Tirado *et al.*, 1996). In addition, we find for the first time several absorption lines which allow us to roughly identify the donor in GRS 1915+ 105. The lower panel clearly shows ^{12}CO absorption band heads characteristic of a low temperature (T < 7000 K) star (e.g. Kleinmann and Hall, 1986). Though weak, we also identify the ^{13}CO (2,0) and ^{13}CO (3,1) transitions, indicating a luminosity class III or brighter (e.g. Wallace and Hinkle, 1997). We also identify the Na doublet (2.20624/2.20897 μm), and possibly the Ca triplet (2.26141/2.26311/2.26573 μm), Al I (2.10988 μm) and the Mg I doublet (2.10655/2.10680 μm) in absorption. Note that the CN doublet (2.0910/2.0960 μm), which in supergiants is more prominent than Al/Mg, is not detected. Thus, we conclude that the donor in GRS 1915+ 105 is a late-type K-M giant.

We have tried to confirm the luminosity class more quantitatively by using the veiling-independent indicator

$r = \log[EW(^{12}\mathrm{CO}(2,0))/(EW(\mathrm{Na}) + EW(\mathrm{Ca}))]$

(Ramírez *et al.*, 1997). Because of the low significance of the Ca triplet our measurement has a large error: $r = 0.25 \pm 0.20$. This value falls in between the ranges covered by dwarfs ($-0.2 \lesssim r \lesssim 0.0$) and giants ($0.4 \lesssim r \lesssim 0.6$) (Ramírez *et al.*, 1997). The ratio of equivalent widths of ^{12}CO to ^{13}CO which depends on luminosity class (Campbell *et al.*, 1990), has been measured for the seven transitions covered (lower panel of Figure 1) to $\sim 3 \pm 1$, again supporting a giant classification.

Unfortunately, the H band spectra are of too low S/N ratio to detect OH or CO, thus not allowing us to use the veiling-independent temperature/luminosity discriminants as proposed by Meyer *et al.* (1998). However, the veiling was roughly determined by comparison of our flux-calibrated 2.35 μm spectrum of GRS 1915+ 105 with that of a K2 III standard star, observed with the same settings. We adopted a 13 mag flat continuum and added the K2 III star spectrum scaled to a brightness in the range 13.5–17.5 mag (Figure 2). A comparison with the GRS 1915+ 105 spectrum gives a (not extinction-corrected) magnitude of K = 14.5–15.0 mag for the donor. With a distance of $\sim$11 kpc and a $A_\mathrm{K} = 3$ mag extinction correction this implies an absolute magnitude of M_K = –2...–3 mag, consistent with the giant classification.

This identification of the donor of GRS 1915+ 105 as a K-M giant implies a rather narrow mass range of 1.0–1.5 $M_\odot$. Typical spherical mass-loss rates of such stars are much too low to sustain the high accretion luminosity of GRS 1915+ 105 via accretion from the donor's stellar wind. We therefore suggest (not too surprisingly) that accretion should occur via Roche lobe overflow. We finally note that while our identification contradicts the findings of Martí *et al.* (2000) who argue

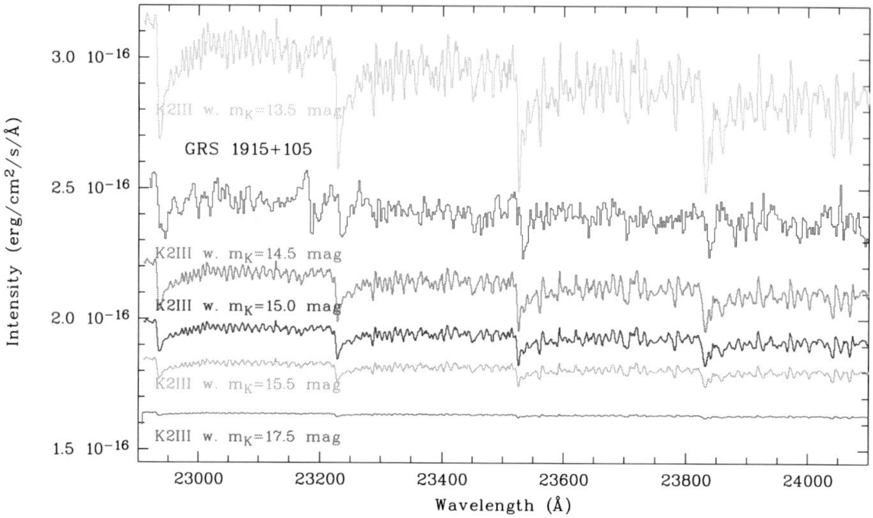

Figure 2. Spectrum of GRS 1915+ 105 in the 2.3–2.4 μm range (second from top, not extinction-corrected), compared to the spectrum of the K2III star HD 202135 scaled to different magnitudes (as marked on each spectrum) which is veiled by a 13 mag flat continuum. This implies that the donor of GRS 1915+ 105 has a magnitude of 14.5–15 mag prior to extinction correction.

for a massive OB-type companion, it is consistent with the constraints derived by Eikenberry *et al.* (2001).

The presence of clear donor absorption features will now allow a period search as well as a determination of the mass of the compact object using radial velocity measurements of the donor.

References

Belloni, T., Mendez, M., King, A.R., *et al.*: 1997, *ApJ* **479**, L145.
Campbell, B., Lambert, D.L. and Maillard, J.-P.: 1990, *PASP* **102**, 79.
Castro-Tirado, A.J., Brandt, S. and Lund, N.: 1992, *IAU* Circ. 5590.
Castro-Tirado, A.J., Brandt, S., Lund, N., *et al.*: 1994, *ApJS* **92**, 469.
Castro-Tirado, A.J., Geballe, T.R. and Lund, N.: 1996, *ApJ* **461**, L99.
Devillard, N.: 2000, Eclipse Users Guide, www.eso.org/projects/aot/eclipse/eug/index.html
Eikenberry, S.S., Matthews, K., Muno, M., *et al.*: 2000, *ApJ* **532**, L33.
Eikenberry, S.S., *et al.*: 2001, these proc.
Greiner, J., Morgan, E.H. and Remillard, R.A.: 1996, *ApJ* **473**, L107.
Greiner, J., Vrba, F.J., Henden, A.A., Guetter, H.H. and Luginbuhl, C.B.: 2001, these proc.
Kleinmann, S.G. and Hall, D.N.B.: 1986, *ApJ Suppl.* **62**, 501.
Martí, J., Mirabel, I.F., Chaty, S. and Rodriguez, L.F.: 2000, *AA* **356**, 943.
Meyer, M.R., Edwards, S., Hinkle, K.H. and Strom, S.: 1998, *ApJ* **508**, 397.
Mirabel, I.F. and Rodriguez, L.F.: 1994, *Nature* **371**, 46.
Ramírez, S.., Depoy, D.L., Frogel, J.A., Sellgren, K. and Blum R.D.: *AJ* **113**, 1411.
Wallace, L. and Hinkle, K.: 1997, *ApJ Suppl.* **111**, 445.

UKIRT IR SPECTRA OF THE MICROQUASAR GRS 1915+105

E. HARLAFTIS
Institute of Astronomy and Astrophysics, National Observatory of Athens, P.O. Box 20048, Athens 118 10, Greece

VIK S. DHILLON
Department of Physics and Astronomy, Univ. of Sheffield, Sheffield S3 7RH, U.K.

ALBERTO J. CASTRO-TIRADO
Instituto de Astrofísica de Andalucía (IAA-CSIC), P.O. Box 03004, E-18080 Granada, Spain; and Laboratorio de Astrofísica Espacial y Física Fundamental (LAEFF-INTA), P.O. Box 50727, E-28080 Madrid, Spain

Abstract. The UKIRT JHK-band spectrum of GRS 1915+105 shows strong He I (2.059 μm) and Brγ (2.166 μm) emission lines. Eikenberry *et al.* (1998) found flux variations of a factor of 5 in these lines on a 5 min timescale. Based on the detection or not of the He II emission line, it was suggested that the companion could be a LMXB or a HMXB, respectively. Since the spectral type of the companion star of GRS 1915+105 is still an open question we attempt to clarify the issue by revisiting the object and comparing the JHK spectrum with that of the Be star DM +49 3718 and the prototype X-ray binary Sco X-1.

1. Introduction

We obtained JHK-band spectra of the X-ray binary GRS 1915+105 (Castro-Tirado *et al.*, 1994) at the 3.8 m UKIRT (+ CGS4) on Mauna Kea during the night of 1999 July 8. We also obtained JHK spectra of Sco X-1 and the Be star DM +49 3718 in order to compare them with the spectral slope and emission line properties of GRS 1915+105. The spectra were bracketed by observations of nearby F-type stars, used during the data reduction to remove the telluric atmospheric features.

2. Observations and Results

GRS 1905+105 is not a Wolf-Rayet star, like Cyg X-3 (van Kerkwijk *et al.*, 1992), since we detect Brγ in emission (see also discussion in Castro-Tirado, Geballe and Lund, 1996) but no absorption features from the companion star (neutral lines or molecular bands), as seen in GX 1+4 (Bandyopadhyay *et al.*, 1997). There is no trace of He II emission line at a time when the radio spectrum was flat and thus the disc was close to quiescence. The last flaring activity was observed on 1999 July 2. The radio and X-ray continua were in their lowest states which indicate that

Astrophysics and Space Science is the original source of publication of this article. It is recommended that this article is cited as: *Astrophysics and Space Science* **276** (suppl.): 35–38, 2001.

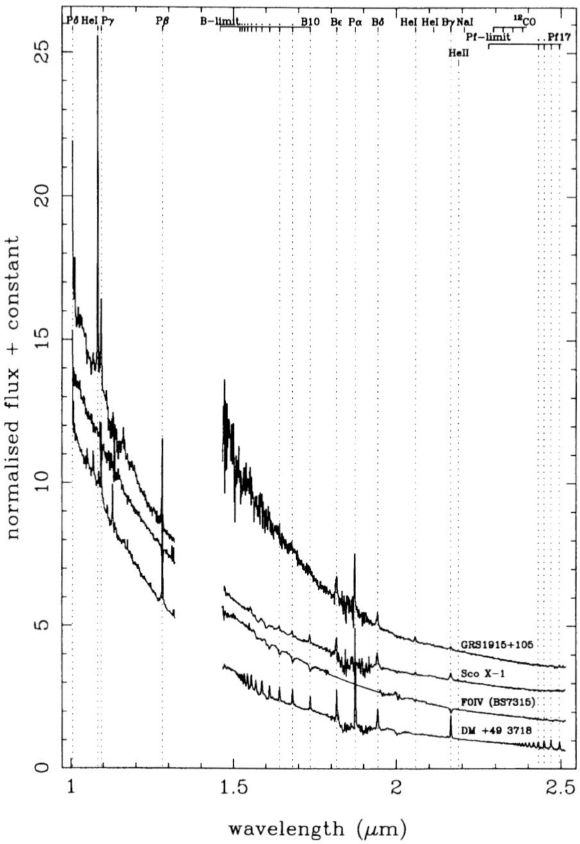

Figure 1. The deredenned HK spectrum of the microquasar GRS 1915+105 and the JHK spectra of Sco X-1 and the Be star DM +49 3718. Also plotted is the JHK spectrum of the F0IV star BS 7315. The spectra have been normalised by dividing by the flux at 2.250 μm and then offset on the y-axis by adding a multiple of 1 to each spectrum.

the accretion disc was very quiet on 1999 July 8 (Martí *et al.*, 2000). It is unclear if the He II emission is produced along the jet, in the accretion disc or on the companion star. However, the X-ray burst activity is related to the jet/disc system centred on the compact object (Belloni *et al.*, 1997), rather than the companion star, and thereafter is the most likely cause for pumping up the transient He II emission. This hypothesis is supported by the absence of the He II emission during quiescence, which is the time that the inner disc is being replenished by gas, suggesting that the X-ray luminosity is not sufficient to pump up the He II line through photoionization of the accretion disc. There is Na I doublet emission (2.206 and 2.209 μm) with an equivalent width of -2.6 ± 1.3 Å – which may be affected by telluric features – and a trace of He I emission at 2.113 μm. However, the spectrum of GRS 1915+105 is different from that of the Be star DM +49 3718 in that only

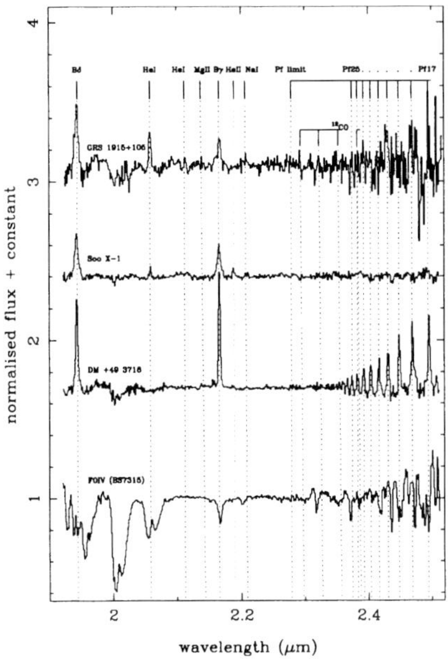

Figure 2. The K-band spectra of the microquasar GRS 1915+105, Sco X-1, and the Be star DM +49 3718. Also plotted is the K-band spectrum of the F0IV star BS 7315 which indicates the location of telluric absorption features. Note the weak He II emission line in Sco X-1. The spectra have been normalized by dividing by a spline fit to their continua and then offset on the *y*-axis by adding a multiple of 0.7 to each spectrum.

a few H lines are visible, which have larger velocity widths and smaller equivalent widths, in addition to the helium emission. We dereddened the GRS 1915+105 spectrum using $E(B-V) = 9.6$ (Boër *et al.*, 1996), and $A_V/E(B-V) = 2.76$ so $A_V = 26.5$ (Chaty *et al.*, 1996) and the extinction law from Howarth (1983). There was insufficient signal in the J-band, given the large extinction, though after smoothing it was possible to combine the J-band spectrum with the H and K-band spectra to derive a power law index of -5.15 ± 0.02. The power-law suggests that the radiation increases towards to J. The steeper slope in GRS 1915+105 than in the Be star DM +49 3718 indicates that either the dereddening is wrong or that there is a significant J-band emission: from the binary components, disc or a star, or from an unknown contributor (or from both). Moreover, during flares the dominant IR component is due to synchrotron emission (Fender and Pooley, 1998; Mirabel *et al.*, 1998) and it is possible that such emission could persist at a much lower level during the quiescence.

3. Conclusions

Our comparison of the IR spectra of the microquasar GRS 1915+105, the prototype X-ray binary Sco X-1 and the Be star DM +49 3718 give the following results:

- The spectral characteristics of the Be star differ from those of Sco X-1 and GRS 1905+105.
- The main difference between the Sco X-1 and GRS 1915+105 spectra is the He II line at 2.189 μm. In Sco X-1, the He II line is persistent and is most likely related to irradiation of the accretion disc around the compact object. In GRS 1915+105, the He II emission line is transient and is observed only during the X-ray bursting activity.
- The other difference between the Sco X-1 and GRS 1915+105 K-band spectra is the weaker He I line at 2.058 μm, which indicates a higher luminosity class for the donor star to GRS 1915+105.

A more detailed discussion can be seen in Harlaftis, Dhillon and Castro-Tirado (2001). These results are also consistent with the identification of the donor as a late-type star by Greiner *et al.* (2001).

References

Bandyopadhyay, R., Shahbaz, T., Charles, P.A., van Kerkwijk, M.H. and Naylor, T.: 1997, *Mon. Not. R. Astron. Soc.* **285**, 718.

Belloni, T., Méndez, M., King, A.R., van der Klis, M. and van Paradijs, J.: 1997, *Astrophys. J.* **479**, L145.

Boër, M., Greiner, J. and Motch, C.: 1996, *Astron. Astrophys.* **305**, 835.

Castro-Tirado, A.J., Brandt, S. and Lund, N.: 1994, *Astrophys. J. Suppl.* **92**, 469.

Castro-Tirado, A.J., Geballe, T.R., Lund, N., *et al.*: 1996, *Astrophys. J.* **461**, L99.

Chaty, S., Mirabel, I.F., Duc, P.-A., Wink, J.E. and Rodriguez, L.F.: 1996, *Astron. Astrophys.* **310**, 825.

Eikenberry, S.S., Matthews, K., Morgan, E.H., Remillard, R.A. and Nelson, R.W.: 1998, *Astrophys. J.* **494**, L61.

Fender, R.P. and Pooley, G.G.: 1998, *Mon. Not. R. Astron. Soc.* **300**, 573.

Greiner, J., McCaughrean, M.J., Cuby, J.G., Castro-Tirado, A.J. and Mennickent, R.: 2001, These Proceedings.

Harlaftis, E., Dhillon, V.S. and Castro-Tirado, A.J.: 2001, *Astron. Astrophys.*, in press.

Howarth, I.D.: 1983, *Mon. Not. R. Astron. Soc.* **203**, 301.

Martí, J., Mirabel, I.F., Chaty, S. and Rodríguez, L.F.: 2000, *Astron. Astrophys.* **356**, 943 .

Mirabel, I.F., Dhawan, V., Chaty, S., Rodríguez, L.F., Martí, J., Robinson, C.R., Swank, J. and Geballe, T.R.: 1998, *Astron. Astrophys.* **330**, L9.

van Kerkwijk, M.H., Charles, P.A., Geballe, T.R., *et al.*: 1992, *Nature* **355**, 703.

ON THE MASS-DONATING STAR IN GRS 1915+105

STEPHEN S. EIKENBERRY
Cornell University

REBA M. BANDYOPADHYAY
NRL/NRC

Abstract. Geometric constraints and the lack of eclipses in GRS 1915+105 are used to place limits on the mass of the companion star. We rule out any companion with $M_s > 19.4 M_{\odot}$ for a black hole mass of $M_{BH} \simeq 30 M_{\odot}$

Keywords: X-ray:stars, black hole physics, GRS 1915+105, close binaries

We present an analysis of constraints on the companion mass for the Galactic microquasar GRS 1915+105. Using the known inclination angle and stability of the jet axis, we can rule out massive ($M_s > 3M_{\odot}$) companions with orbital axes not aligned with the jet axis. For an aligned orbital axis, we constrain the ratio of the stellar radius to the binary semi-major axis to be $R_s/a < 0.342$, based on the lack of X-ray eclipses. We then show that these constraints together with the X-ray luminosity, approximate black hole mass, and observations of the X-ray absorbing column density towards GRS 1915+105 rule out accretion via stellar wind from a massive companion, implying that accretion occurs via Roche-lobe overflow. In that case, the constraint on R_s/a rules out any companion with $M_s > 19.4 M_{\odot}$ for a black hole mass of $M_{BH} \simeq 30 M_{\odot}$. The lack of significant X-ray reprocessing in the infrared from GRS 1915+105 also provides constraints on the mass and temperature range of a Roche-lobe-filling mass donor.

These results are inconsistent with the massive companion claimed by Martí *et al.* (2001), but match well with the results of Greiner *et al.* (2001).

References

Martí, J., *et al.*: 2001, these Proceedings.
Greiner, J., *et al.*: 2001, these Proceedings.

Astrophysics and Space Science is the original source of publication of this article. It is recommended that this article is cited as: *Astrophysics and Space Science* **276** (suppl.): 39–40, 2001.

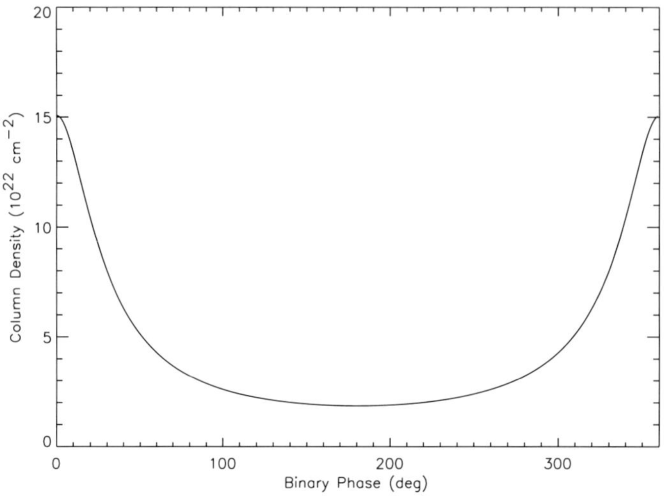

Figure 1. Orbital variation of observed X-ray absorption column density for wind-fed accretion from an LBV-type star. Assumes $\dot{M} = 2.65 \times 10^{-5}\ M_{\odot}\ \text{yr}^{-1}$, $a = 2.9$ AU, and $v_{wind} = 10^3 \text{kms}^{-1}$. Note that both the maximum amplitude of the absorption and its strong variations are inconsistent with observations of GRS 1915+105.

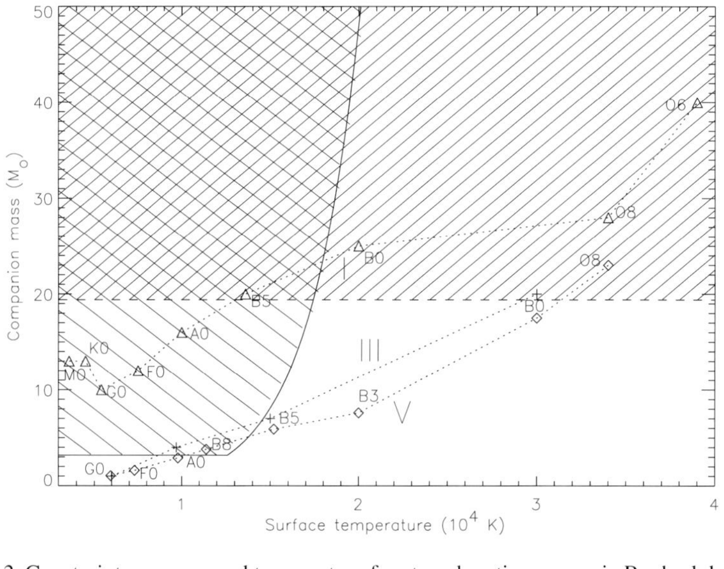

Figure 2. Constraints on mass and temperature for stars donating mass via Roche-lobe overflow. The hatched areas are regions which are ruled out by: (i) the limit on R_s/a (bounded by horizontal dashed line); (ii) limits on X-ray reprocessing (bounded by solid curve); (iii) the limit on $q > 0.1$ (allowing all stars with $M < 3M_{\odot}$). Roman numerals indicate stellar luminosity class for supergiant (I), giant (III), and main-sequence (V) stars. Assumes $M_{BH} = 30M_{\odot}$.

SUPERHUMPS IN BLACK HOLE X-RAY TRANSIENTS *

C.A. HASWELL and D.J. ROLFE
Department of Physics and Astronomy, The Open University
E-mail: C.A.Haswell@open.ac.uk

A.R. KING and J.R. MURRAY
Department of Physics and Astronomy, Leicester University

P.A. CHARLES
Department of Physics and Astronomy, University of Southampton

Abstract. The superhumps in black hole X-ray transients result from a modulation in the area of the accretion disc. We discuss implications of this for parameter determinations in LMXBs.

1. Introduction

Superhumps are a periodic photometric modulation with period $P_{\rm sh}$ slightly longer than $P_{\rm orb}$. They were discovered in extreme mass-ratio cataclysmic variables (CVs) harbouring eccentric accretion discs performing apsidal precession (Warner, 1995). CV superhumps are powered by tidally-driven modulations in viscous dissipation (Murray, 2000). Many black hole X-ray transients (BHXRTs) satisfy the extreme mass-ratio requirement for the development of an eccentric disc (Haswell *et al.*, 2000; Whitehurst and King, 1991) and several (e.g. GS 1124–68, GRO J0422+32) have shown superhumps (O'Donoghue and Charles, 1996; Uemura *et al.*, 2000).

The optical light in X-ray bright LMXBs is dominated by reprocessing: the optical emission from viscous dissipation is smaller by a factor of $\sim 10^{-3}$. This means superhumps in BHXRTs cannot be powered in the same way as those in CVs (Haswell *et al.*, 2000).

2. The Mechanism Powering LMXB Superhumps

Recently, eclipse mapping of V348 Pup, a superhumping CV, revealed clear evidence that the spatial distribution of the disc emission is modulated on $P_{\rm sh}$ (Rolfe, Haswell and Patterson, 2000). Numerical simulations of a precessing eccentric disc in an extreme mass ratio binary (Murray, 2000) reveal a modulation in shape on $P_{\rm sh}$ (Haswell *et al.*, 2000). Figure 1 shows a $\sim 10\%$ modulation in the area of the disc. This will clearly modulate $L_{\rm opt}$, as $L_{\rm opt} \propto A_{\rm eff} L_X$.

* This work was supported by grant F/00-180/A from the Leverhulme Trust.

Astrophysics and Space Science is the original source of publication of this article. It is recommended that this article is cited as: *Astrophysics and Space Science* **276** (suppl.): 41–42, 2001.

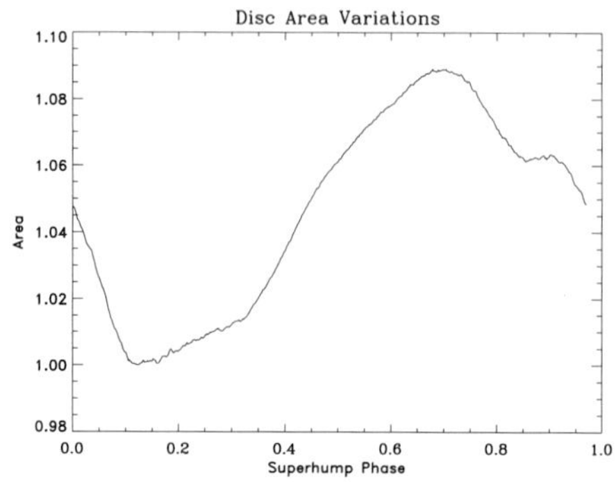

Figure 1. Disc surface area as it varies over the course of a single superhump cycle.

3. Implications

Superhumps occur only if the disc extends out to the radius where $\Omega_{\text{Kep}} = 3\Omega_{\text{orb}}$; this requires an extreme mass ratio. For neutron star LMXBs, superhumps consequently require $P_{\text{orb}} < 4.2\text{hr}$. For a black hole accretor there is no clear restriction on mass, so superhumps are not restricted to short P_{orb}. Thus we can infer black hole primaries for superhumping LMXBs with $P_{\text{orb}} > 4.2\text{hr}$ and main sequence mass donors (Haswell *et al.*, 2000).

The mass donor star also reprocess X-rays. Consequently, the modulation in disc shape is likely to cause the most prominent superhump modulation when the disc is viewed face-on. At higher inclinations the disc reprocessing area, A_{eff}, is foreshortened, diminishing the superhump modulation, while the orbital modulation due to the changing aspect of the mass donor will be increased. The disc shape modulation probably includes changes in vertical structure. For systems which are viewed edge-on, this may lead to modulations in occultation of the central emitting regions. X-ray dips attributed to such occultation are observed in several LMXBs. These dips have been used to infer P_{orb}, but they may be modulated at P_{sh} or the disc precession period.

References

Haswell, C.A., King, A.R., Murray, J.R. and Charles, P.A.: 2000, *MNRAS*, in press (astro-ph/0008367).

Murray, J.R.: 2000, *MNRAS* **314**, L1.

O'Donoghue D. and Charles P.A.: 1996, *MNRAS* **282**, 191.

Rolfe, D.J., Haswell, C.A. and Patterson, J.: 2000, *MNRAS* **317**, 759.

Warner, B.: 1995, in: *Cataclysmic Variable Stars*, Cambridge Univ. Press, Cambridge

Whitehurst, R. and King A.R.: 1991, *MNRAS* **249**, 25.

Uemura, M. *et al.*: 2000, *IAU Circ.* 7418.

NO
FUMAR

XTE J1550−564: A SUPERLUMINAL EJECTION DURING THE SEPTEMBER 1998 OUTBURST

DIANA HANNIKAINEN
Department of Physics & Astronomy, University of Southampton, Southampton SO17 1BJ, UK

DUNCAN CAMPBELL-WILSON, RICHARD HUNSTEAD and VINCE MCINTYRE*
School of Physics, University of Sydney, NSW 2006, Australia

JIM LOVELL, JOHN REYNOLDS and TASSO TZIOUMIS
ATNF, CSIRO, PO Box 76, Epping, NSW 1710, Australia

KINWAH WU**
MSSL, University College London, Holmbury St. Mary, Dorking, Surrey, RH5 6NT, UK

Abstract. In 1998 September, the X-ray transient XTE J1550−564 underwent a major outburst in soft and hard X-rays, followed by a radio flare. Australian Long Baseline Array images obtained shortly after the peak in the radio flare showed evolving structure. The components observed have an apparent separation velocity of $> 2c$.

Keywords: XTE J1550−564, radio continuum:stars

1. Introduction

The X-ray transient XTE J1550−564 was discovered simultaneously by the All-Sky Monitor on the Rossi X-ray Timing Explorer (Smith, 1998) and by the Burst and Transient Source Experiment on the Compton Gamma-Ray Observatory (Wilson *et al.*, 1998) on MJD 51063 (1998 Sept 7; MJD=JD−240000.5). Following the discovery of the radio counterpart (Campbell-Wilson *et al.*, 1998), XTE J1550−564 was monitored with the Molonglo Observatory Synthesis Telescope (MOST) at 843 MHz between MJD 51065 and 51092. A flare was recorded, peaking with an estimated flux density of 380 mJy on MJD 51078. Two Very Long Baseline Interferometry (VLBI) observations using the Australian Long Baseline Array (LBA) were performed 2–3 days after the MOST flare. In this paper we present the resulting images from these observations, and discuss briefly the implications of the evolving structure observed.

* Present address: ATNF, CSIRO, PO Box 76, Epping, NSW 1710, Australia
** Permanent address: School of Physics, University of Sydney, NSW 2006, Australia

Astrophysics and Space Science is the original source of publication of this article. It is recommended that this article is cited as: *Astrophysics and Space Science* **276** (suppl.): 45–48, 2001.

TABLE I
VLBI Observations

Date	T_{start} (UT)	T_{end} (UT)	Antenna	Frequency (GHz)
1998 Sept 24	02:17	06:08	DSS45 (34m)	2.29 & 8.4
Mid-point (MJD):	51080.7		Hobart (26m)	2.29 & 8.4
			Mopra (22m)	2.29
1998 Sept 25	22:30	09:00	DSS43 (70m)	2.29 & 8.4
Mid-point (MJD):	51081.7		Hobart (26m)	2.29 & 8.4
			Mopra (22m)	2.29
			ATCA (6×22m)	2.29

TABLE II
Flux densities at 2.29 GHz

	Flux density (mJy)			
Date	East	Central	West	Total
1998 Sept 24	71	–	20	91
1998 Sept 25	19	25	8	52

2. Observations and Results

The VLBI observations were carried out on 1998 Sept 24 and 25 and are summarized in Table I. The data post-correlation used the AIPS task FRING to find fringes and calibrate the fringe amplitude. Due to the small number of baselines, the Caltech program DIFMAP was used to model and source fit the data, and measure the flux densities.

Figure 1 shows the resulting VLBI images at 2.29 GHz from 1998 Sept 24 in the top panel and from 1998 Sept 25 in the bottom panel. Phase-referenced observations were not performed on either day so the zero coordinate in the images is arbitrary. The flux densities (with uncertainties of 10–20%) of the components are summarized in Table II.

The image from 1998 Sept 25 clearly shows an additional component compared to the image on 1998 Sept 24. As the image center is arbitrary, it is not obvious how to relate the components seen in the two images. However, two-point spectral indices derived from Australia Telescope Compact Array data at the time of the radio flare show a flattening just prior to the first VLBI observation from about $\alpha = -0.45$ to -0.1 ($S_\nu \propto \nu^\alpha$) over the interval 4.8–8.6 GHz (Hannikainen *et al.*, 2001).

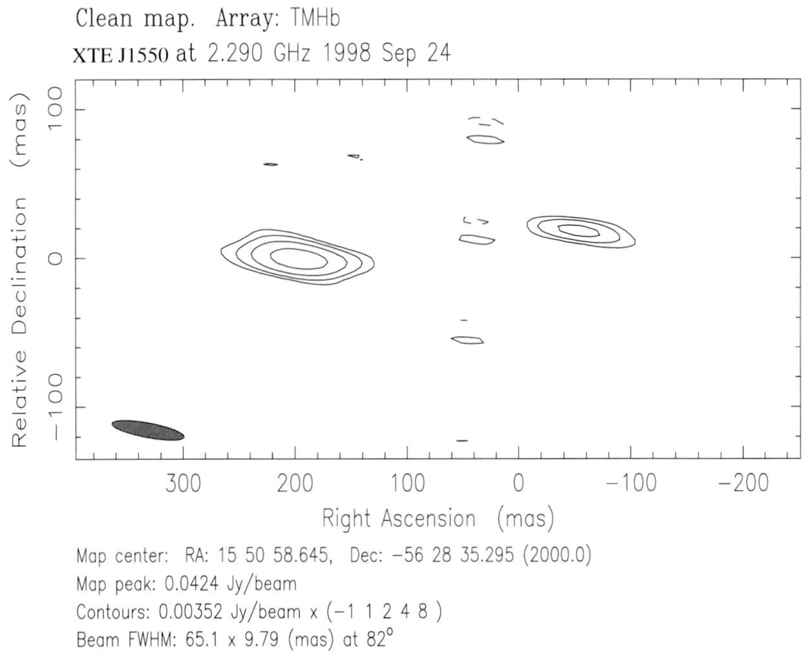

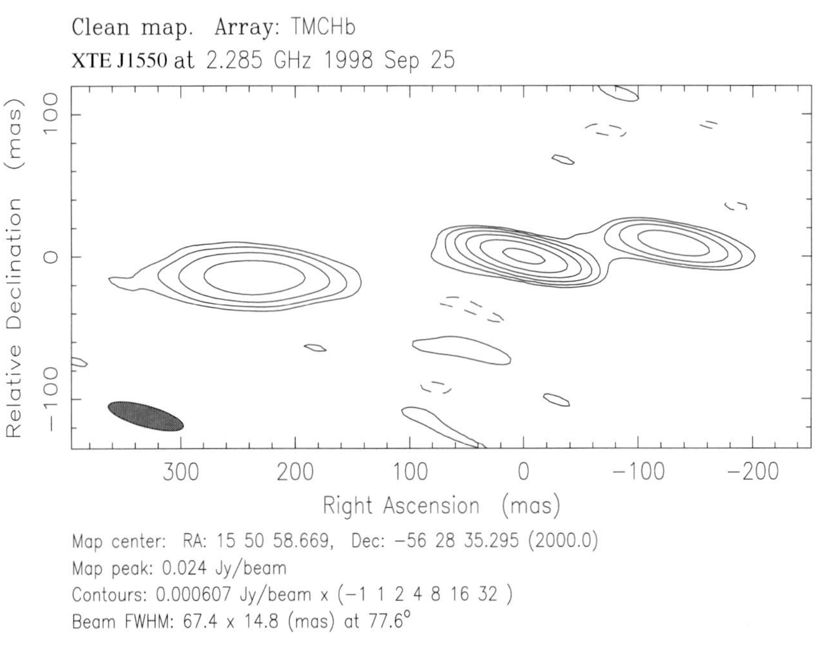

Figure 1. VLBI images at 2.29 GHz from 1998 Sept 24 (MJD 51080.7, top) and 1998 Sept 25 (MJD 51081.7, bottom). The zero coordinate is arbitrary. The beam is shown in the lower left-hand corner in each image.

This is very similar to the behavior seen in the 4.8–8.6 GHz two-point spectral index of GRO J1655−40 during the 1994 jet ejection episodes (Hannikainen *et al.*, 2000) and strongly suggests that a new optically thick outburst occurred. If this interpretation is correct, then we associate the central component in the bottom panel with the flaring core and thus the two components in the top panel, the ejecta, can be associated with the two outermost components in the bottom panel. This interpretation is reinforced by the 1998 Sept 25 single-baseline 8.4 GHz data which shows a new unresolved component not present in the 1998 Sept 24 data. This means that in one day the ejecta have moved apart by approximately 115 mas. At an estimated distance of 3.5–5 kpc (McIntyre, in preparation) this implies a separation velocity $> 2c$ for the ejecta, which makes XTE J1550−564 the fifth Galactic source to exhibit apparent superluminal motion.

3. Conclusions

We have presented two VLBI images of XTE J1550−564 from the time of the 1998 September outburst and showed that the source evolved over a period of 24 hours. The two outermost components seen in the 1998 Sept 25 image are associated with the two components in the 1998 Sept 24 image, implying an apparent separation velocity of $> 2c$.

Acknowledgements

The Australia Telescope Compact Array and the Mopra Telescope are part of the Australia Telescope which is funded by the Commonwealth of Australia for operation as a National Facility managed by CSIRO. We thank the staff of both the University of Tasmania, Hobart, and the Canberra Deep Space Communications Complex, Tidbinbilla. DH acknowledges the support of a PPARC postdoctoral research grant to the University of Southampton.

References

Campbell-Wilson, D., McIntyre, V., Hunstead, R. and Green, A.: 1998, IAUC 7010.

Hannikainen, D., Wu, K., Campbell-Wilson, D., Hunstead, R., Lovell, J., McIntyre, V., Reynolds, J. and Tzioumis, T.: 2001, *Proceedings of the 4th INTEGRAL Workshop*, ESA-SP, in press.

Hannikainen, D.C., Hunstead, R.W., Campbell-Wilson, D., Wu., K., McKay, D.J., Smits, D.P. and Sault, R.J.: 2000, *ApJ*, **540**, 521.

Smith, D.A.: 1998, IAUC 7008

Wilson, C.A., Harmon, B.A., Paciesas, W.S. and McCollough, M.: 1998, IAUC 7010

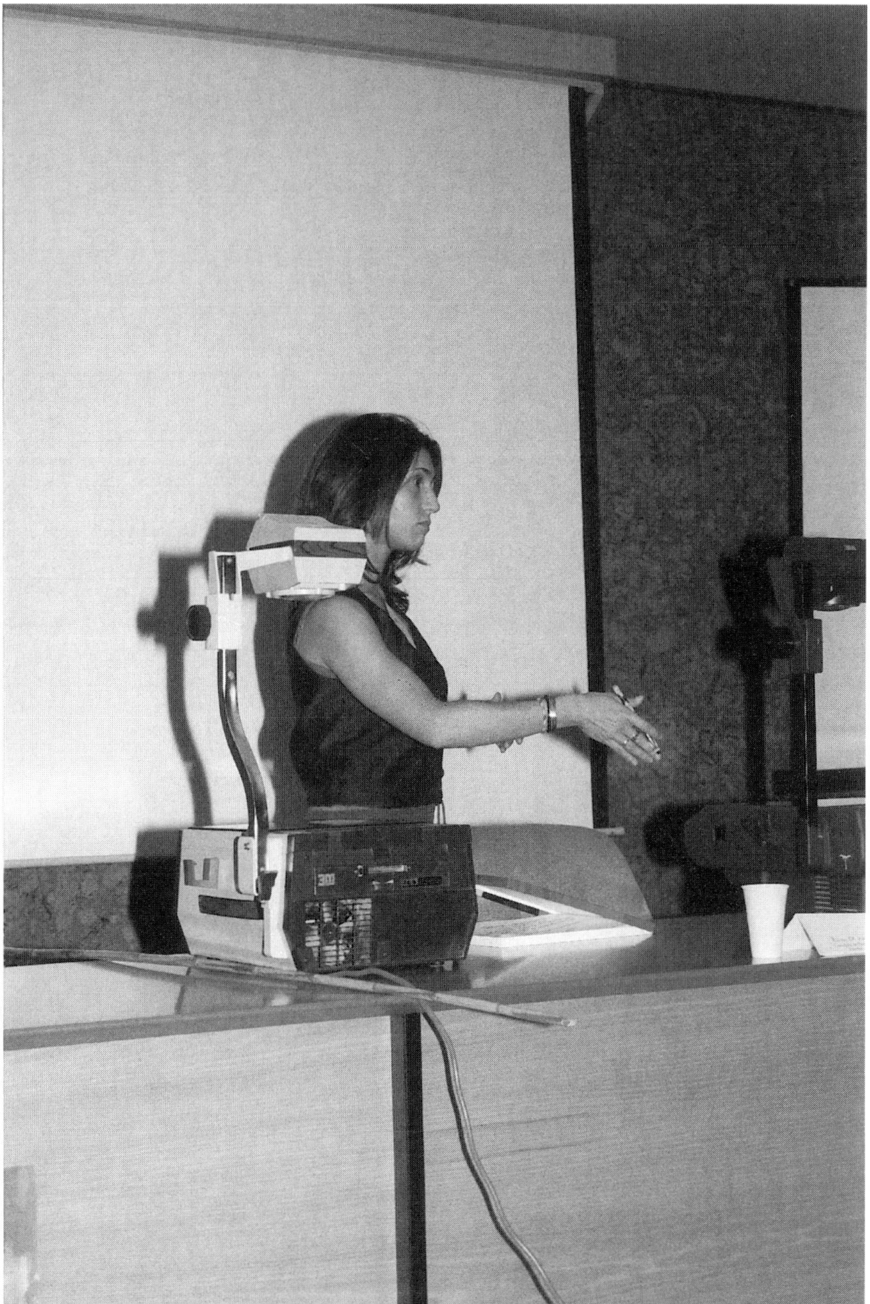

OPTICAL BEHAVIOUR OF XTE J1550-564 AND XTE J1859+226 FROM OUTBURST TO QUIESCENCE

C. SÁNCHEZ-FERNÁNDEZ
LAEFF-INTA, P.O. Box 50727, E-28080 Madrid, Spain; E-mail: celia@laeff.esa.es

A. J. CASTRO-TIRADO and A. GIMÉNEZ
IAA-CSIC, Granada (Spain) and LAEFF-INTA, Madrid, Spain

C. ZURITA and J. CASARES
IAC, Tenerife, Spain

N. LUND
DSRI, Kobenhavn, Denmark

Abstract. We present here the optical light curves of the Soft X-ray Transients XTE J1550-564 and XTEJ1859+226 from outburst to quiescence

Keywords: X-ray Transients, black hole candidates, optical observations

1. XTE J1550-564

XTE J1550-564 was first detected on Sep. 7.09 UT 1998 by the All-Sky Monitor on RXTE (Smith and Remillard, 1998). This object is the brightest SXT yet detected by RXTE. The source was also detected in hard X-rays (20-100 keV) by BATSE/CGRO (Wilson *et al.*, 1998). On the basis of its timing and spectral properties in X-rays (Cui *et al.*, 1999; Wijnands *et al.*, 1999) the source was classified as a black hole candidate. An optical counterpart with V$\sim$ 16 mag was proposed by Orosz *et al.* (1998). This object was confirmed as the optical counterpart by means of a 300-s optical spectrum which displayed prominent H_α emission as well as broad and weaker emission lines from H_β and HeII (Castro-Tirado *et al.*, 1998, Sánchez-Férnandez *et al.* 1999). Radio emission from the source was also detected (Campbell-Wilson *et al.*, 1998).

1.1. Observations and Discussion

Figure 1 displays the optical light curve of XTE J1550-564 based on our own measurements. The X-ray light curve from ASM public data as well as optical data from IAU Circulars have been included for completeness.

Astrophysics and Space Science is the original source of publication of this article. It is recommended that this article is cited as: *Astrophysics and Space Science* **276** (suppl.): 51–54, 2001.

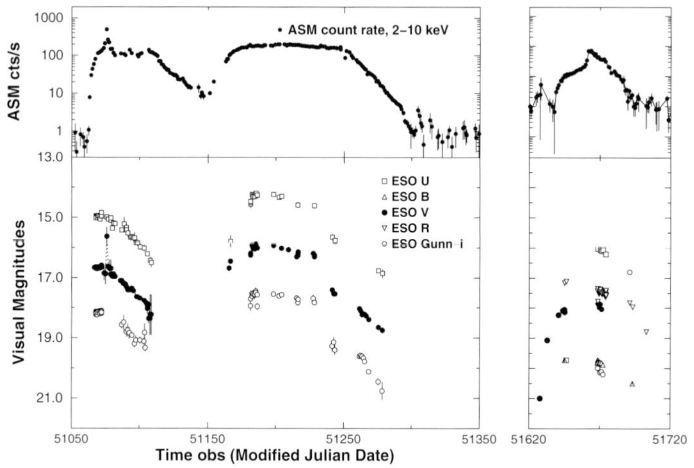

Figure 1. Optical/X-rays light curves of XTE J1550-564 showing the three outbursts referred in the text.The quiescent phase between the second and third outbursts has been omitted for clarity.

1.1.1. *The October 1998 outburst*

The soft X-rays light curve is characterized by a gradual brightening, a giant flare, which is also detected in optical wavelengths, a post-flare plateau ($\sim$ 25 days) and a subsequent decay. Amazingly, the optical emission does not follow the same pattern and the plateau phase is not present. The decay phase in the optical light curves takes place just after the flare occurrence, with decay rate $\sim$ 0.045 mag/day (Sánchez-Fernández *et al.*, 1999).

1.1.2. *The 1998/1999 outburst*

By the end of 1998, when the source appeared to be heading towards quiescence, it experienced a prolonged reflare which was observed at X-ray and optical wavelengths, and which lasted for about 6 months (Jain *et al.*, 2000, Sánchez-Fernández et al., 2000; see Figure 1). In this reflare, a new plateau appeared, both in soft X-rays and optical wavelengths, though with different durations ($\sim$ 75 days in soft X-rays; $\sim$ 50 days in optical). The optical decay rates are $\sim$ 0.053 mag/day.

1.1.3. *The April 2000 outburst*

After this secondary outburst, the source stayed in quiescence for about one year until it experienced a minor outburst in April, 2000 (Smith *et al.*, 2000). Stochastic flaring activity was observed during this outburst but no convincing evidence for the orbital period was found. An optical spectrum displaying emission lines characteristic of SXTs in outburst was obtained during this same epoch (Sánchez-Fernández *et al.*, 2001).

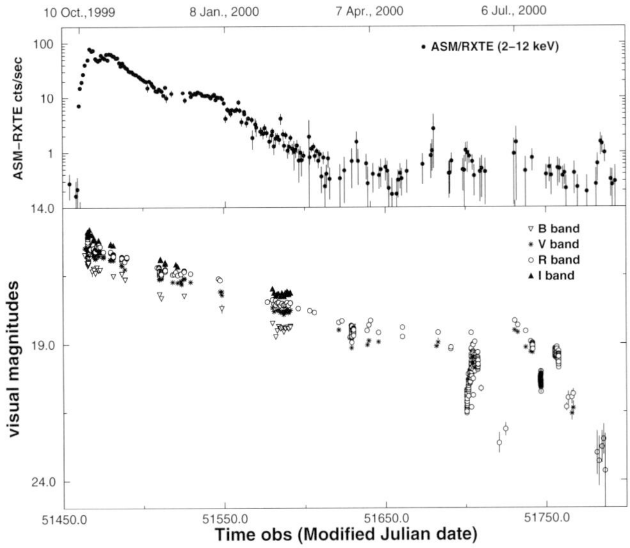

Figure 2. Optical/X-rays light curves of XTE J1859+226 from outburst to quiescence. Soft X-ray data have been obtained from RXTE public archive.

2. XTE J1859+226

XTE J1859+226 was first detected on Oct. 9.658 by ASM/RXTE (Wood *et al.*, 1999) at an intensity of 160 $\pm$ 15 mCrab (2-12 keV). On the basis of its timing and spectral properties in X-rays, the source was classified as a black hole candidate (see for example Markwardt *et al.*, 1999). The optical counterpart was soon discovered by Garnavich *et al.* (1999). A likely radio counterpart to the X-ray transient was also detected (Pooley *et al.*, 1999).

2.1. Observations and discussion

We present on Figure 2 the optical light curve of XTE J1859+226 based on our own measurements. X-ray light curve from ASM public data as well as optical data from IAU Circulars have been included for completeness.

The optical and soft X-rays light curves display the typical Fast Rise + Exponential Decay (*FRED*) shape, with rise-time (10%-90% peak) of $\sim$ 5 days in ASM data and an e-fold time of $\sim$ 23 days. The initial decay rate observed in optical is $\sim$ 0.018 mag/day. A secondary maxima is detected both in ASM and optical light curves $\sim$ 55 days after primary outburst. This secondary maxima should be due to a new injection of matter in the accretion disk as a response from the secondary star to x-ray heating. In addition, the optical light curve shows some additional peculiarities. Namely, a fast decay from R$\sim$19 to 21.5 mag 235 days after the extrapolated onset of outburst (Casares *et al.*, 2000), followed by 3 minioutbursts (see Zurita *et al.*, 2001a). There is no evidence of minioutuburst in the ASM light curve so an explanation based on X-ray reprocessing to these phenomena should be ruled out (see Zurita *et al.* 2001b for a further discussion).

2.1.1. *Evidences of an outside-in outburst*

By carefully examining the light curve at the begining of outburst we find evidences that the optical light curve had already peaked while the soft X-rays light curve was still in the rising phase. This behaviour has also been observed in some other SXTs (Aql X-1, GRO J1655-40, XN 2023+338, GRS 1124+68 and GRO J0422+32) and is thought to be the signature of an 'outside-in' outburst, in which a thermal instability in the disk propagates inward.

References

Campbell-Wilson, D., McIntyre, V., Hunstead, R., Green, A., Wilson, R.B. and Wilson, C.A.: 1998, IAU Circular 7010.

Casares, J., Rodrigues, P., Zurita, C., Shahbaz, T., Charles, P., Hynes, R., Abbott, T. and Hakala, P.: 2000, IAU Circular 7451.

Castro-Tirado, A.J., Duerbeck, H.W., Hook, I. and Yan, L.: 1998, IAU Circular 7013.

Cui, W., Zhang, S.N., Chen, W. and Morgan, E.H.: 1999, *ApJ* **512**, 793.

Garnavich, P.M., Stanek, K.Z. and Berlind, P.: 1999, IAU Circular 7269.

Jain, R.K., Bailyn, C., Orosz, J.A., McClintock, J.E., Sobczak, J.E. and Remillard, R.: 2000, *ApJ*, in press.

Markwardt, C.B., Marshall, F.E. and Swank, J.H.: 1999, IAU Circular 7274.

Orosz, J., Bailyn, C. and Jain, R.: 1998. IAU Circular 7009.

Pooley, G.G. and Hjellming, R.M.: 1999, IAU Circular 7278.

Remillard, R., Morgan, E., McClintock, J. and Sobczak, G.: 1998, IAU Circular 7219.

Sánchez-Fernández, C., Castro-Tirado, A.J., Duerbeck, H.W., Mantegazza, L., Beckmann, V. *et al.*: 1999, *AandA* **348**, L9.

Sánchez-Fernández, C., Zurita, C., Castro-Tirado, A.J. *et al.*: 2001, *AandA*, in press.

Smith, D.A. and Remillard R.: 1998, IAU Circular 7008.

Smith, D.A. *et al.*: 2000, IAU Circular 7399.

Wijnands, R., Homan, J. and van der Klis, M.: 1999, *ApJ* **526**, 33.

Wilson, C.A., Harmon, B.A., Paciesas, W.S. and McCollough, M.L.: 1998, *ApJ* **517**, 121.

Wood, A., Smith, D.A., Marshall, F.E. and Swank, J.: 1999, IAU Circular 7274.

Zurita, C., Casares, J., Charles, P., Rodriguez Gil, P., Sánchez-Fernández, C., Castro-Tirado, A., Shahbaz, T., Abbot, T. and Hakala, P.: 2001a, These Proceedings.

Zurita, C., Sánchez-Fernández, C., Casares, J. *et al.*: 2001b, *MNRAS*, in press.

MINIOUTBURSTS IN XTE J1859+226

C. ZURITA, J. CASARES, P. RODRÍGUEZ–GIL and T. SHAHBAZ
Instituto de Astrofísica de Canarias (IAC)

P.A. CHARLES
University of Southampton

C. SÁNCHEZ–FERNÁNDEZ
Laboratorio de Astrofísica Espacial y Física Fundamental (LAEFF-INTA)

A. J. CASTRO–TIRADO
Instituto de Astrofísica de Andalucía (IAA-CSIC) y LAEFF-INTA

T. ABBOT
Nordic Optical Telescope (NOT)

P. HAKALA
Turku University

Abstract. Photometry of the X-ray transient XTE J1859+226, obtained during outburst, showed at least 3 minioutbursts. During the second one, the source exhibited a 0.3 mag modulation with QPO flaring activity during the minimum.

Keywords: black hole physics, close binaries, XTE J1859+226

We show on Figure 1 the overall lightcurve of XTE J1859+226 in R-band and X-rays (2–12 keV), since the Oct 1999 outburst. The R light curve can be classified as *FRED*: 'fast–rise exponential–decay'. A 'glitch' is detected in the exponential decay phase $\sim 160\ d$ after the peak, exactly as in A0620-00 (Tsunemi *et al.* 1977). We have also detected at least 3 minioutbursts when XTE J1859-226 was approaching quiescence. *Minioutbursts* have only been detected in GRO J0422+32 (Chevalier and Ilovaiski, 1995) and GRS 1009-45 (Bailyn and Orosz, 1995). The source may have reached the quiescent state since August 20, 2000. B-band photometry on July 5, 2000, during the second minioutburst, showed a decay of about 0.3 mag for 3 hr. During the minimum, the source exhibited *QPO* flaring activity with 20% amplitude (Figure 2). We find a period of 21.7 ± 0.6 min, very close to that reported by Hynes *et al.* (1999). We note that flare activity was also found in an individual night during a *minioutburst* in GRO J0422+32 (Chevalier and Ilovaiski, 1995). The flaring ceased when the source returned to its predip level. On July 6, the *QPOs* disappeared and instead the source exhibited a smooth sinusoidal modulation with 0.10 mag amplitude. Assuming that the dip in the July 5 light curve is the primary minimum, it occurs exactly 1.3 d before a previous reported minimum (McClintock *et al.*, 2000). Therefore the orbital period must be

Astrophysics and Space Science is the original source of publication of this article. It is recommended that this article is cited as: *Astrophysics and Space Science* **276** (suppl.): 55–56, 2001.

a submultiple. We favor an orbital period of 0.65 *d* although more data is needed for confirmation.

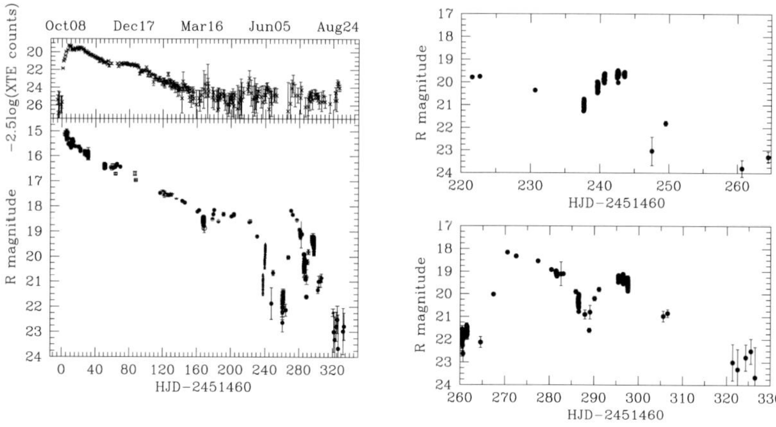

Figure 1. X-Ray and R-band long term light curve of XTE J1859+226 and minioutbursts.

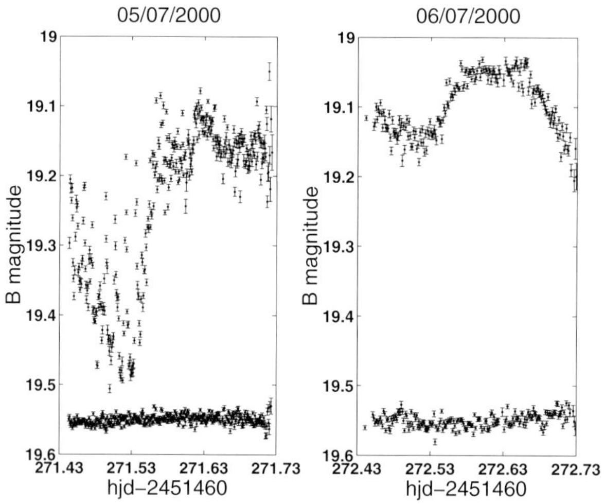

Figure 2. B-band light curves of XTE J1859+226 and comparison star during the second minioutburst.

References

Bailyn, C. D. and Orosz, J.A.: 1995, *ApJ* **440**, L73.

Chevalier, C. and Ilovaiski, S.A.: 1995, *A&A* **297**, 103.

Hynes, R., Haswell, C., Norton, A. and Chaty, S.: 1999, IAU Circ. 7294.

McClintock, J., Remillard, R., Heindl, W. and Tomsick, J.: 2000, IAU Circ. 7466.

Tsunemi, H., Matsuoka, M. and Takagishi, K.: 1977, *ApJ* **211**, L15.

OPTICAL OBSERVATIONS OF THE X-RAY TRANSIENTS XTE J1859 +226 AND XTE J1118+480

SERGEI S. GUZIY and ALEKSEY S. SHLYAPNIKOV
Kalinenkov Astronomical Observatory, Nikolaev State University, Nikolskaya 24, Nikolaev, 54030, Ukraine; E-mail: root@aok.mk.ua

Abstract. We present some results of observations on the X-ray transients XTE J1859+226 and XTE J1118+480 by means of the 0.7-metre telescope at the Kalinenkov Astronomical Observatory of Nikolaev State University. The observations of XTE J1859+226 were obtained in November 1999 and April 2000. The observations of the XTE J1118+480 started in April 2000. The light curves are presented here.

Keywords: X-ray transient, black hole candidate, XTE J1859+226, XTE J1118+480

1. Introduction

XTE J1859+226 was detected by RXTE All-Sky Monitor (ASM) on Oct 9, 1999. Extrapolation indicates that the outburst began around Oct 8.55 UT. The ASM spectrum was roughly Crab-like and did not show evolution (Wood *et al.*, 1999). An optical counterpart was detected on October 12 (Garnavich *et al.*, 1999).

XTE J1118+480 were detected by RXTE ASM on March 29, 2000. Retrospective ASM analysis indicated that the x-ray flux was slowly rising since March 5. The X-ray spectrum did not vary significantly during intense flares. Preliminary spectral analysis indicated a power law visible to 30 keV. The photon index was roughly 1.8, which is similar to Cyg X-1 in its hard state (Remillard *et al.*, 2000). The optical counterpart was detected on March 30 (Uemura *et al.*, 2000).

2. Observations and Results

We used the 0.7 m telescope of the Kalinenkov Astronomical Observatory of Nikolaev State University and a SBIG ST-7 CCD. The observations were obtained in the R-band, with a scale of the $0.73''$/pixel.

2.1. XTE J1859+226

Optical observations of XTE J1859+226 were started on November 3, 1999, 28 days after the discovery by Garnavich *et al.* (1999). CCD photometry of XTE J1859+226 was obtained on 2 nights in November 1999 (120 frames) and April

Astrophysics and Space Science is the original source of publication of this article. It is recommended that this article is cited as: *Astrophysics and Space Science* **276** (suppl.): 57–58, 2001.

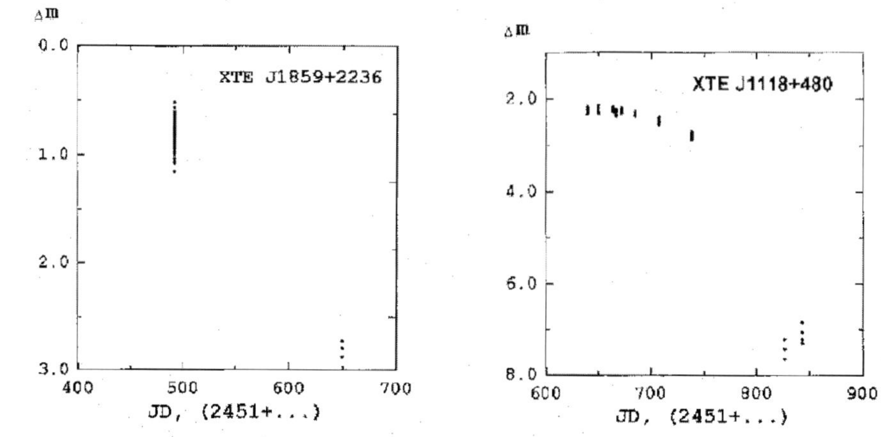

Figure 1. The light curve of XTE J1859+226 and XTE J1118+480 in the R-band.

2000 (3 frames). Exposure times were 50 s (November 1999) and 300 s (April 2000). After a standard reduction (correction for a flat field, hot pixels and a background), magnitudes were measured by means of aperture photometry. For comparison we used the reference star at R.A. = $18^h54^m41.^s59$ and December = $+22°38'57''$ (2000) with R = 15.36.

2.2. XTE J1118+480

Optical observations of XTE J1118+480 were started on April 3, 2000, 4 days after the discovery (Uemura *et al.*, 2000). CCD photometry in the R-band was obtained on 8 nights from April to July 2000. Exposure times of 30 s (April–May) and 60 s (June–July).We present here the results of optical observations of two X-ray transients, For comparison we used the star USNO-A2.0 1350-07926028 (R.A. = $11^h18^m40.^s75$, December = $+48°03'25.''9$) with R = 10.66.

Acknowledgements

We are grateful to the LOC and to the Instituto de Astrofísica de Andalucía for the support given in order to attend the Workshop.

References

Garnavich P. M, Stanek, K. Z. and Berlind, P.: 1999, IAU Circ. 7276.
Remillard, R. *et al.*: 2000, IAU Circ. 7389.
Uemura M., Kato, T. and Yamaoka, H.: 2000, IAU Circ. 7390.
Wood A. *et al.*: 2000, IAU Circ. 7274.

RECENT HUBBLE SPACE TELESCOPE RESULTS ON X-RAY NOVAE

R.I. HYNES
Department of Physics and Astronomy, University of Southampton, Southampton, SO17 1BJ, UK
E-mail: rih@astro.soton.ac.uk

C.A. HASWELL and S. CHATY
Department of Physics and Astronomy, The Open University, Walton Hall, Milton Keynes, MK7 6AA, UK

C.R. SHRADER
Laboratory for High-Energy Astrophysics, NASA Goddard Space Flight Center, Greenbelt, MD 20771, USA

W. CUI
Department of Physics, Purdue University, West Lafayette, IN 47907, USA

C.W. MAUCHE
Lawrence Livermore National Laboratory, L-43, 7000 East Avenue, Livermore, CA 94550, USA

Abstract. *RXTE* has recently discovered two new optically bright black hole candidates (BHCs): XTE J1859+226 and XTE J1118+480. We observed both with *HST*/STIS, the first outburst observations of X-ray novae with this instrument. We followed broad band spectral evolution, detected correlated X-ray/UV rapid variability and performed high resolution UV spectroscopy. We present a summary of initial results on the spectral energy distributions observed.

XTE J1859+226 is a classical X-ray nova showing an exponential decay lightcurve and quasi-periodic oscillations typical of BHCs (Cui *et al.*, 2000). The STIS/G230L spectrum defines the 2175 Å interstellar absorption feature $\Rightarrow E(B-V) = 0.58 \pm 0.07$. Hence we deredden the spectra to obtain the intrinsic SED. Early in the outburst SEDs are well fitted by a simple irradiated disc model ($T \propto R^{-3/7}$, Vrtilek *et al.*, 1990). The edge temperature, $T_{\rm out}$, remains constant so the edge of the hot region must move inward as the X-rays decline, possibly due to a cooling wave or self-shielding by the disc. The last spectrum is better fitted by a viscously heated disc model (Shakura and Sunyaev, 1973) with $T_{\rm out} \sim 8000$ K as predicted by disc instability models.

XTE J1118+480 lies at high latitude and suffers very little extinction making *EUVE* observations possible. Its unusual outburst only reached the canonical low/hard state of black hole candidates with a very low X-ray to optical flux ratio. Its IR–UV SED shows roughly constant f_ν (Hynes *et al.*, 2000). A weak Balmer jump in absorption implies some thermal emission, likely the disc. The spectrum of a disc extending to last stable orbit (A) clearly ruled out but the exact interpretation depends on an uncertain $N_{\rm H}$: Model B has the inner disc at a few hundred $R_{\rm s}$,

Astrophysics and Space Science is the original source of publication of this article. It is recommended that this article is cited as: *Astrophysics and Space Science* **276** (suppl.): 61–62, 2001.

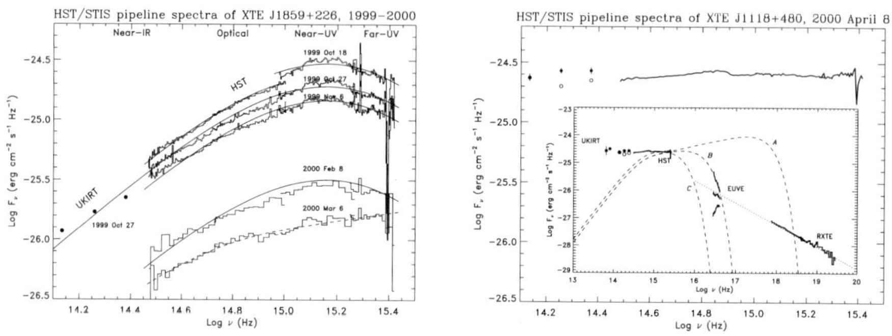

Figure 1. Left: *HST* and UKIRT observations of XTE J1859+226 showing the evolution of the SED through the outburst. Right: SED of XTE J1118+480 on 2000 April 8 from *HST* and UKIRT. Inset also shows simultaneous *RXTE* and *EUVE* data. Note that both plots use the same axes, so the shape of the SEDs can be directly compared.

consistent with thermal EUV if $N_{\rm H} = 1.0 \times 10^{20}\,{\rm cm}^{-2}$. Model C has the inner disc at a few thousand $R_{\rm s}$ and works if $N_{\rm H} = 0.75 \times 10^{20}\,{\rm cm}^{-2}$ and the EUV is non-thermal. The X-ray power-law is likely non-thermal, and has a spectral index typical of the low/hard state. A disc model cannot explain the flatness of the IR spectrum so another low energy component, likely synchrotron emission, is needed; see also contributions by Fender *et al.* and Markoff *et al.* for discussion of this component.

Acknowledgements

This work was supported by grant F/00-180/A from the Leverhulme Trust and is based on observations with the NASA/ESA *Hubble Space Telescope*, obtained at STScI, operated by AURA under NASA contract NAS5-26555. *EUVE* observations were made possible by a generous grant of Director's Discretionary Time. UKIRT is operated by the JAC on behalf of the UK PPARC. Thanks to the staff at these facilities.

References

Cui, W., Shrader, C.R., Haswell, C.A. and Hynes, R.I.: 2000, *ApJ* **535**, L123.

Hynes, R.I., Mauche, C.W., Haswell, C.A., Shrader, C.R., Cui, W. and Chaty, S.: 2000, *ApJ* **539**, L37.

Shakura, N.I. and Sunyaev, R.A.: 1973, *AandA* **24**, 337.

Vrtilek, S.D., Raymond, J.C., Garcia, M.R., Verbunt, F., Hasinger, G. and Kurster, M.: 1990, *AandA* **235**, 162.

THE MODERATE AND HIGH-SPEED PHOTOMETRY OF THE BLACK HOLE CANDIDATE XTE J1118+480

ELENA P. PAVLENKO, ELENA S. DMITRIENKO and NIKOLAI M. SHAKHOVSKOI
Crimean Astrophysical Observatory, Naychny , Crimea, Ukraine
E-mail: pavlenko@crao.crimea.ua

SERGEI YU. SHUGAROV, NATALY A. KATYSHEVA and IGOR M. VOLKOV
Sternberg Astronomical Institute, Moscow State University, Universitetskii pr. 13, Moscow, 119899 Russia; E-mail: shugarov@sai.msu.ru

Abstract. Our March–May photometry of the optical counterpart of the black hole candidate XTE J1118+480 during the plateau of outburst confirmed the existence of the known 0.17085 *d* variations in the R-band with amplitude $0^m.06$ and the 10-s QPOs, which are stronger in the I-band.

Keywords: close binaries, X-ray novae, XTE J1118+480

1. Introduction

We observed the optical counterpart of XTE J1118+480 (Remillard *et al.*, 2000; Uemura *et al.*, 2000) starting in late March 2000 for 23 nights in the R-band at the SAI Zeiss-600 and CrAO 38-cm Cassegrain telescopes in Crimea with a SBIG CCD ST-7 camera, and at the Zeiss-300 cm telescope of the Students SAI Observatory in Moscow. These observations have been carried out with exposure times of several tens of seconds. High-speed UBRI photometry was made at the CrAO AZT-11 1.25m telescope on April, 30 with 2-s and May, 15 with 5-s exposure times.

2. Observational Results

During the first 45 days after the March outburst, XTE J1118+480 has shown no prominent brightness decline (Figure 1, top panel of left part). The plateau-like shape of the outburst maximum lasted $\sim$ 100 *d* and then changed to a rapid fading. A similar behaviour was seen in A0620–00 (Shugarov 1976) and GRO J0422+32 (Chevalier and Ilovaisky 1995).

We confirm the existence of 0.17085 *d* light variations (Cook *et al.*, 2000; see Figure 1, middle and bottom panels of left part), and also confirm the 10-s QPOs (Patterson *et al.*, 2000) which we found to be more prominent in the I band. On the right part of Figure 1 the power spectra for the two runs of of the observations are

Astrophysics and Space Science is the original source of publication of this article. It is recommended that this article is cited as: *Astrophysics and Space Science* **276** (suppl.): 63–64, 2001.

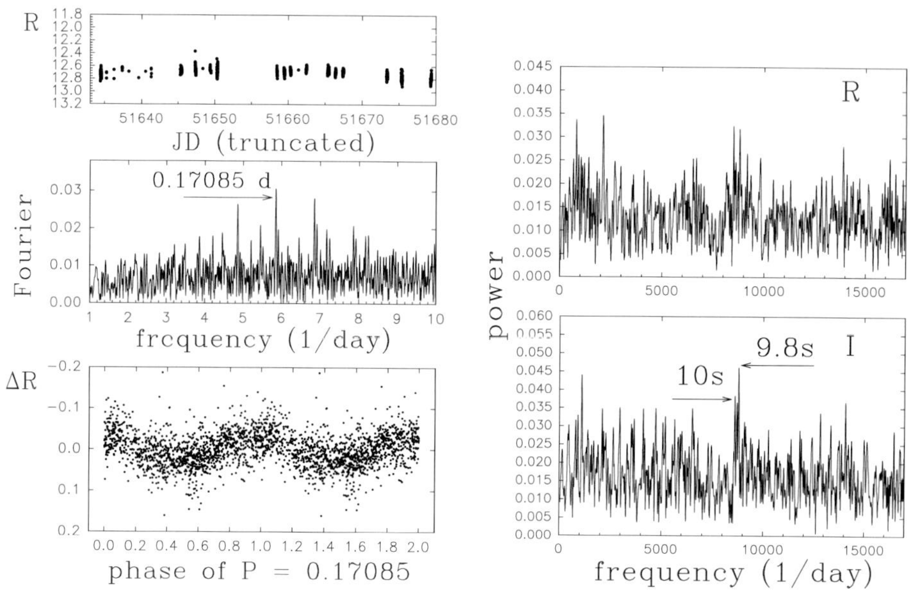

Figure 1. **Left:** Top panel: The behaviour of XTE J1118+480 at outburst maximum; middle panel: The power spectrum for all data; the arrow indicates the most significant period at 0.17085 days; bottom panel: Data, folded on this period. **Right:** The summed power spectra in R and I for data obtained on Apr 30.04 and May 15.05, 2000

given. The main peak is split between 9.08 s and 10 s. This may be caused by a change of these QPOs during two weeks.

3. Acknowledgements

The work is partially supported by the RFBR grants RFBR NN 00-15-96553, 00–02–16471 and 99-02-17589. EPP is thankful to the LOC and the Instituto de Astrofísica de Andalucía for supporting her visit to Granada.

References

Chevalier, C. and Ilovaisky, S.A.: 1995, *Astron. Astrophys.* **297**, 103.
Cook, L., Patterson, J., Buczinski, D. *et al.*: 2000, IAU Circ., 7397.
Patterson, J., Fried, R., Harvey, D. *et al.*: 2000, IAU Circ., 7412.
Remillard, R., Morgan, E., Smith, D. and Smith, E.: 2000, IAU Circ., 7389.
Shugarov, S.: 1976, *Var. Star* **20**, 251.
Uemura, M., Kato, T. and Yamaoka, H.: 2000, IAU Circ., 7390.

V404 CYG: 10 YEARS IN QUIESCENCE

ELENA P. PAVLENKO
Crimean Astrophysical Observatory, Naychny , 98409 Crimea, Ukraine
E-mail: pavlenko@crao.crimea.ua

YULIANA G. KUZNETSOVA
Kiev University, Physical Faculty, 64 Volodymyrska Str., Kiev 252017, Ukraine

SERGEI YU. SHUGAROV and VLADISLAV S. PETROV
Sternberg Astronomical Institute, Moscow State University, Universitetskii pr. 13, Moscow, 119899 Russia; E-mail: shugarov@sai.msu.ru

Abstract. We present the preliminary results of a long-term photometric study of V404 Cyg in quiescence. Its ellipsoidal light curves show the variations in both maxima and minima from year to year over 1992–2000. The most prominent are variations of the depth of minimum at phase 0.0 in V (1992–1997) and at phase 0.5 in R (1998–2000). The $\sim$ 6-*hr* QPOs persist but with variable amplitude.

Keywords: close binaries, X-ray novae, V404 Cyg

V404 Cyg is one of the brightest objects ($V \sim 18^m$) among known X-ray novae in the quiescent state. It is a black hole X-ray binary (Casares and Charles, 1994), where the mass of the black hole was estimated as $14 - -18M_\odot$ (Antokhina *et al.*, 1993) and $\sim 12M_\odot$ (Shahbaz *et al.*, 1994). V404 Cyg has a relatively long orbital period of 6.4714 days and 6-hour QPOs of variable amplitude (up to $0^m.3$) (Pavlenko *et al.*, 1996 and references therein).

Observations of V404 Cyg were carried out during 138 nights: In 1991–1997 at the 0.5 m telescope of the Crimean astrophysical observatory (CrAO) in a broad 'B + V' system; in 1998–2000 mostly at 0.38-m Cassegrain telescope (CrAO), also at 1.25 m, 0.6 m and 0.7 m telescopes of the Sternberg Astronomical Institute in a broad 'R + I' system (1998–1999) and in R (2000). All data were reduced to the standard V and R system.

The double-humped light curves are believed to be a result of the ellipsoidal effect. We folded our data on the orbital period using the ephemeris (Casares and Charles, 1994) separately in V and R. Phase 0.5 corresponds to the inferior conjunction of the black hole.

For each year we calculated a mean ellipsoidal light curve by least squares fitting the first two harmonics of the orbital period to the data (Figure 1). Both minimum and maximum magnitude varies from year to year in both spectral band. In V for 1992–1997 the maximum at phase 0.25 is more stable than that at phase

Astrophysics and Space Science is the original source of publication of this article. It is recommended that this article is cited as: *Astrophysics and Space Science* **276** (suppl.): 65–66, 2001.

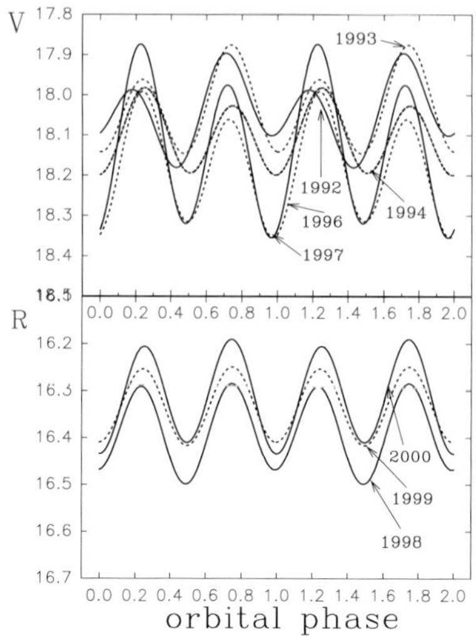

Figure 1. The V404 Cyg ellipsoidal light curves evolution in the period 1992–2000.

0.75, and the depth of minimum at phase 0.0 is more variable than that at phase 0.5. In R for 1998–2000 the minimal magnitude variations occur at the minimum at phase 0.0. Since V and R observations were not simultaneous, we cannot conclude whether this behaviour shows dependence on time or on the spectral band. We suggest that changes in accretion disk geometry and brightness with time could produce such lightcurves variations. We also studied the QPO evolution and found that these light oscillations are persistent over all quiescent epochs but display variable amplitude. In 2000 the typical time of the QPOs was 6.26 hours and mean amplitude $0^m.06$.

Acknowledgements

The work is partially supported by the RFBR grants RFBR NN 00-15-96553, 00-02-16471 and 99-02-17589. EPP is thankful to the LOC and the Instituto de Astrofísica de Andalucía for supporting her visit to Granada.

References

Antokhina, E., Pavlenko, E., Cherepaschuk, A. and Shugarov, S.: 1993, *Astronomy Reports*, **37**, 407.
Casares, J. and Charles, P. A.: 1994, *MNRAS*, **271**, L5.
Pavlenko, E. P., Martin, A. C. and Casares, J. *et al.*: 1996, *MNRAS*, **281**, 1094.
Shahbaz, T., Ringwald, F. A. and Bunn, J. C. *et al.*: 1994, *MNRAS*, **271**, L10.

Chapter II
PERSISTENT AND RECURRENT JETS AND OUTFLOWS

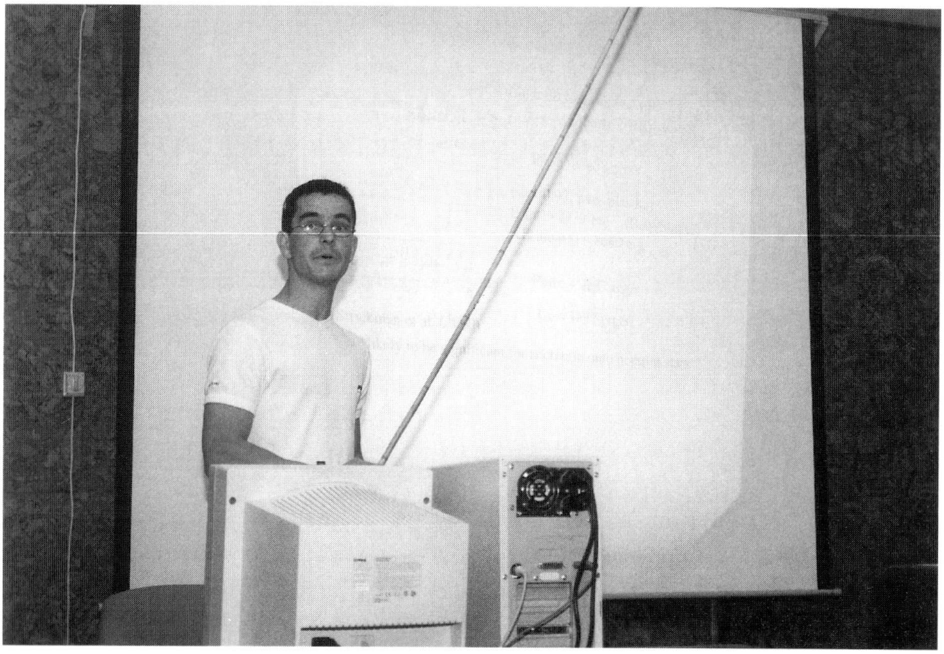

ENERGETICS OF JETS FROM X-RAY BINARIES

ROB FENDER
Astronomical Institute 'Anton Pannekoek' and Center for High-Energy Astrophysics, University of Amsterdam, Kruislaan 403, 1098 SJ Amsterdam, The Netherlands

Abstract. I discuss the energetics of synchrotron-emitting outflows, increasingly found to be present in many different classes of X-ray binary systems. It is shown that the outflow is likely to be comparable in power to the integrated X-ray luminosity, traditionally taken to be an indicator of the global mass-transfer rate. This is especially found to be the case in the (low/)hard states of black hole candidate systems. I conclude that jets are extremely important, energetically and dynamically, for the accretion process in the majority of known X-ray binary systems.

Keywords: black hole physics, neutron star, close binaries, ISM, jet, outflow

1. Introduction – the Ubiquity of Jets from X-ray Binaries

Radio emission as a property of X-ray binary systems (XRBs) has developed in recent years from being a rare and unusual facet of their broadband spectrum to being recognised as a ubiquitous property of several classes of XRBs. In the early 1990s, it was still (almost) possible to discuss individually radio-emitting X-ray binaries (e.g. Hjellming and Han, 1995), and numbers of systems in which collimated outflows, or jets, were directly observed could be counted on the fingers of one hand. As the 1990s drew on, more and more of these systems became resolved into such jets, and in the middle of the decade the apparent superluminal motions observed from GRS 1915+105 (Mirabel and Rodriguez, 1994; see also Fender *et al.*, 1999) and GRO J1655-40 (Tingay *et al.*, 1995; Hjellming and Rupen, 1995) established beyond doubt that XRBs could accelerate powerful flows to bulk velocities in excess of $0.9c$. By 2000 we are beginning to realise that jets may be rather more ubiquitous than previously imagined (e.g. Mirabel and Rodriguez, 1999; Fender and Hendry, 2000; Fender, 2001), and in Figures 1(a),(b) I summarise our (or at least my) current understanding of the empirical relation of radio emission to X-ray 'state' of the neutron star (NS) and black hole candidate (BHC) systems respectively. In addition to the relations indicated in these figures, it seems that discrete, radio-emitting ejection events (in which the radio spectrum rapidly becomes optically thin) are associated with state transitions (including outbursts of transients).

While some areas of the coupling between accretion flow and jet are still empirically uncertain (the Atoll sources, the Intermediate/Very High state for BHCs), it seems that all XRBs except the high-field X-ray pulsars will, under the right

Astrophysics and Space Science is the original source of publication of this article. It is recommended that this article is cited as: *Astrophysics and Space Science* **276** (suppl.): 69–77, 2001.

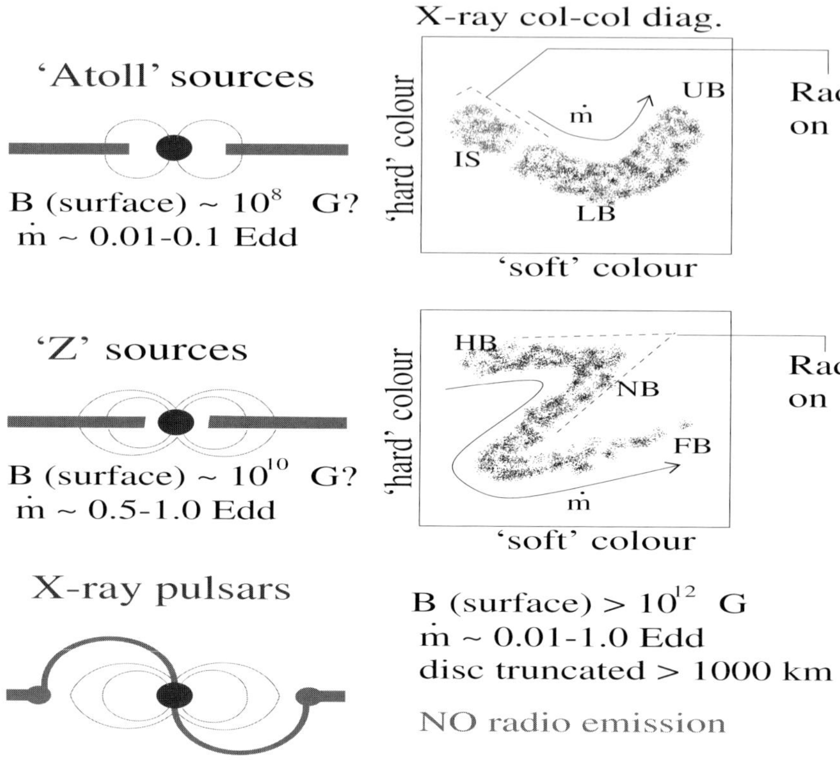

Figure 1. (a): A qualitative sketch of the relation of jets to accretion in the three 'types' of neutron star XRB. In the low-field 'Atoll' sources the accretion rate is believed to be < 10% Eddington, except possibly during rare transient outbursts (e.g. Aql X-1); the evidence is marginal so far but it appears such sources are 'radio on' when in the 'Island State' (IS) in the X-ray colour-colour diagram (CD). The Z sources are believed to be accreting at a much higher rate, near Eddington, and are 'radio on' when on the 'Horizontal Branch' and maybe also, at a lower level, the 'Normal Branch' in the CD. Note that for both Atoll and Z sources the estimates of surface magnetic field are very uncertain. Finally, in the high-field X-ray pulsars no radio synchrotron emission has ever been detected; possibly this is due to truncation of the accretion disc a long way from the neutron star.

conditions, produce a synchrotron-emitting jet. It is therefore an important question to address the significance of this jet, energetically and dynamically, for the process of accretion onto compact objects as a whole. One thing obvious from inspection of Figs 1(a),(b) is the apparent anti-correlation between the mass accretion rate, $\dot{m}$, traditionally estimated from X-ray studies alone, and the presence of a jet, in both NS and BHC systems (see also Belloni, Migliari and Fender, 2000 and Homan *et al.*, 2001 for further discussion).

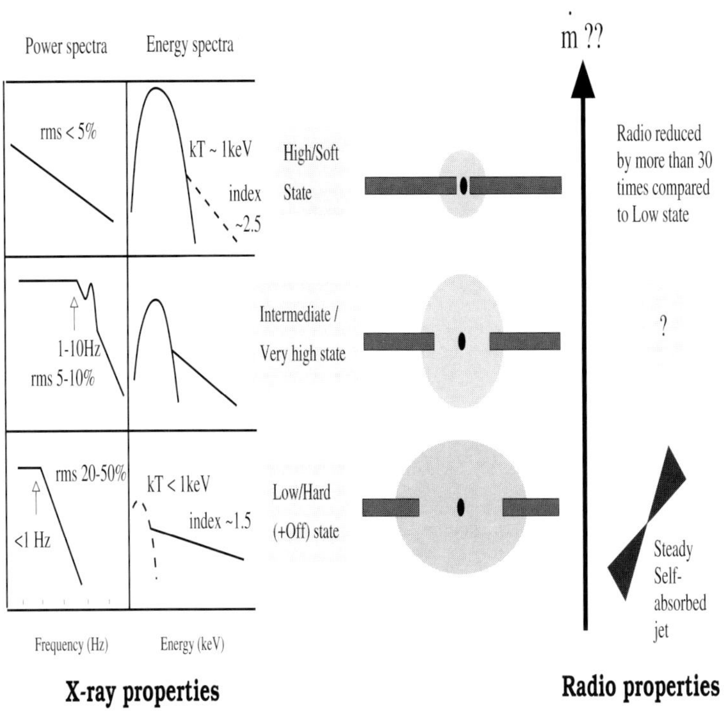

Figure 1. (b): The qualitative relation of radio emission to X-ray 'state' in BHC XRBs: the low/ hard state is found to produce a steady, flat- (or inverted-)spectrum jet, the soft state produces no detectable radio emission. The relation of $\dot{m}$ to these states is not certain (e.g. Homan *et al.*, 2001); nor are the radio characteristics of the relatively rare 'Intermediate/Very High' states well determined.

2. Energetics from 'Equipartition'

In cases where it is possible to associate a given synchrotron luminosity with a given volume, the minimum energy associated with the event can be estimated assuming approximate equipartition. In this situtation, the energy associated with the magnetic field is approximately equal to that associated with the electrons, and there is a corresponding equipartition magnetic field, B_{eq}. Below the equipartition field, the energy in particles ($\propto B^{-3/2}$) dominates; above equipartition, the energy in the field ($\propto B^2$) dominates. While there is little obvious physical justification for assuming equipartition, it is useful in that it represents the case of maximal radiative efficiency for a synchrotron-emitting plasma.

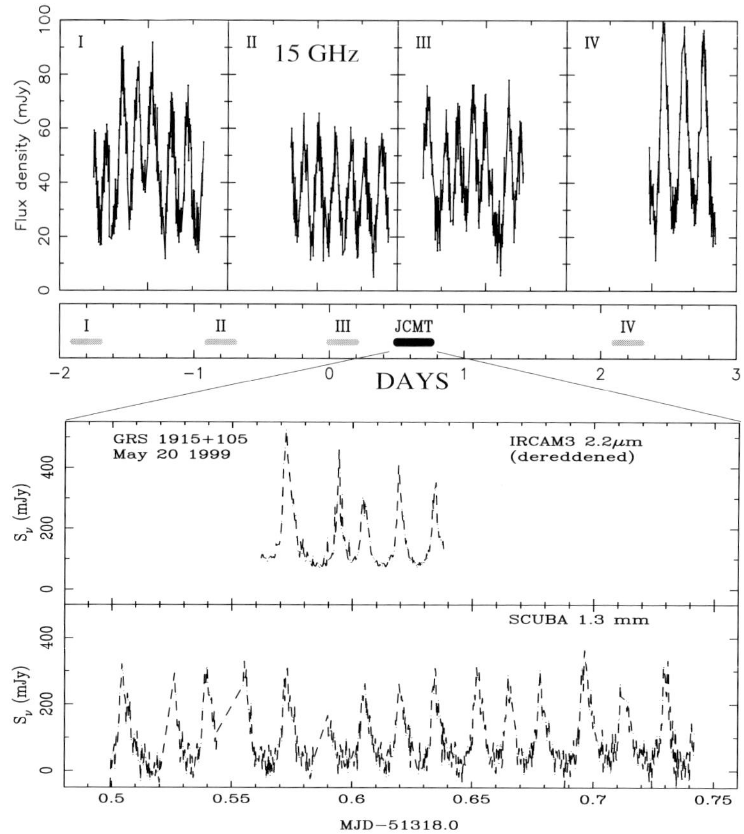

Figure 2. Simultaneous mm and infrared observations of synchrotron oscillations GRS 1915+105 (lower panel). Each event is believed to be associated with a discrete ejection, and the observations came in the middle of a period of more than a week of continuous oscillations as observed at 15 GHz (top panel); see Fender and Pooley (2000). Note that each of the four 15 GHz observations lasts ~ 0.1 hr.

One obvious way to associate luminosity and volume is by observing discrete ejection events sufficiently well that the rise time (and therefore maximum volume) and peak flux can be accurately measured. In Fig 2 we see an example of a sequence of repeated discrete ejection events, observed from radio through the mm band to the near-infrared band. Full details of the observations, and the application of equipartition estimations to these data, are found in Fender and Pooley (2000). The results of the equipartition calculations are presented in Table I. A more basic introduction to minimum energy requirements via equipartition can be found in e.g. Longair (1994).

From the estimates in Table I, it is clear that even in the minimum energy case, ie. a pair plasma with no significant bulk kinetic energy, the power required for the jet is 3×10^{38} erg s^{-1}. The integrated X-ray luminosity (presumably *not* coming from the jet ?) at this time was $\sim 10^{39}$ erg s^{-1}, so we find that even in this case

TABLE I

Calculation of (synchrotron) radiative luminosity, equipartition magnetic field, total energy, jet power and mass-flow rate for the oscillations reported here, given different physical assumptions. Γ is the bulk motion Lorentz factor; a filling factor of unity is assumed, otherwise the equipartition magnetic field would be high enough that radiative losses would be significant in the infrared. In these calculations a distance of 11 kpc and Doppler factors for relativistic bulk motion which are the same as those reported in Fender *et al.* (1999) are all assumed. Mass flow rate $\dot{M}_{\rm jet}$ and jet power P are based upon one ejection every 20 min. For more details, see Fender and Pooley (2000).

Case	Γ	L(erg s^{-1})	B$_{\rm eq}$(G)	$E_{\rm min}$(erg)	P (erg s^{-1})	$\dot{M}_{\rm jet}$(g s^{-1})	η
e^+:e^-	1	3×10^{37}	40	4×10^{41}	3×10^{38}	–	< 0.1
e^+:e^-	5	4×10^{39}	115	3×10^{43}	3×10^{40}	–	~ 0.15
p^+:e^-	1	3×10^{37}	40	4×10^{41}	3×10^{38}	2×10^{20}	< 0.1
p^+:e^-	5	5×10^{39}	115	1×10^{46}	8×10^{42}	4×10^{21}	< 0.02

the ratio of jet power to total X-ray luminosity, $P_{\rm J}/L_{\rm X} \sim 0.3$. If we assume that the ejections share the bulk relativistic motion of the major ejections, then it seems $P_{\rm J} > L_{\rm X}$.

Spencer (1996) presents a summary of energy estimates based on the same application of equipartition to radio varibility, for five XRB systems. In all cases the equipartition magnetic field lies in the range 0.01–1 Gauss, lower than that derived for the events from GRS 1915+105 considered above; however this may not be unreasonable considering that the peak emission from the longer events considered by Spencer would have originated in larger volumes. In all the five cases considered by Spencer (1996), the apparent jet power was at least equal to the integrated X-ray luminosity of the system.

In cases where the radio structure can be clearly resolved, the volume responsible for a certain luminosity can be estimated directly from images, and again equipartition applied. Spencer *et al.* (these proceedings) apply this method directly to the steady, resolved jets in Cygnus X-1, and Paredes *et al.* (2000) apply it to jets from the unusual system LS 5039, finding equipartition magnetic fields of 0.3 and 0.2 Gauss respectively. Both works further conclude that the energy associated with the resolved outflows is large, $> 10^{39}$ erg; however a major uncertainty in estimating the power in the outflow is associated with how continuously it must be generated.

3. Steady Jets – Energetics from Radiative Efficiency

Without large amplitude variability, or directly resolved jets, it is no longer possible to associate a given luminosity with a certain volume, and we must apply other arguments in order to estimate the total jet power. This is the situation we face in trying to estimate the power in jets from BHCs in the Low/Hard X-ray state, which produce a flat spectrum in the radio band with no large-amplitude variability (Fender, 2001). In this case we may estimate the total jet power by (a) carefully measuring the extent of the synchrotron spectrum which it produces, and (b) introducing a radiative efficiency, η, which is the ratio of total to radiated power (in the jet's rest frame). From this we can estimate the jet power as

$$P_{\rm J} \sim L_{\rm J}\eta^{-1}F(\Gamma, i)$$

where $L_{\rm J}$ is the total radiative luminosity of the jet, η is the radiative efficiency, and $F(\Gamma, i)$ is a correction factor for bulk relativistic motion with Lorentz factor Γ and Doppler factor δ, ($F(\Gamma, i) \sim \Gamma\delta^{-3}$ – see Fender, 2001).

3.1. Spectral extent and radiative luminosity

Starting from the reasonable assumption that all the emission observed in the radio band is synchrotron in origin, we can try to see how far this spectrum extends to other wavelengths. Firstly, it should be made clear that most systems have not been observed at $\nu < 1$ GHz, and while some low-frequency turnovers have occasionally been observed, there are no reported cases of a complete cut-off to the synchrotron emission at low radio frequencies. In any case, while a low-frequency cut-off is important for estimating the mass of the ejecta in the (by no means certain) case that there is a proton for each emitting electron, the radiative luminosity is dominated by the high-frequency extent of the synchrotron spectrum. Since the spectrum is unlikely to have a spectral index $\alpha < -1$ (where $S_\nu \propto \nu^\alpha$), then the most important observation is simply determination of the highest frequency at which synchrotron emission is observed.

In Figure 3 the broadband radio–optical spectra of eight XRB systems in (low/) hard X-ray states is presented; in all cases the spectra hint at an extension of the synchrotron spectrum to the near-infrared or even optical bands. There is a large amount of additional evidence for such a high-frequency extent to the spectrum, each piece of which may be considered 'circumstantial' but when gathered together presents a compelling case (a lot of this evidence is presented in Fender, 2001). Note also that observations of GRS 1915+105 (admittedly not a 'typical' system!), such as those shown in Figure 2, unambiguously establish that the synchrotron spectrum can extend to the near-infrared band.

Our best recent case is the low/ hard state transient XTE J1118+480, which clearly shows excess emission at near-infrared and probably also optical wavelengths (Hynes *et al.*, 2000) and whose radio spectrum smoothly connects to a sub-mm detection at 850 μm (Fender *et al.*, 2001). In Fender *et al.* (2001) it is argued that in this case the synchrotron radiative luminosity is already $\geq 1\%$ of the bolometric X-ray luminosity. How important the total jet power is then depends on our estimates for the radiative efficiency, η.

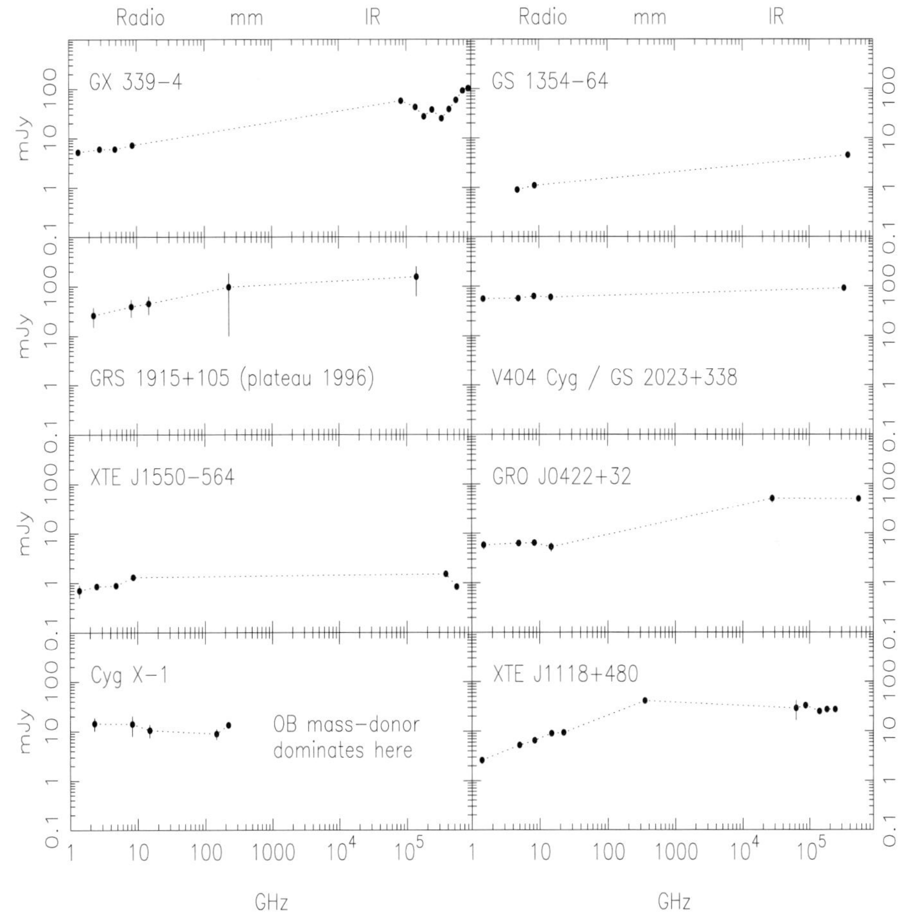

Figure 3. Broadband radio–mm–infrared–optical spectra of eight black hole candidate XRBs in hard X-ray states (apart from GRS 1915+105, all of these can be considered to be the canonical 'low/hard' X-ray state). There seems to be clear evidence in all cases, where observable (ie. not directly in Cyg X-1) for a continuation of the radio–mm synchrotron spectrum to at least the near-infrared band. Data and/or references for all systems in Fender (2001), except GX 339-4 (infrared–optical data from Corbel and Fender 2001), XTE J1550-564 (Corbel *et al.*, 2001) and XTE J1118+480 (Fender *et al.*, 2001).

3.2. Radiative efficiency

In Fender and Pooley (2000) a detailed study was made of the power required to maintain the repeated ejection events shown in Figure 2. In Table I, we also tabulate this radiative efficiency for the different cases considered; its maximum value is ~ 0.15. In one of the original works on the power in steady jets, Blandford and Königl (1979) determined that $\eta \sim \frac{1}{2}k_e(1 + \frac{2}{3}k_e\Lambda)^{-1}$, where $k_e \leq 1$ and $\Lambda = \ln(\gamma_{\rm e,max})/(\gamma_{\rm e,min})$, $\gamma_{\rm e,max/min}$ being the upper and lower bounds of the electron energy spectrum respectively. Since it seems likely that $(\gamma_{\rm e,max}/(\gamma_{\rm e,min})$ is in the range 10^2–10^4, then $\eta < 0.15$. Celotti and Ghisellini (2001) find a comparable value for the radiative efficiency of AGN jets.

4. Comparison of Jet Power to Accretion Luminosity

The best-studied cases of clearly resolved (in time) ejection events, such as those presented in Fig 2, seem to require that the ratio of jet power to X-ray luminosity, $P_J/L_X > 0.3$. A more general study of large outbursts from five different systems shows that, via the simple equipartition arguments outlined above, $P_J/L_X > 1$ (Spencer, 1996). Other applications of equipartition to XRB ejection events exist, and in all cases it seems that they imply the jet power is extremely significant.

Using a different, but not altogther unrelated, method of measuring the high-frequency extent of the synchrotron spectrum we find that for the BHC transient XTE J1118+480, $P_J/L_X > 0.01\eta^{-1}$, where $\eta < 0.15$ for BHC systems in the canonical low/ hard X-ray state. Since it seems that the relation of X-ray to radio flux in this state is comparable for all the systems (i.e. the similarities in the broadband spectra of the sources presented in Figure 3 extend to at least the X-ray band), then it seems reasonable to conclude that $P_J/L_X \geq 0.1$ for all BHC XRBs in the low/ hard X-ray state. Again it seems that the jet must be highly significant for the process of accretion.

5. Conclusions

Two different (but not entirely independent) approaches have revealed that a large fraction of the entire luminosity (radiative and mechanical) of some *classes* (this is important – we are no longer talking about individual 'oddballs') of X-ray binaries may be carried by the jet. This fraction is likely to be > 0.1 and may be > 0.5, ie. the jet may even be the *dominant* power output channel. Since it is clear from studies of e.g. GRS 1915+105 that the disc–jet coupling is occurring in the inner few 100 km of the accretion flow, the extraction of such a large amount of energy (and presumably angular momentum) must have a dramatic effect on the accretion flow close to the compact object.

Based on this, it seems that models for accretion in X-ray binaries which at least parameterise such an outflow are needed. They are not entirely non-existent – for GRS 1915+105 (again!) both Nayakshin, Rappaport and Melia (2000) and Janiuk, Czerny and Siemiginowska (2000) have begun to consider the effect of a moderately powerful outflow on the accretion process. A more radical step has been taken by Markoff, Falcke and Fender (2001; see also these proceedings) in which almost the entire broadband spectrum of the BHC low state (excepting thermal emission in the optical – UV range) is explained by jet emission, and in which it is suggested that the X-ray power-law may even be *synchrotron* emission. This would be contrary to the generally accepted view that the X-ray power law in the low state arises in Comptonisation of 'seed' photons in a hot, thermal, corona. It is further worth noting that once mass-loss via jets is considered, and the empirical connections between X-ray 'state' and jet formation, as outlined in Figs 1(a),(b) recognised, then it is no longer clear that the global mass transfer rate can be inferred from X-ray observations alone.

Acknowledgements

I would like to thank many of the participants at the Granada workshop for stimulating conversations, and Annalisa Celotti and Kieran O'Brien for reading through a draft of this work. In particular I would like to acknowledge many relevant conversations, over several years, with Ralph Spencer, Guy Pooley and the late Bob Hjellming.

References

Belloni, T., Migliari, S. and Fender, R.P.: 2000, *AandA* **358**, L29.
Blandford, R. and Königl, A.: 1979, *ApJ* **232**, 34.
Celotti, A. and Ghisellini, G.: 2001, in prep.
Corbel, S. and Fender, R.P.: 2001, in prep.
Corbel, S. *et al.*: 2001, *ApJ*, in press.
Fender, R.P.: 2001, *MNRAS*, **322**, 31.
Fender, R.P. and Hendry, M.A.: 2000, *MNRAS* **317**, 1.
Fender, R.P., Garrington, S.T., McKay, D.J., Muxlow, T.W.B., Pooley, G.G., Spencer, R.E., Stirling, A.M. and Waltman, E.B.: 1999, *MNRAS* **304**, 865.
Fender, R.P., Hjellming, R.M., Tilanus, R.P.J., Pooley, G.G., Deane, J.R., Ogley, R.N. and Spencer, R.E.: 2001, *MNRAS*, **322**, L23.
Fender, R.P. and Pooley, G.G.: 2000, *MNRAS* **318**, L1.
Hjellming, R.M. and Han, X.: 1995, in: *X-ray binaries*, Lewin, W.H.G., van Paradijs, J. and van der Heuvel, E.P.J. (eds.), Cambridge University Press, Cambridge, p. 308.
Hjellming, R.M. and Rupen, M.P.: 1995, *Nature* **375**, 464.
Homan, J., Wijnands, R., van der Klis, M., Belloni, T., van Paradijs, J., Klein-Wolt, M., Fender, R. and Mendez, M.: 2001, *ApJS*, **132**, 377.
Hynes, R.I., Mauche, C.W., Haswell, C.A., Shrader, C.A., Cui, W. and Chaty, S.: 2000, *ApJ* **539**, L37.
Janiuk, A., Czerny, B. and Siemiginowska, A.: 2000, *ApJ* **542**, L33.
Longair, M.S.: 1994, in: *High Energy Astrophysics*, Vol. 2, Cambridge University Press, 2nd edition.
Markoff, S., Falcke, H. and Fender, R.P.: 2000, *AandA*, in press.
Mirabel, I.F. and Rodríguez, L.F.: 1994, *Nature* **371**, 46.
Mirabel, I.F. and Rodríguez, L.F.: 1999, *ARAandA* **37**, 409.
Nayakshin, S., Rappaport, S. and Melia, F.: 2000, *ApJ* **535**, 798.
Paredes, J.M., Martí, J., Ribó, M. and Massi, M.: 2000, *Science* **288**, 2340.
Spencer, R.E.: 1996, in: *Radio Emission from the Stars and the Sun*, Taylor, A.R. and Paredes, J.M. (eds.), *ASP Conf. Ser.* **93**, p. 252.
Tingay, S.J. *et al.*: 1995, *Nature* **374**, 141.

RADIO JETS IN LS 5039

JOSEP M. PAREDES and MARC RIBÓ
Departament d'Astronomia i Meteorologia, Universitat de Barcelona, Av. Diagonal 647, E-08028 Barcelona, Spain

JOSEP MARTÍ
Departamento de Física, Universidad de Jaén, Calle Virgen de la Cabeza 2, E-23071 Jaén, Spain

MARIA MASSI
Max Planck Institut für Radioastronomie, Auf dem Hügel 69, D-53121 Bonn, Germany

Abstract. LS 5039 is a high mass X-ray binary whose optical counterpart is a bright (V = 11) star of O7 V((f)) spectral type, at an estimated distance of 3.1 kpc. The radio emission from LS 5039 is persistent and its radio spectrum is clearly non-thermal. Here we report VLBA and VLA observations which reveal that LS 5039 is resolved into bipolar radio jets emanating from a central core. In addition, this object appears to be associated with one of the enigmatic γ-ray sources detected by EGRET on board the COMPTON-GRO satellite.

Keywords: LS 5039, 3EG J1824-1514, gamma-ray, jet

1. Introduction

It is known that only a small fraction ($\sim$ 10%) of the catalogued X-ray binaries are radio emitters. Some of them are members of the selected class of galactic microquasars, i.e., systems where the radio emission has been resolved into collimated jet structures. During the last years we have been involved in a search for new radio emitting X-ray binaries, looking specially for those with radio jet morphology. The method used for the identification of potential candidates is based on the cross-identification of different radio and X-ray catalogues under very strict selection criteria (Ribó, 2001). The main tools of this search are the NRAO VLA Sky Survey (NVSS) (Condon *et al.*, 1998) and the ROSAT All Sky Survey (RASS) (Voges *et al.*, 1999). After a careful selection process, several candidates were identified. One of them, the star LS 5039, soon focused most of our attention. In a series of VLA observations, LS 5039 appeared as a bright and unresolved radio source, with non-thermal synchrotron spectrum (spectral index $\alpha = -0.5$) and showed some evidences of variability, although with moderate amplitude (Martí et al., 1998). In the X-ray domain, Ribó *et al.* (1999) reported the absence of any periodic modulation on time scales of 0.02$-$2000 s and 2$-$200 d. In this paper, we present high resolution interferometric results on LS 5039 and propose its possible association with the EGRET source 3EG J1824-1514 (Paredes *et al.*, 2000).

Astrophysics and Space Science is the original source of publication of this article. It is recommended that this article is cited as: *Astrophysics and Space Science* **276** (suppl.): 79–82, 2001.

2. Milliarc Second Radio Jets

The interferometric observations of LS 5039 were carried out with the VLBA plus the VLA* at the 6 cm wavelength on 1999 May 8th. The data were calibrated using the standard procedures. The resulting map, uniformly weighted, is shown in Figure 1. This map clearly indicates that LS 5039 is a source of bipolar radio jets emanating from a bright central core. The jets extend over 6 milliarc second on the sky and account for 20% of the total 16 mJy flux density. There is a brightness asymmetry between the jet components, being the south-east component about two times stronger than the north-west component. If we assume this asymmetry to be due to Doppler boosting effects, then a lower limit for the jet velocity ($v \geq (0.15 \pm 0.04)c$) and an upper limit for the ejection angle ($\theta \leq 81^\circ \pm 2^\circ$) can be derived. All these results agree with LS 5039 being a new microquasar.

The radio emission of LS 5039 was at its typical persistent level at the time of our observation and during the weeks before, as inferred from the Green Bank Interferometer (GBI)** radio monitoring. The lack of a precursor flaring event suggests us that the jets may be a permanent feature of the source. The size of the emitting region and the observed flux density allow us to estimate a brightness temperature of $\sim 9.5 \times 10^7$ K, indicating synchrotron radiation. By assuming energy equipartition, we estimate a total energy content in relativistic electrons of $\sim 4.8 \times 10^{39}$ erg and a magnetic field of ~ 0.2 G.

3. High-Energy (> 100 MeV) γ-Ray Source

The Energetic Gamma Ray Experiment Telescope (EGRET) instrument on board the COMPTON Gamma Ray Observatory (CGRO) has discovered 271 persistent high energy γ-ray sources (Hartman *et al.*, 1999). However, about 65% of them remain unidentified. We explored the possibility that LS 5039 is related to one of these unidentified high energy γ-ray source by searching in the Third Egret Catalog. The position of LS 5039 is well inside the 95% confidence contour of 3EG J1824-1514, a persistent source of high energy γ-rays in the Galactic Plane. In addition to the position agreement, LS 5039 is the only X-ray source from the RASS Bright Source Catalog (Voges *et al.*, 1999) within this region of the sky, and is also the only one simultaneous X-ray and radio emitter. The combination of all these clues leads us to propose that the microquasar LS 5039 and the EGRET source 3EG J1824-1514 are very likely to be associated. The supernova remnant SNR G16.8–1.10 is also consistent in position with 3EG J1824-1514 (Romero *et*

* The VLBA and VLA are operated by the National Radio Astronomy Observatory (NRAO) with Associated Universities Inc. and are funded by the NSF.

** The GBI is a facility of the NSF operated by NRAO with support from the NASA High Energy Astrophysics program.

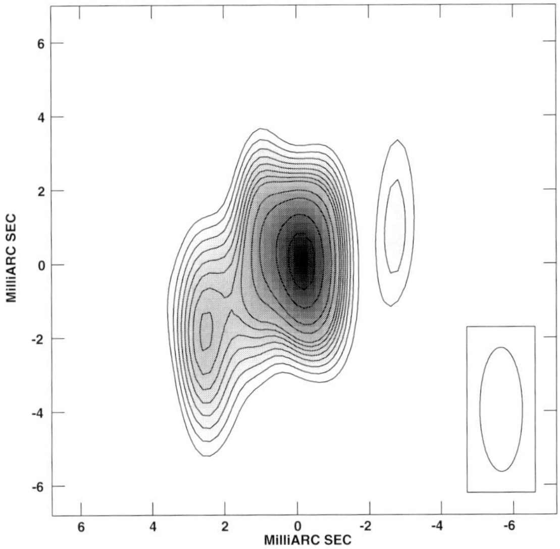

Figure 1. The milliarc sec radio structure of LS 5039 observed at 6 cm with the VLBA+VLA. The contour levels correspond to 6, 8, 10, 12, 14, 16, 18, 20, 25, 30, 40 and 50 times 0.085 mJy beam^{-1}, the map rms noise. The synthesized beam at the bottom right corner is 3.3×1.2 milliarc second^2.

al., 1999). In order to study any possible connection between LS 5039 and G16.8–1.10, we have estimated the proper motion of LS 5039 from several databases and our own observations. Our results are consistent with LS 5039 being a runaway star originated from this supernova remnant in a 10^5 yr old explosion.

Figure 2 shows the radio and γ-ray light curves of LS 5039 and 3EG J1824-1514 for the last ten years. Both, the γ-ray and radio emission observed reveal a rather persistent flux. Concerning the origin of the γ-ray photons, it has been proposed that they are created by inverse Compton scattering of the ultraviolet photons from the hot optical star and the radio emitting electrons. A detailed energetic study can be found in Paredes *et al.* (2000).

Acknowledgements

This project has been partly supported by DGESIC (PB97-0903) in Spain. MR acknowledges receipt of a fellowship from Generalitat de Catalunya (ref. 1999 FI 00199). MM and JM acknowledge support received from the European Commission's TMR program and Junta de Andalucía, respectively.

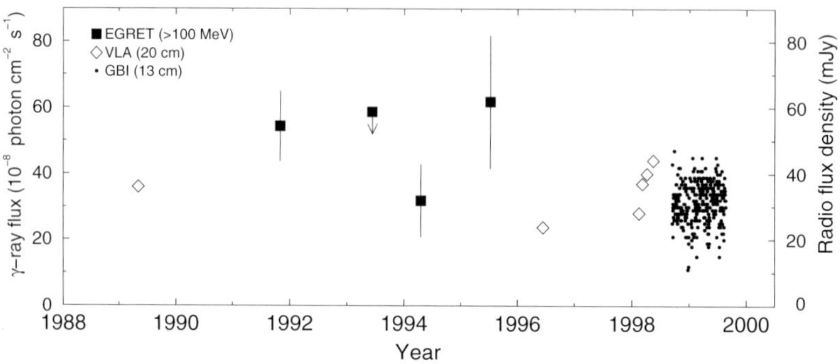

Figure 2. γ-ray and radio light curves of 3EG J1824-1514 and LS 5039. Both sources, with a persistent level of emission, are proposed to be the same object.

References

Condon, J.J., Cotton, W.D. and Greisen, E.W., *et al.*: 1998, *AJ*, **115**, 1693.
Hartman, R.C., Bertsch, D.L. and Boom, S.D., *et al.*: 1999, *ApJS*, **123**, 79.
Martí, J., Paredes, J.M. and Ribó, M.: 1998, *A&A*, **338**, L71.
Paredes, J.M., Martí, J., Ribó, M. and Massi, M.: 2000, *Science*, **288**, 2340.
Ribó, M., Reig, P., Martí, J. and Paredes, J.M.: 1999, *A&A*, **347**, 518.
Ribó, M. PhD thesis, University of Barcelona, 2001.
Romero, G.E., Benaglia, P. and Torres, D.F.: 1999, *A&A*, **348**, 868.
Voges, W., Aschenbach, B. and Boller, Th., *et al.*: 1999, *A&A*, **349**, 389.

E-BOPP

INTERNAL SHOCK MODEL FOR THE RADIO EMISSION OF MICROQUASARS

CHRISTIAN R. KAISER*
Max-Planck-Institut für Astrophysik, Garching, Germany

Abstract. The superluminal radio components observed in microquasars are usually interpreted as individual ejection events. We discuss some of the shortcomings of this model and propose the internal shock model in quasi-continuous relativistic jets as an alternative. This model can resolve the problems with the single ejection model and is in agreement with radio observations. We outline some further testable model predictions some of which are already confirmed by observations.

1. The Classical 'Blob' Picture *vs.* the Internal Shock Model

In this paper we concentrate on the most luminous events observed; events A in the scheme of Eikenberry (this proceedings). Since the discovery of superluminal motion in GRS 1915+105 (Mirabel and Rodríguez, 1994), the moving radio components were thought of as discrete ejections filled with a relativistic and magnetised plasma. The 'blob' picture is the simplest interpretation of the observations. However, in a detailed comparison of observational data with the predictions of this model several shortcomings become apparent. Firstly, unless it is confined by an external pressure, a blob of plasma will expand freely and in the process its radio synchrotron emission will dim rapidly. Instead, a rather gradual decline of the lightcurves is usually observed. It is of course possible to invoke a confinement mechanism, like the ram pressure of the external medium being pushed aside by the blob, to fit the observations. Nevertheless, this appears to be inconsistent with the observed ballistic trajectories of the blobs. Also, if the blobs are filled with a single population of relativistic electrons, then radiative losses will quickly lead to the formation of a steep spectral break (e.g. Atoyan and Aharonian, 1999). With time this break moves to lower frequencies. At a fixed observing frequency we would therefore expect to measure an almost constant spectral index for some time before the spectral break moves below this frequency and the blob vanishes altogether. Instead, a gradual steepening of the radio spectrum is observed (Mirabel and Rodríguez, 1994). Finally, the flux ratio of approaching and receding blobs is often found to be inconsistent with the blob model.

We developed a relativistic version of the internal shock model to apply to the radio observations of microquasars (Kaiser, Sunyaev and Spruit, 2000). This model

* Present address: Department of Physics and Astronomy, University of Southampton
E-mail: crk@astro.soton.ac.uk

Astrophysics and Space Science is the original source of publication of this article. It is recommended that this article is cited as: *Astrophysics and Space Science* **276** (suppl.): 85–88, 2001.

was originally invoked to explain the radio knots in the jets of M87 (Rees, 1978) and was later 'rediscovered' for Gamma Ray Bursts (Rees and Meszaros, 1994) and blazars (Ghisellini, 1999). The basic mechanism is that of a shock, caused by a fluctuation in the bulk speed of the jet material, which travels along a quasi-continuous, conical jet and accelerates electrons to relativistic velocities. In other words, the jet material is moving at a somewhat slower velocity than the shock which overtakes it and 'lights up' the jet.

2. The Model in Comparison with Observations

The shock continuously replenishes the supply of relativistic electrons to the emitting region. The region immediately behind the shock will be brightest. For this region radiative losses are yet unimportant and its luminosity, as it progresses down the jet, decreases because the strength of the magnetic field decreases away from the source centre. Also, the kinetic energy of the material driving the shock is limited and so the shock will eventually weaken. This way, the decline of the predicted radio lightcurve is rather gentle in agreement with observations. The final fading of the radio emission is caused by exhausting the kinetic energy of the material driving the shock.

The accelerated electrons are left behind by the shock and start to 'age' because of energy losses due to adiabatic expansion, synchrotron radiation and inverse Compton scattering of the CMB. In most radio observations the emission regions are not resolved and so a superposition of the radiation from a wide range of electron populations is observed. Starting from the current position of the shock along the jet, these populations were accelerated at successively earlier times and have developed spectral breaks at progressively lower frequencies. Superposition of these spectra results in a composite spectrum which is flat at low frequencies, steepens at higher frequencies where a larger jet region has 'dropped out' and finally cuts off at the frequency corresponding to the highest energies electrons are accelerated to by the shock. This process is illustrated in Figure 1. The whole pattern moves to lower frequencies as time passes and so the model can explain the observed gradual steepening of the radio spectrum.

The region just behind the shock will usually be brightest in this model as here a fresh supply of accelerated electrons is available. So most of the emission of the observed blobs will come from here. This implies that the measured advance speed is that of the shock rather than the bulk velocity of the jet material which is overtaken by the shock. However, the flux ratio of the approaching to the receding jet is set by the bulk velocity of the jet material. This explains the inconsistencies between the observed expansion speeds and the flux ratio.

In Kaiser *et al.* (2000) we show that the model successfully explains the radio observations of GRS 1915+105 during its radio flare in 1994 (Mirabel and Rodríguez 1994).

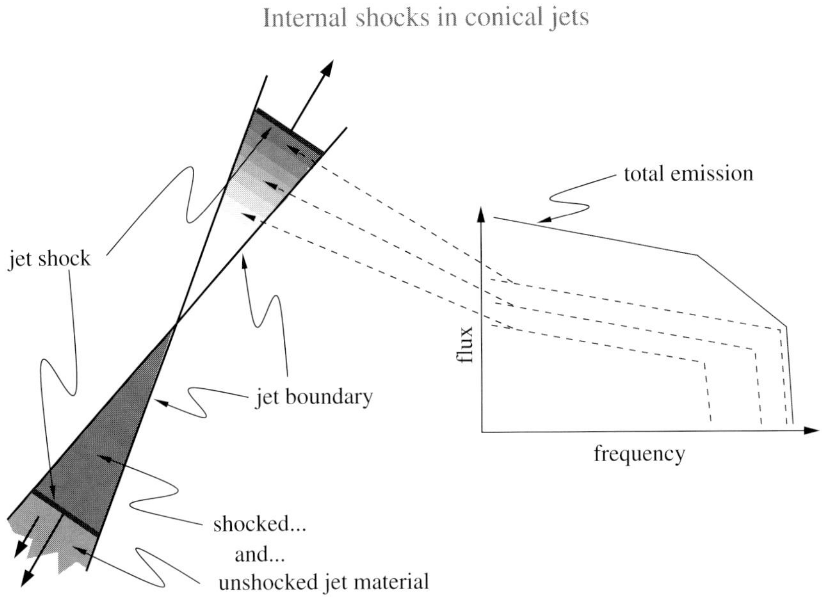

Figure 1. Illustration of the superposition of emission from jet regions of different 'age' (see text).

The internal shock model can be tested further in various ways. Any moving radio component should show steepening of its radio spectrum towards high frequencies before sharply cutting off. This may be detected in multi-frequency observations. However, limited resolution particularly at low frequencies will always lead to uncertainties whether or not the detected emission is caused by a single shock or is a blend of consecutive shocks in the jet.

The final fading of the radio emission is caused by the shock 'running out of steam'. Less and less electrons are accelerated and so the lightcurve steepens considerably. Because of relativistic effects, this appears to happen earlier in the approaching jet than in the receding jet. The model thus predicts that the flux ratio of the two should change significantly when the lightcurves begin to steepen. This effect may have been observed in a flare of GRS 1915+105 following the famous 1994 event (Rodríguez and Mirabel, 1999).

The jet in our model is expanding sideways. Thus the density of the jet material decreases and it should eventually be stopped by the ram pressure of the surrounding ISM. Although the bulk velocity of the jet material is slower than inferred from the measurement of superluminal motions, it is almost certainly supersonic. Therefore, at the end of the jet a strong shock should be formed and the former jet material should inflate a detectable radio lobe as in the extragalactic radio sources of type FRII (Fanaroff and Riley, 1974). It has been suggested that the diffuse radio emission region W50 around SS433 is just this lobe (Begelman *et al.* 1980).

Similar structures are detected around other objects (Mirabel and Rodríguez, 1999 and references therein).

Simple momentum balance arguments imply that the shock velocity along the jet should be constant (Kaiser *et al.*, 2000). However, in cases when we can observe the shock reaching the end of the jet we would expect a sudden deceleration. For the 'single blob picture' any deceleration should be rather gradual as material is swept up by the blob, unless there is a sudden dramatic increase in the density of the surrounding gas. In the only case any deceleration was ever observed in a microquasar (XTE J1748-288, Hjellming *et al.*, 1999), it was very rapid.

The internal shock model is based on a 'quasi-continuous' jet. In other words, the central engine must produce a jet for some time before the ejection of the material causing the shock. Even without the shock lighting up the jet it is likely that some low level radio emission from at least the inner regions of the jet is detectable. Mirabel and Rodríguez (1999) present high resolution radio maps of GRS 1915+105 during radio quiescence showing an elongated emission region aligned with the axis of the previously observed superluminal components.

References

Atoyan, A.M. and Aharonian, F.A.: 1999, *MNRAS* **302**, 253.
Begelman, M.C., Sarazin, C.L., Hatchett, S.P., McKee, C.F., and Arons, J.: 1980, *ApJ* **238**, 722.
Fanaroff, B.L. and Riley, J.M.: 1974, *MNRAS* **167**, 31.
Ghisellini, G.: 1999, *Astron. Nach.* **4**, 232.
Hjellming, R.M., Rupen, M.P., Mioduszewski, A.J., Smith, D.A., Harmon, B.A., Waltman, E.B., Ghigo, F.D., and Pooley, G.G.: 1998, *BAAS* **193**, 103.08.
Kaiser, C.R., Sunyaev, R., and Spruit, H.C.: 2000, *AandA* **356**, 975.
Mirabel, I.F. and Rodríguez, L.F.: 1994, *Nature* **371**, 46.
Mirabel, I.F. and Rodríguez, L.F.: 1999, *ARAandA* **37**, 409.
Rees, M.J.: 1978, *MNRAS* **184**, 61.
Rees, M.J. and P. Meszaros: 1994, *ApJ* **430**, L93.
Rodríguez, L.F. and Mirabel, I.F.: 1999, *ApJ* **511**, 398.

REVIEW ON PERSISTENT AND RECURRENT JET SOURCES

PH. DUROUCHOUX
CEA Saclay, DSM/DAPNIA/Service D'Astrophysique, 91191 Gif sur Yvette Cedex, France
E-mail: pdurouchoux@cea.fr

Abstract. In this paper I will review the different categories of Jet Sources. These sources can be separated into persistent and recurrent sources, and into extended and compact sources. GRS 1915+105 exhibits properties of both an extended and a compact source. Typically, new results obtained after the Microquasar Workshop in Paris in 1998 will be presented. A large number of these results are a consequence of multi-wavelength observations carried out since the Paris Workshop.

1. 1E 1740.7-2942, GRS 1758-258

1E 1740.7–2942 and GRS 1758-258 are the only two persistent bright sources and persistent hard ($\alpha^{-1.5}$) sources in the Galactic Bulge. Both sources exhibit a core and jet structure in the radio. No optical counterpart has been identified for either source, due to the high extinction, although an F dwarf and an early K giant have been found in the VLA error cycle for GRS 1758-258 (Martí *et al.*, 2000a).

Smith, Heindl and Swank (2000) weekly monitored 1E 1740.7–2942 and and GRS 1758-258, and found that each source has a long period (of the order of 500 days) and a short period (over 15 days) which can be interpreted as disk precession and orbital periods respectively. Such long orbital periods imply the need for a giant companion to fill the Roche lobe which could be, for GRS 1758-258 one of the stars seen by Martí *et al.* (2000).

The method used by Smith *et al.* (2000) to extract the short period is to substract a running average from the light curve and perform a Lomb-Scargle periodogram on the residuals. For 1E 1740.7–2942 they derive a slightly broad peak at 12.7 days allowing for a small period derivative. For the GRS 1758-258 the peak is centered at 18.45 days.

Another new interesting result is the recent analysis of a *Chandra* observation of 1E 1740.7–2942 (Cui *et al.*, 2000). Using HRGT/ACIS, they discovered a peculiar extended X-ray emission ($3''$ in length), roughly perpendicular to the radio lobe, but surprisingly, no X-ray jets were detected in the 0th order image. Moreover, VLA observations in 1997 of GRS 1758-258 in the hard state show a flat radio spectrum, which could indicate a possible conical jet geometry, related to synchrotron emission of relativistic electrons in an optically thick medium (which will be discussed below) and which can be a signature of a micro-jet.

Astrophysics and Space Science is the original source of publication of this article. It is recommended that this article is cited as: *Astrophysics and Space Science* **276** (suppl.): 91–96, 2001.

2. GRS 1915+105

This first galactic superluminal source was recently observed with the VLBA, while the BH binary was in varying X-ray states. It exhibited two different radio states (Dhawan, Mirabel and Rodriguez, 2000): (a) Plateau State, (b) Flare State. The plateau state is characterized by:

- Weak Soft and Strong Hard X-rays (called Hard Low State)
- Radio flux density in the range 10–100 mJy.
- Flat radio spectrum, usually associated with a compact jet (see below).

VLT observations in May 1999 and the use of the ISAAC spectro-imager in July 1999 (Martí *et al.*, 2000b) showed that GRS 1915+105 is a high mass X-ray system with an early type primary, as proposed by Chaty *et al.*, 1996 and Mirabel *et al.*, 1997. Moreover, due to its P Cygni profiles characterised by blue emission wings when the system is bursting, GRS 1915+105 must be surrounded by an expanding envelope. During the flare state, optically thin ejecta (up to 1 Jy) are expelled to thousands of AU (Dhawan *et al.*, 2000), whereas during the plateau state, optically thick ejecta, corresponding to a compact jet are expelled from the surroundings of the compact object. GRS 1915+105 is therefore the first source having both: compact and extended jets. This is a major result for the understanding of the compact objects and accretion disk instabilities. We must point out that absorption lines have been seen in GRS 1915+105 spectra (Doppler shifted He-like Fe edge) as well as in GRO J1655-40 (Kotani *et al.*, 2000) and could be a signature of superluminal sources.

Concerning the QPOs, GRS 1915+105 exhibits high frequency QPO (67 Hz), usually interpreted as severe effects due to general relativity in the inner accretion disk (Morgan *et al.*, 1999) and a low frequency QPO ($\sim 1 - 10$ Hz) interpreted by Tagger and Pellat (1999) as an instability occuring in the inner part of the disk. This accretion-ejection instability is responsible in GRO J16554-40 and GRS 1915+105 for the correlation between the spectral parameters of the inner disk and the QPO frequency (see Psaltis *et al.*, 1999; Muno *et al.*, 1999; Varniere *et al.*, 2000 and Rodriguez *et al.*, 2000).

3. XTE J1748-288

XTE J1748-288 was discovered by Smith *et al.* (1998) and its superluminal nature established by Hjellming *et al.* (1998). It presents an interesting example of jet interaction with the ambient medium. The jet slows down considerably at a later stage of evolution and at the same time, the tip of the jet brightens. The enhanced radio emission is probably produced by a shock resulting from the interaction between the jet and the surrounding medium. VLA observations in February 2000 showed a two sided ejection event (Hjellming *et al.*, 2000, in prep). XTE J1748-288 has been observed to brighten and decelarate from 0.93c to 0.23c, probably crashing into the circumstellar matter. Recent observations by ASCA of XTE J1748-288 in a quiescent state (Kotani *et al.*, 2000) showed that the jet blob colliding with the environmental matter was supposedly not related to the X-ray source or to the possible presence of a K line from highly ionized iron.

4. Elongated Jets versus Compact Jets

Optically thin medium: Relativistic electrons in a magnetic field in an optically thin medium produce coherent synchrotron radiation, with a radio spectrum with a negative spectral index. The radio emission is usually associated with a high degree of linear polarization. This scenario characterizes extended jets.
Optically thick medium: On the other hand, relativistic electrons in an optically thick medium with a magnetic field produce incoherent synchrotron radiation, whose radio spectral index is flat or inverted, and is associated with a low degree of linear polarization. This is characteristic of compact jets.

The persistent X-ray source GX 339-4 presents X-rays properties usually associated with BHCs. Is it a jet source? Although Fender *et al.* (1997) claimed they could have detected an elongation in the western direction, with the ATCA, there is still a question mark on whether GX 339-4 is a jet source or not. Corbel *et al.* (2000) have shown that there is a clear correlation between the radio and hard X-ray behavior. Moreover, they measured an inverted radio spectrum, with the ATCA, with a radio spectral index = +0.15, corresponding to an optically thick medium and a very small amount of linear polarization, leading to the conclusion that GX 339-4 is a compact plasma jet source (Corbel *et al.*, 2000; Fender *et al.*, 1999).

5. Cyg X-1 and Cyg X-3

Cyg X-1: GX 339-4 is not the first potential compact source. Cyg X-1 has a flat radio spectrum (Fender *et al.*, 2000a), and a similar hard X-ray-radio correlation. VLBA observations of Cyg X-1 (de La Force *et al.*, 2000) clearly show a compact jet of about 20 m.a.s extension.

Cyg X-3: Cyg X-3 is also a jet source and VLA observations at 4.995 GHz (Dhawan *et al.*, 2000) on April 3, 2000 and April 21, 2000 show a clear jet propagation.

6. SS433

W50 is one of the largest Galactic SNRs and the only one confirmed to harbor a twin jet binary system: SS433. In radio it is elongated in the east-west direction forming the radio ‘ears’.

Safi-Harb *et al.* (2000) pin down the non thermal nature of the X-ray emission of the eastern lobe as synchrotron radiation from highly energetic particles resulting from the interaction between the eastern jet and the SNR shell. Moreover, new X-ray observations of the western jet with RXTE and radio millimeter observations with MOPRA have probed the interaction of the western jet with a dense molecular cloud (Durouchoux *et al.*, 2000). Finally, extended features, though much smaller than for 1E1740-2942, have also been observed in SS433 at radio wavelengths (Paragi *et al.*, 1999), as well as circularly polarized radio emission (Fender *et al.*, 2000b).

7. Potential New Jet Sources

XTE J1859+226 is a possible eclipsing system due to its long term variation (Chaty *et al.*, 2000), with a periodicity of the order of 9 hours. Its spectral energy distribution is best fitted by a (i) simple X-ray heated disk model (ii) viscously heated disk model with an edge temperature of 8000 K. Its hard X-ray spectrum and high frequency QPOs make XTE J1859+226 a good BHC.

XTE J1118+48 exhibits a periodicity of 4.1 hours. Its spectral energy distribution and mini outbursts in the Low State, make this source an unusual object (Chaty *et al.*, 2000). ASCA observations in May 2000 (Yamaoka *et al.*, 2000) show: (i) rapid variability in X-rays, (ii) low frequency (0.11 Hz) QPO with a frequency shift (Revnivtsev reported 0.085 Hz a week earlier). A high-brightness core of 1 m.a.s size with an inverted radio spectrum (Dhawan, private communication) makes XTE J1118+48 a potential compact jet source.

XTE J1550-564: Multi frequency radio observations of XTE J1550-564 with MOST and ATCA during the Sept 1998 outburst and associated VLBI observations shortly after a radio flare showed a triple radio structure corresponding to ejections from a jet source (Hannikainen *et al.*, 2000). Moreover, a negative radio spectrum index and a small percentage of polarization (about 6 percent) are good arguments for XTE J1550-564 to be a compact jet source.

SAX J1819.3-2525 is an X-ray transient likely to be associated with the variable star V4641 Sgr. XTE observations of SAX J1819.3-2525 in Sept 1999 showed a flux of the order of 1.5 crab, and a huge H alpha line (6700 km s^{-1}) with a blue wing. VLA images obtained on Sept 16, 1999 (Hjellming, 1999) showed an extension of about $0.3''$ to the North. Moreover superluminal motion was detected after several observations (IAUC 7265). A major question remains: what is the distance of V4641/SAX J1819.3-2525? From radio measurements a distance lower than 1 kpc is derived (making this source the closest radio jet source) whereas a distance greater than 8 kpc is deduced from optical measurements.

LS 5039 is a HMXB with an O7 companion. This source, assumed to be a new micro-quasar system with apparent continuous ejection of relativistic plasma (Parades *et al.*, 2000), may be associated with the EGRET source 3EG J1824-1514.

VLA and VLBA observations resolved LS5039 into bipolar radio jets emanating from a central core; The jets extend over 6 m.a.s. on the sky and the spectral radio index is negative ($\alpha^{-0.5}$) which corresponds to a non thermal optically thin emission mechanism, implying that LS5039 is a new compact jet source.

8. Elongated SNRs

SS433 is not the only SNR associated with a jet source. A few good examples are listed in SNR catalogs. It is interesting to point out that G54.1+0.3, among few others, has an elongated radio map which fits quite well with the elongated X-ray map (Lu, 2000). We have found (Durouchoux *et al.*, 2000) a molecular cloud in the direction of G54.1+0.3, and CO(1-0) line enhancement at the location of the X-ray hot spot, which could be the signature of the interaction of the jets with a dense medium.

9. Conclusions

Since the last microquasar workshop in Paris in 1998, many new possible jet sources have been discovered. In addition it is now clear that there are two different categories of jets: extended jets, and compact jets. A major result is the radio observations of GRS 1915+105 with simultaneous compact and extended jets. Is GRS 1915+105 a special case or do all jet sources (or superluminal sources) begin with a compact ejection which, in some cases, is tranformed into extended jets? An accurate balance of the jet's energetics associated with numerous new observations could answer this question.

Acknowledgements

It is a pleasure to thank S. Chaty, V. Dhawan, D. Smith, F. Mirabel, W. Cui, and D. Hannikainen for permission to include in this review unpublished results, and R. Sood for useful comments on the original manuscript.

References

Chaty, S., *et al.*: 1996, *Astron. Astrophys.* **310**, 825.
Chaty, S., *et al.*: 2000, these Proceedings.
Corbel, S., Fender, R.P., Tzioumis, A.K., Nowak, M., McIntyre, V., Durouchoux, P. and Sood, R.: 2000, *Astron. Astrophys.* **359**, 251.
Cui, W., *et al.*: 2000, *Astrophys. J.*, in press (astro-ph/0009380).

de La Force, C., *et al.*: 2000, these Proceedings.
Dhawan, V., Mirabel, I.F. and Rodriguez, L.F.: 2000, these Proceedings.
Durouchoux, Ph., *et al.*: 2000, these Proceedings.
Fender, R., *et al.*: 1997, *Mon. Not. R. Astron. Soc.* **286**, L29.
Fender, R., *et al.*: 1999, *Astrophys. J.* **519**, L65.
Fender, R.P., Pooley, G.G., Durouchoux, P., Tilanus, R.P.J. and Brocksopp, C.: 2000a, *Mon. Not. R. Astron. Soc.* **312**, 853.
Fender, R., *et al.*: 2000b, *Astrophys. J.* **530**, L29.
Hannikainen, D., *et al.*: 2000, these Proceedings.
Hjellming, R., *et al.*: 1998, IAU Circ. 6934
Kotani, T., *et al.*: 2000, *Astrophys. J.* **543**, 133.
Lu, F.J., Aschenbach, B., Song, L.M. and Durouchoux, P.: 2000, these Proceedings.
Martí, J., *et al.*: 2000a, *Astron. Astrophys.* **356**, 943.
Martí, J., *et al.*: 2000b, *Astron. Astrophys.*, in press (astro-ph/0009442).
Morgan, E.H., Remillard, R.A. and Greiner, J.: 1997, *Astrophys. J.* **482**, 993.
Mirabel, F., *et al.*: 1997, *Astrophys. J.* **477**, L45.
Muno, M.P., Morgan, E.H. and Remillard, R.A.: 1999, *Astrophys. J.* **527**, 321.
Parades, J.M., Martí, J., Ribó, M. and Massi, M.: 2000, *Science* **288**, 2340.
Paragi, Z., *et al.*: 1999, *Astron. Astrophys.* **348**, 910.
Psaltis, D., *et al.*: 1999, *Astrophys. J.* **13**, L37.
Safi-Harb, S., Durouchoux, Ph. and Petre, T.: 2000, these Proceedings.
Smith, D., Heindl, W. and Swank, J.: 2000, in: *Proceedings of the* 4th *Integral Workshop*, ESA-SP, in press.
Smith, D., Levine, A. and Wood, A.: 1998, IAU Circ. 6932.
Rodriguez, J., Varnière, P., Tagger, M. and Durouchoux, P.: 2000, *Astron. Astrophys.*, in press.
Tagger, M. and Pellat, R.: 1999, *Astron. Astrophys.* **349**, 1003.
Varnière, P., Rodriguez, J. and Tagger, M.: 2000, *Astron. Astrophys.*, in press.
Yamaoka, K., *et al.*: 2000, these Proceedings.

ISO OBSERVATIONS OF THE GALACTIC CENTER BLACK HOLE CANDIDATES 1E 1740.7-2942 AND GRS 1758-258

PHILIP MALONEY
CASA, University of Colorado

SEAN COLGAN, ANGELA COTERA and DAVID HOLLENBACH
NASA Ames Research Center

Abstract. The Galactic Center black hole candidates 1E 1740.7-2942 and GRS 1758 are luminous hard X-ray sources, with $L_X > 10^{37}$ erg s^{-1}. It has been suggested that both of these sources are powered by Bondi-Hoyle accretion from the ambient dense interstellar medium. If this is the case, luminous far-infrared fine-structure line emission should be detectable from both sources. We have obtained *Infrared Space Observatory* observations of both sources and failed to detect emission at the predicted levels, implying that either (1) the hard X-ray luminosities averaged over the last ~ century are considerably lower (by factors of $\gtrsim 3-5$) than presently, or (2) the mass of gas in immediate proximity to the sources is smaller than in the models (which assumed the X-ray sources were embedded in dense ISM) by about the same factor. Option (2) seems more likely, implying that these sources are *not* fueled by Bondi-Hoyle accretion.

The most luminous hard X-ray sources in the Galactic center are 1E 1740.7-2942 and the nearly identical source GRS 1758-258, with 1–100 keV luminosities above 10^{37} erg s^{-1}. Both are also 'microquasars', with associated time-variable triple radio sources. In addition, 1E 1740.7 was suggested to be the 'Great Annihilator' source of narrow, time-variable annihilation-line radiation at 511 keV, a source whose existence now seems unlikely (Malet *et al.*, 1995; Purcell *et al.*, 1997).

The projected position of 1E 1740.7-2942 is directly atop a Galactic center molecular cloud (Bally and Leventhal, 1991), while GRS 1758 lies within a few hundred parsecs of the center of the Galaxy. It has been suggested that the prominence of both sources in the X-ray and radio is a result of their embedding in the dense interstellar medium (ISM) of the Galactic Center (Phinney, 1992; Chen, Gehrels and Leventhal, 1994), which would also account for the rapid annihilation of the positrons from the Great Annihilator source. This same medium might also power the sources through Bondi-Hoyle accretion, which would require in addition a low velocity of the sources (presumably stellar-mass black holes) relative to the ISM.

If these hard X-ray sources are embedded in dense ISM, then the irradiation of the surrounding gas should produce luminous far-infrared fine-structure line emission, which can be observed in spite of the large visual extinction towards the sources (Maloney, Colgan and Hollenbach, 1997). We obtained *Infrared Space*

Astrophysics and Space Science is the original source of publication of this article. It is recommended that this article is cited as: *Astrophysics and Space Science* **276** (suppl.): 97–98, 2001.

Observatory (ISO) observations of both sources (Colgan *et al.*, in preparation) which we compare with the model predictions of Maloney *et al.* (1997).

The sources were observed with *ISO* on 22 February and 1 March 1997, using both the Short Wavelength Spectrometer and the Long Wavelength Spectrometer. A number of positions were observed on and around both sources, 10 for 1E 1740.7-2942 and 18 for GRS 1758-258 (Colgan *et al.* discuss the data reduction in detail). The interpretation is complicated by the clear presence of unrelated emission in the vicinity of the sources, presumably from photdissociation regions. However, no strong line emission was detected from either source. The upper limits to line emission associated with the compact objects are typically factors of 3 – 5 below the model predictions.

This implies either (1) the hard X-ray luminosities of the sources, averaged over the thermal response time of the gas (of order 100 years) are considerably lower (by factors of $\gtrsim 3 - 5$) than presently, or (2) the mass of gas in immediate proximity to the sources is smaller than in the models (which assumed the X-ray sources were embedded in dense ISM) by about the same factor. (This limit on the gas mass is more stringent than that imposed by the failure to detect fluorescent iron $K\alpha$ emission at 6.4 keV: Churazov, Gilfanov and Sunyaev, 1996). The latter possibility seems more likely, implying that these sources are not fueled by Bondi-Hoyle accretion from the ISM.

Acknowledgements

This research was supported by the NSF under grant AST-9900871 and by *ISO* grant JPL 961636 to SETI.

References

Bally, J. and Leventhal, M.: 1991, *Nature* **353**, 234.
Chen, W., Gehrels, N. and Leventhal, M.: 1994, *Astrophys. J.* **426**, 586.
Churazov, E., Gilfanov, M. and Sunyaev, R.: 1996, *Astrophys. J.* **464**, L71.
Malet, I., *et al.*: 1995, *Astrophys. J.* **444**, 222.
Maloney, P.R., Colgan, S.W.J. and Hollenbach, D.J.: 1997, *Astrophys. J.* **482**, L41.
Phinney, E.S.: 1992, *Nature* **358**, 189.
Purcell, W.R., *et al.*: 1997, *Astrophys. J.* **491**, 725.

MID-INFRARED OBSERVATIONS OF GRS 1915+105 AND SS433

YAËL FUCHS[1], I. FÉLIX MIRABEL[1,2] and RICHARD N. OGLEY[1]
[1]*DSM/DAPNIA/Service d'Astrophysique, CEA/Saclay, Orme des Merisiers Bât. 709, 91191 Gif sur Yvette cedex, France;* [2]*IAFE/CONICET, Argentina*
E-mail: yfuchs@cea.fr

Abstract. We have observed in the mid-infrared (4–18 μm) the counterpart of the compact object GRS 1915+105, and the western jet of SS433. The images were carried out with the ISOCAM infrared camera on board of the Infrared Space Observatory (ISO). The mid-infrared photometry of GRS 1915+105 shows the presence of an additional contribution besides the synchrotron emission. The 15 μm images of the large-scale western lobe of the SS433/W50 nebula are compared to the radio and X-ray ones. They show infrared synchrotron emission on the western edge of SS433/W50 lobe.

1. GRS 1915+105

We have imaged the close environment of GRS 1915+105 with ISOCAM at two epochs, on 1996 April 28 and on 1997 October 24. The binary system appears in mid-infrared wavelengths as a faint point source, globally brighter in our 1997 observations than in the 1996 ones. No elongated structure was observed with 1.5″ pixel field of view, the best ISOCAM resolution. We obtained the photometry with several large-band filters between 4 and 18 μm, and have compared it to simultaneous observations at near-infrared, radio and X-ray wavelengths.

On 1996 April 28 GRS 1915+105 was in a High/Soft state, with nearly constant RXTE/ASM flux ($\sim$ 90 counts/s), no significant BATSE flux, and $\sim$ 3 mJy at 15 GHz (Ryle Telescope) few days before and after. On 1997 October 24, 5 days before a major radio and X-ray flare, the compact source was in a plateau state with $\sim$ 50 mJy at 2.25 & 8 GHz (GBI) giving a radio spectral index ~ -0.08, with low constant RXTE/ASM flux ($\sim$ 20 counts/s) and BASTE photon flux $\sim$ 0.06.

The IR data have been de-reddened according to the law from Lutz *et al.* (1996). We used very simple models combining a hot black body emission or a hot synchrotron one to fit the near infrared data, with a cold black body emission (for a dust envelope heated by the binary) and a synchrotron emission extrapolated from radio wavelengths. We conclude that synchrotron emission alone is not sufficient to explain the mid-infrared flux of GRS 1915+105, and that an additional contribution, which could be dust emission around the binary, is needed.

Astrophysics and Space Science is the original source of publication of this article. It is recommended that this article is cited as: *Astrophysics and Space Science* **276** (suppl.): 99–100, 2001.

2. SS433 / W50

SS433 is a high mass X-ray binary surrounded by the supernova remnant W50, a bright radio nebula with a $\sim 1^\circ \times 2^\circ$ unusual ellipsoidal morphology. SS433 produces relativistic jets at subarcsec scales and large scale lobes ($\sim$ 90 pc at a distance of 5 kpc). We have imaged the large scale west radio lobe with the ISOCAM 14–16 μm filter and a $6''$ pixel field of view resolution in September–October 1997.

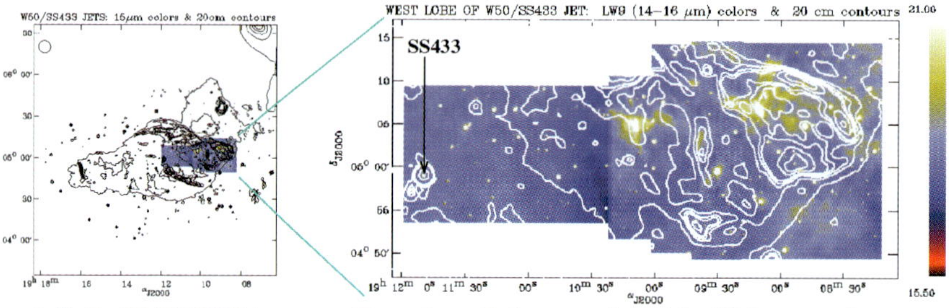

Figure 1. **Left:** SS433/W50 image at 20 cm from Dubner *et al.* (1998). **Right:** superimposing radio 20 cm contours with ISOCAM 15 μm image shows that the IR emission matches the radio one along the western edge of the radio nebula. **This IR emission is likely to be due to synchrotron radiation.**

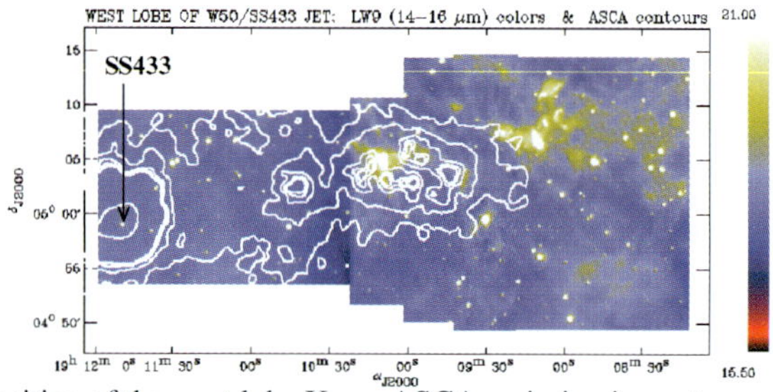

Figure 2. Superimposition of the west lobe X-ray ASCA emission in contours with the 15 μm image in colors. It shows that the IR emission lies along the same axis as the X-ray western lobe. Thus both emissions are the result of the interaction of the mass outflow with the interstellar medium. Most of the IR emission occurs at the western edge of W50 whereas the X-ray lobe fills the space between SS433 and this edge, which is an indication that the jet is present between these two components without being seen (Mirabel & Rodríguez, 1999).

References

Dubner, G.M., *et al.*: 1998, *ApJ* **116**, 1842.
Lutz, D., *et al.*, 1996: *A&A* **315**, L269.
Mahoney, W.A., *et al.*: 1997, *Proc. of the Fourth Compton Symposium*, AIP Conf. Proc. **410**, 912.
Margon, B.: 1984, *ARA&A* **22**, 507.
Mirabel, I.F. and Rodríguez, L.F., 1999, *ARA&A* **37**, 409.

THREE MODES OF LONG WAVELENGTH FLARING FROM GRS 1915+105: IMPLICATIONS FOR JET FORMATION

STEPHEN S. EIKENBERRY
Cornell University

Abstract. GRS 1915+105 has exhibited at least three modes of long-wavelength (infrared to radio) flares. Class A flares are the bright ($\sim$ 1 Jy) radio events whose apparent superluminal motion in many ways defines the microquasars. Class B flares have intermediate ($\sim$ 50–300 mJy) flux densities in the IR and radio, and are associated with hard X-ray dipping behaviors. Class C flares are faint ($\sim$ 5 – 50 mJy) IR/radio events associated with soft X-ray dips. We discuss each of these classes and their inter-relationships.

1. Class A – Large Radio Flares

The discovery of superluminal motions in a radio flare from GRS 1915 +105 by Mirabel and Rodríguez (1994) in many ways launched the field of microquasar astrophysics. Such flares have been seen since then (e.g. Fender *et al.*, 1999), and always seem to have peak flux densities of $F_\nu \sim 1$ Jy. The resolved jets with $v = 0.92c$ show spectral evolution from optically thick to optically thin, and seem to recur every few days during an ejection episode. The ejection episodes themselves recur on timescales of a few years. While Harmon *et al.* (1997) observed an association between the Class A flares and hard X-ray activity, no simultaneous time-resolved X-ray coverage has yet been obtained.

2. Class B – Mid-size IR/Radio Flares

The second class of long-wavelength events were first clearly characterized by Fender *et al.* (1997), whose same-day IR/radio observations showed flares with flux densities of $F_\nu \sim 50 - 100$ mJy and decay timescales of $\sim$ 10 minutes. The brightness temperature in the radio implied a synchrotron origin for the flares and the constancy of the decay timescale over 4 decades of frequency implied that adiabatic expansion dominates the cooling process.

Eikenberry *et al.* (1998a,b) observed a long sequence of Class B flares in the IR over two nights in August 1997. These flares had peak flux densities $F_\nu \sim 200 - 250$ mJy and recurrence timescales of $\sim 30 - 60$ minutes. Simultaneous X-ray observations with RXTE revealed that these flares are associated with complex

Astrophysics and Space Science is the original source of publication of this article. It is recommended that this article is cited as: *Astrophysics and Space Science* **276** (suppl.): 101–104, 2001.

X-ray behaviors (Figure 1). In particular, the flares come after the hard X-ray dips, during which the blackbody temperature in the inner disk drops while the power-law index hardens. Belloni *et al.* (1997) first interpreted dips such as these as the *disappearance* of the inner accretion disk, and combined with the observations of Fender *et al.* (1997) and Eikenberry *et al.* (1998a,b), we recognize that *these events provide the first direct time-resolved observations of relativistic jet formation in a black hole system.*

Mirabel *et al.* (1998) made simultaneous IR/radio/X-ray observations of several Class B flares, finding wavelength-dependent time delays, with the longer wavelengths peaking later. Such behavior is roughly consistent with that expected from an expanding van der Laan blob, and extrapolations based on this timing would indicate that the ejection occurs near the end of the hard X-ray dip.

3. Class C – Faint Flares

The existence of a third class of long-wavelength flares was first clarified through IR/X-ray observations by Eikenberry *et al.* (2000). These flares have peak flux densities in the IR of $\sim 4-8$ mJy and are associated with rapid soft X-ray dips (Figure 2), during which the blackbody temperature increases and the power-law index softens (as opposed to the hard Class B dips). These flares also seem to explain the quasi-steady IR excess during similar X-ray behaviors in 1997 (Eikenberry *et al.*, 2000).

It is interesting to note that the IR flares appear to begin *before* the X-ray activity, implying an 'outside-in' propagation of the disturbance – opposite of the Class B behavior. The spotty coverage from RXTE makes 'aliasing' a problem for verifying this. However, Eikenberry *et al.* (2000) showed that no constant delay fits with the X-rays preceeding the IR flares, and even a non-steady delay would have to reach up to $\sim$ 3000 seconds. Thus, the 'IR-first' explanation seems the least objectionable for the observations.

4. Three Separate Behaviors?

A key issue here is the reality of the distinction between the three classes presented above. The Class C flares seem distinguishable from Class B due to the marked differences in their corresponding X-ray behaviors (soft versus hard dips, IR-first versus X-ray-first), as well as differences in recurrence times and amplitudes. Class B flares, on the other hand, behave in many ways like 'baby jet' versions of Class A ejections. In particular, the relative amplitudes and decay timescales are roughly consistent with identical ejections, but with $\sim$ 1000 times less mass in the Class B events. However, Dhawan *et al.* (2000) have used VLBA to resolve the Class B events, and find that their axis of motion differs significantly from the Class

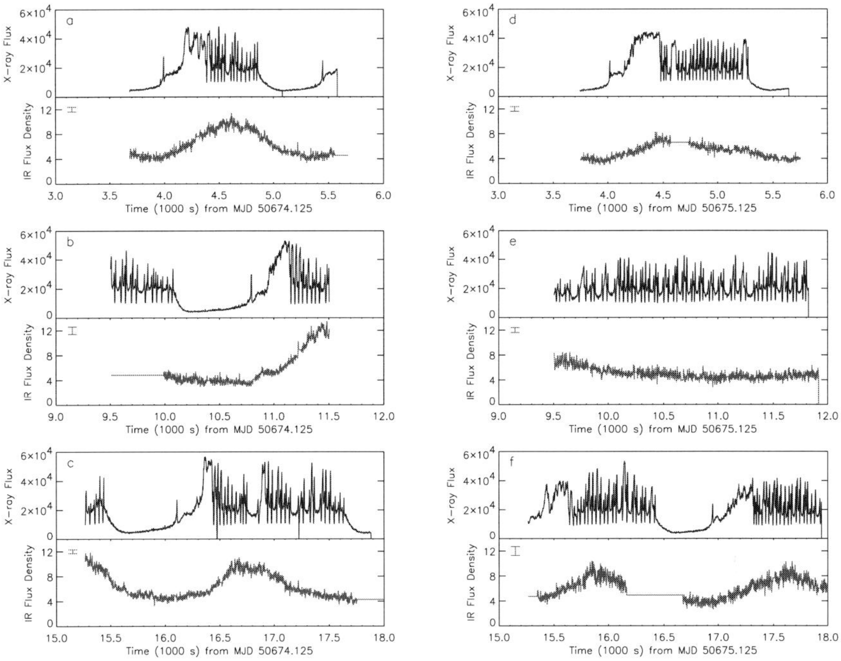

Figure 1. Simultaneous IR/X-ray observations of Class B flares from Eikenberry *et al.* (1998a).

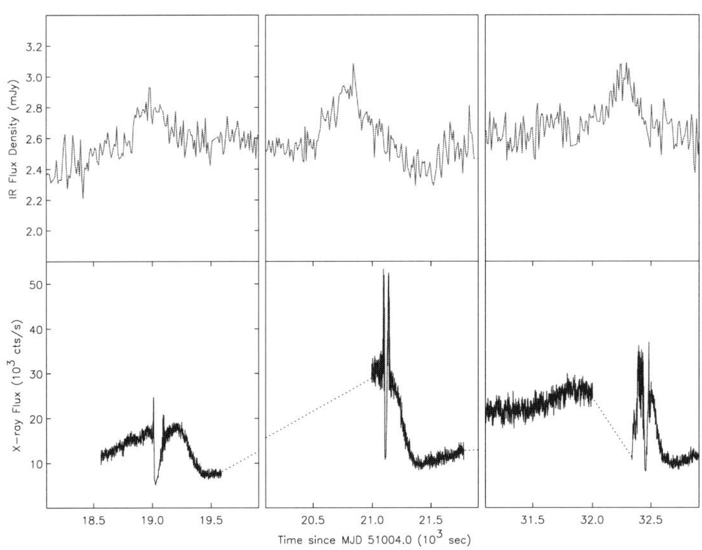

Figure 2. Simultaneous IR/X-ray observations of Class C flares from Eikenberry *et al.* (2000).

A events. Thus, it appears that there are in fact at least 3 distinct modes of jet production in GRS 1915+105.

References

Belloni, T., Mendez, M., King, A.R., van der Klis, M. and van Paradijs, J.: 1997, *Astrophys. J.* **488**, L109.

Dhawan, V., *et al.*: 2001, These Proceedings.

Eikenberry, S.S., Matthews, K., Morgan, E.H., Remillard, R.A. and Nelson, R.W.: 1998a, *Astrophys. J.* **494**, L61.

Eikenberry, S.S., Matthews, K., Muno, M., Blanco, P.R., Morgan, E.H. and Remillard, R.A.: 2000, *Astrophys. J.* **532**, L33.

Eikenberry, S.S., Matthews, K., Murphy, T.W., Nelson, R.W., Morgan, E.H. and Remillard, R.A.: 1998b, *Astrophys. J.* **506**, L31.

Fender, R.P., Garrington, S.T., McKay, D.J., Muxlow, T.W.B., Pooley, G.G., Spencer, R.E., Stirling, A.M. and Waltman, E.B.: 1999, *Mon. Not. R. Astron. Soc.* **304**, 865.

Fender, R.P., Pooley, G.G., Brocksopp, C. and Newell, S.J.: 1997, *Mon. Not. R. Astron. Soc.* **290**, L65.

Harmon, B.A., Deal, K.J., Paciesas, W.S., Zhang, S.N., Robinson, C.R., Gerard, E., Rodríguez, L.F. and Mirabel, I.F.: 1997, *Astrophys. J.* **477**, L85.

Mirabel, I.F., Dhawan, V., Chaty, S., Rodríguez, L.F., Martí, J., Robinson, C.R., Swank, J. and Geballe, T.: 1998, *Astron. Astrophys.* **330**, L9.

Mirabel, I.F. and Rodríguez, L.F.: 1994, *Nature* **371**, 46.

K BAND MONITORING OF GRS 1915+105 DURING 1999–2000

J. GREINER
Astrophysical Institute Potsdam, 14482 Potsdam, Germany

F.J. VRBA, A.A. HENDEN, H.H. GUETTER and C.B. LUGINBUHL
US Naval Observatory, Box 1149, Flagstaff Station, Flagstaff, AZ 86002, USA

Abstract. We report on long-term K band photometry of the galactic microquasar GRS 1915+105 since March 1999. We see variations of more than 1 mag, which are seemingly correlated with variations in X-ray intensity and spectral slope.

Keywords: X-ray binary, GRS 1915+105, infrared observations

1. Introduction

GRS 1915+105 is the prototypical microquasar, a galactic X-ray binary ejecting plasma clouds at v≈0.92 c (Mirabel & Rodriguez 1994). Three different sources of infrared (IR) radiation are expected from a X-ray binary, namely thermal emission from the outer part of accretion disk, thermal emission from the companion star or synchrotron emission from the jet(s). Previous IR observations of GRS 1915+105 include, besides a few single measurements since 1993, some few hrs photometry in conjunction with X-ray/radio measurements (Eikenberry *et al.* 2000), and a 2 month monitoring in 1996 (Bandyopadhyay *et al.* 1998).

2. Observations and Results

K band observations have been performed with the 1.55-m telescope of the US Naval Observatory Flagstaff equipped with the 256 × 256 pixel HgCdTe IRCAM camera during 8–10 day intervals around full Moon. Because the main interest was in the long-term behaviour, 'only' about 1–3 observations per month were done, consisting of groups of 3 or 4 individual observations per night. Each group represents a set of coadded integrations with an exposure of 400 sec. In total, observations were performed on 34 nights between 05 May 1999 and 15 June 2000.

These observations show GRS 1915+105 to be variable between $12.1 \lesssim K \lesssim 13.7$ mag (not extinction-corrected; Figure 1). Our long coverage allows to search for correlations with other emission characteristics, e.g. the RXTE All-Sky Monitor (ASM) quick-look results provided by the ASM/RXTE team. After selecting those K band observations which were done within 4 hrs of an ASM scan, we find that

Astrophysics and Space Science is the original source of publication of this article. It is recommended that this article is cited as: *Astrophysics and Space Science* **276** (suppl.): 105–106, 2001.

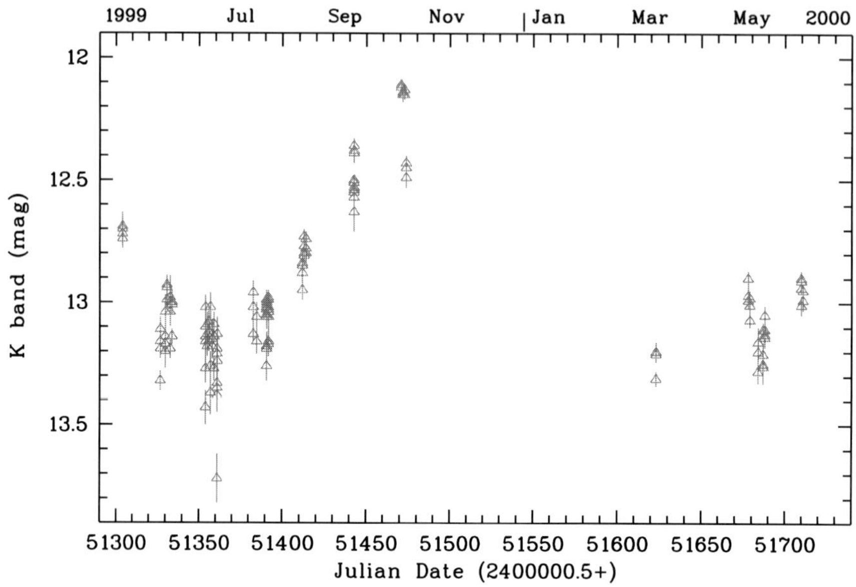

Figure 1. K band light curve of GRS 1915+105, showing the mean of all 'groups'.

both the X-ray flux as well as the X-ray hardness ratio (flux ratio of 5–12 keV vs. 3–5 keV band) are positively correlated with the K band flux.

A regression analysis suggests that a doubling of the X-ray count rate (flux) leads to a 33% increase in the K band flux. This argues against an explanation due to illumination of the secondary. Instead, it rather could be synchrotron radiation related to (small) ejections/flares as seen also at shorter time scales (Eikenberry *et al.*, 2000). In this picture the correlation of the K band flux with the X-ray hardness ratio is compatible with the correlation of X-ray spectral slope and radio flux found by Rau and Greiner (2001). A large (small) value of the hardness ratio corresponds to a flat (steep) X-ray power law slope, and therefore a high (low) temperature of the electrons in the corona if the general comptonization paradigm is adopted. These correlations then imply that strong K band flux and weak radio flux occur during times of a flat X-ray power law, i.e. most of the synchrotron power is radiated either in the infrared or radio band depending only on the temperature of electrons in the corona.

References

Bandyopadhyay, R., Martini, P. and Gerard, E., *et al.* : 1998, *MNRAS* **295**, 623.
Eikenberry, S.S., Matthews, K. and Muno, M., *et al.* , 2000, *ApJ* **532**, L33.
Mirabel, I.F. and Rodriguez, L.F.: 1994, *Nature* **371**, 46.
Rau, A. and Greiner, J., 2001, these Proceedings.

GRS 1915+105: VLBA IMAGES OF THE AU-SCALE JET *

V. DHAWAN**
NRAO, U.S.A.; E-mail: vdhawan@aoc.nrao.edu

I.F. MIRABEL ***
CEA Saclay, France; E-mail: mirabel@discovery.saclay.cea.fr

L.F. RODRIGUEZ ‡
UNAM, Mexico; E-mail: l.rodriguez@astrosmo.unam.mx

Abstract. High-resolution observations of GRS 1915+105 are summarized. For full discussion see *Astrophys. J.* **543**, 1 November 2000; astro-ph/0006086.

Radio imaging with the VLBA at radio wavelengths (13 cm to 0.7 cm), in different states of the black hole binary, always resolves the nucleus as a compact jet of length $\sim 10\lambda_{cm}$ AU. The brightness temperature of the nucleus is $T_B \sim 10^9$ K, on scales of 2.5–7 AU (0.2–0.6 milliarcseconds), and its properties are well fit by a conically expanding synchrotron jet model. The nuclear jet varies in $\sim$ 30 min during minor X-ray/radio outbursts, and re-establishes within $\sim$ 8 hours of a major outburst, indicating the robustness of the disk/jet system.

At lower resolution (80–240 AU) more extended ejecta are imaged at $\sim$ 500 AU from the core, with superluminal motion measurable in just hours. The velocity of 1.4 ± 0.1 c ($D/12$ kpc) for the approaching component is consistent with ballistic motion from 500 AU outwards, perhaps even since birth. The axes of the superluminal ejecta and AU-scale jet differ by $\leq 12°$, and both are stable in time for over 3 years.

Astrometry relative to the extragalactic frame has located the black hole to ± 1.5 mas. Its secular parallax, due to Galactic rotation, is 5.8 ± 1.5 mas yr^{-1}, consistent with a 12 kpc distance. A limit of $\leq$ 100 km s^{-1} is placed on its proper-motion with respect to its neighbourhood.

Some black hole systems (e.g. Cyg X-1, 1E 1740-2942, GRS 1758-258, GX 339-4, XTE J1118+48, SgrA*) have compact radio cores with steady, flat-spectrum 'plateau' states, but lack evidence of flares and transient ejecta. GRS 1915+105 is the only source known to exhibit both states.

* We thank the RXTE/ASM team for quick-look results.
** NRAO is a facility of the National Science Foundation operated under cooperative agreement by Associated Universities, Inc.
*** Partial support from CONICET, Argentina.
‡ Partial support from CONACyt, Mexico.

Astrophysics and Space Science is the original source of publication of this article. It is recommended that this article is cited as: *Astrophysics and Space Science* **276** (suppl.): 107–108, 2001.

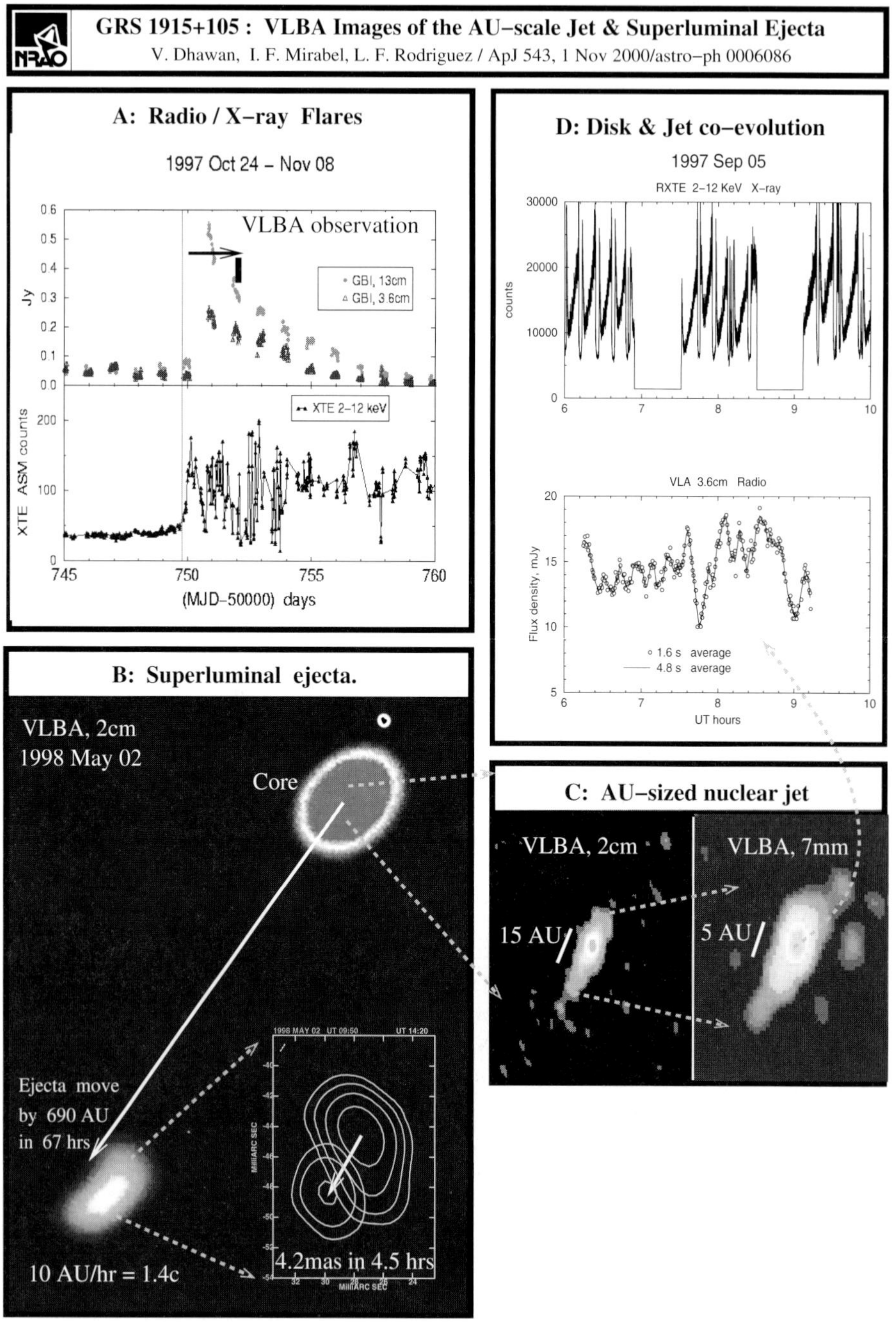

Figure 1. Coordinated X-ray/Radio observations demonstrate Disk/Jet co-evolution.

AN ATTEMPT TO IDENTIFY THE EXTENDED SYNCHROTRON STRUCTURE ASSOCIATED WITH THE MICRO-QUASAR GRS 1915+105

E. FÜRST[1] and M. OSTROWSKI[2]
[1] *Max-Planck-Institut für Radioastronomie, Bonn, Germany*
[2] *Obserwatorium Astronomiczne, Uniwersytet Jagielloński, Kraków, Poland*

Abstract. We used the Effelsberg 100 m telescope at 10.45 GHz in an attempt to identify extended synchrotron structures formed by jets of GRS 1915+105. For this distant galactic plane source a few suspect structures were pointed out.

To plan new observations we analysed a $2^\circ \times 2^\circ$ region centered at the position of GRS 1915+105 ($\alpha_{1950} = 19h\ 12m\ 45s$, $\delta_{1950} = 10^\circ\ 51'\ 27''$). At the map composed of the Effelsberg Galactic plane survey at 11 cm superposed on the IRAS 60 μm emission two complexes of HII regions dominate by their strong emission ($\alpha_{1950} = 19h\ 11m\ 47s$, $\delta_{1950} = 11^\circ\ 07'\ 03''$; $\alpha_{1950} = 19h\ 11m\ 06s$, $\delta_{1950} = 10^\circ\ 48'\ 25''$) and the shell-type emission feature near $\alpha_{1950} = 19h\ 14m$, $\delta_{1950} = 11^\circ\ 4'$ is a supernova remnant. Along the jet PA = 155° determined by Mirabel and Rodríguez (1994) three patchy emission features A, B and C were selected, which were not yet identified. We have conducted new observations of these regions at 10.45 GHz. At the map (Figure 1) the jet position lines given in the literature are plotted (PA = 155° in the middle). In polarization no signal beyond 3 r.m.s was detected.

Source A is a faint extended emission plateau which could be of nonthermal nature. The integrated flux density is about 1.1 Jy. At the bottom of this area a compact source of 99 mJy is visible, coinciding with IRAS 19132+1035 (Rodríguez and Mirabel, 1998; Chaty *et al.*, 2000). The sources in regions B and C, outside the range studied by the above authors, are very likely thermal emission features, probably faint HII emission. The ‘hop-spot’ at the bottom of region B is characterized with a steeper non-thermal spectrum. A map of region C shows an extended source with a thermal-like spectrum. It is characterized with a complicated morphology with a number of brighter spots over the structure. We note that the brightest knots in regions A, B and C are situated close to the GRS 1915+105 jets’ axis.

MO acknowledges support from the *KBN* through the grants PB 179/P03/96/11 and PB 258/P03/99/17.

Astrophysics and Space Science is the original source of publication of this article. It is recommended that this article is cited as: *Astrophysics and Space Science* **276** (suppl.): 109–110, 2001.

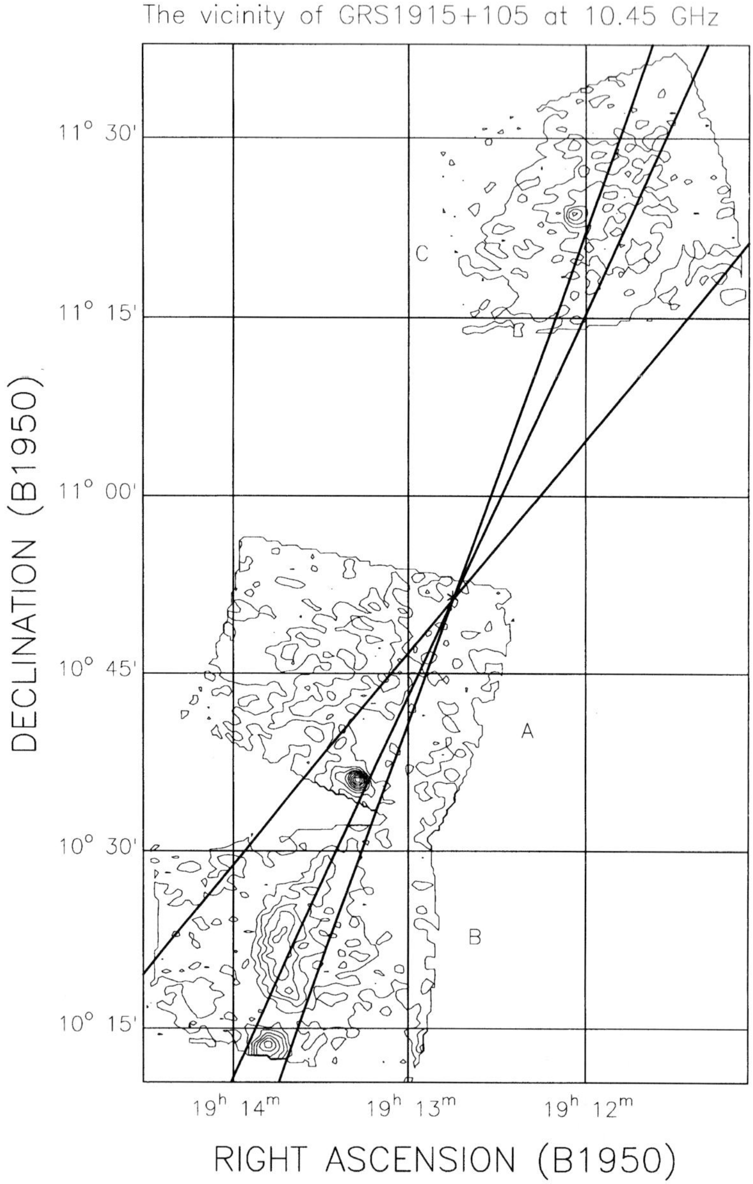

Figure 1. The total power 10.45 GHz Effelsberg maps of regions A, B and C.

References

Chaty, S., *et al.*: 2000, in: *X-ray Astronomy 99*, in press (astro-ph/0007470).
Mirabel, I.F. and Rodríguez, L.F.: 1994, *Nature* **371**, 46.
Rodríguez, L.F. and Mirabel, I.F.: 1998, *Astron. Astrophys.* **340**, 47.

EVN OBSERVATIONS OF GRS 1915+105

G. GIOVANNINI[1], L. FERETTI[1], M. TORDI[1], T. VENTURI[1], S. MASSAGLIA[2], G. BODO[3], E. TRUSSONI[3], M. GLIOZZI[4], M. TAVANI[5], J. CONWAY[6], A. FOLEY[7], D. GRAHAM[8], A. KUS[9], R. SPENCER[10] and C. TRIGILIO[11]

[1] *IRA, via Gobetti 101, 40129 Bologna, Italy;* [2] *Univ. Torino;* [3] *OA. Torino;* [4] *MPE Garching;* [5] *IFCTR Milano;* [6] *Onsala Observatory;* [7] *NFRA Dwingeloo;* [8] *MPifR Bonn;* [9] *Torun Observatory;* [10] *NRAL Jodrell Bank;* [11] *IRA Noto*

Abstract. New European VLBI Network observations are presented.

We present new European VLBI Network (EVN) observations aimed at studying GRS 1915+105 at high resolution. TOO observations were performed during gaps in the official observing schedule on 1998 May 31, June 2 and June 9 at 6 cm, 6 cm and 18 cm, respectively. Integration times were of about 4 h in each observation. The telescopes involved in the observations were Medicina, Noto, Effelsberg, Jodrell, Onsala, Westerbork and Torun. The observations at 6 cm were obtained a few days before a major flare (see Figure 1) while the 20 cm observation was obtained after the radio flare.

In the two images obtained before the major flare, the source is resolved, showing two opposite asymmetric jets (Figure 2). The main jet is in the SE direction in position angle $\sim 160°$, consistent with the orientation of the main jet present in previous MERLIN and VLA data (Fender *et al.*, 1999; Mirabel and Rodriguez, 1994). The total source extent is of $\sim$ 13 mas, corresponding to 2.1×10^{15} cm ($\sim$ 140 AU). The total fluxes are roughly consistent with the Green Bank Interferometry (GBI) fluxes, implying that no large scale structure was missed.

In the third observation, obtained after the flare, the source is unresolved. The total flux is 17 mJy, much lower than the GBI flux. This would imply the presence of extended structure not detected in the present observation, supporting the idea that during the flare strong high velocity components have been ejected.

Assuming that the jets are relativistic and intrinsically symmetric, we can derive an estimate of the velocity from the jet to counter-jet brightness ratio and from the jet length ratio. Using a source orientation to the plane of the sky of 66° (Fender *et al.*, 1999), we derive a range for the velocity of 0.2–0.6c up to a distance of $\sim$ 8 mas ($\sim$ 90 AU). The absence of visible proper motion between our observations could be due to a continuous ejection in a stable jet.

From the present results, it is evident that steady jets exist in the source also before a major flare, in agreement with Dhawan *et al.* (2000). However a flare produces a strong ejection resulting in extended and relativistic symmetric jets while steady jets are short ($\sim$ 90 AU) and only mildly relativistic ($v = 0.2$–0.6c).

Astrophysics and Space Science is the original source of publication of this article. It is recommended that this article is cited as: *Astrophysics and Space Science* **276** (suppl.): 111–112, 2001.

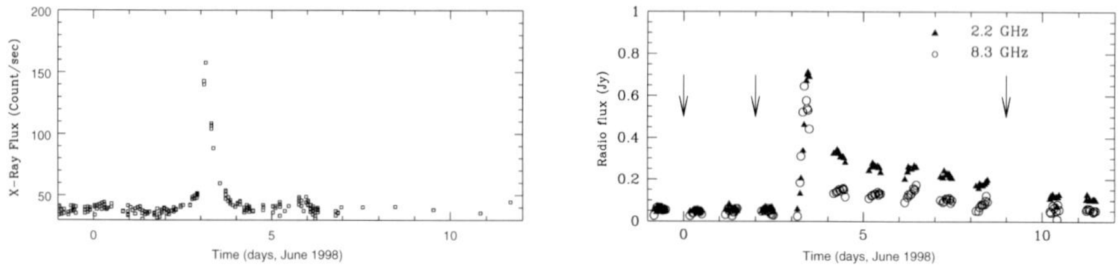

Figure 1. Source light curve in X-rays (RXTE/ASM, left) and in radio (GBI, right), with arrows indicating the epochs of the EVN observations.

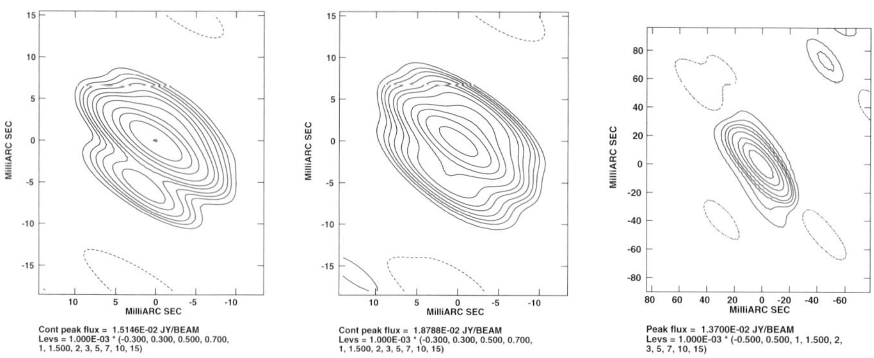

Figure 2. Images of the 3 EVN observations. The first contour is always at the 2σ noise level. Left panel: 1998 May 31, 6 cm, HPBW = 10×3 mas. Central panel: 1998 June 2, 6 cm, HPBW = 10×3 mas. Right panel: 1998 June 9, 18 cm, HPBW = 39×12 mas.

From the present velocity estimate and the high resolution images obtained by Dhawan *et al.* (2000) before a major flare, which show a quite symmetric source on a scale of $\sim$ 50 AU, we find that steady jets appear to accelerate at distance between 50 AU and 90 AU from the black hole as found in a few AGN on the pc scale (see e.g. Cotton *et al.*, 1999; Giovannini *et al.*, 2001).

Acknowledgement

This work has been partially supported by the European Commission's TMR-LSF programme under contract ERBFMGECT950012.

References

Cotton, W.D., Feretti, L., Giovannini, G., *et al.*: 1999, *Astrophys. J.* **519**, 108.
Dhawan, V., Mirabel, I.F. and Rodríguez, L.F.: 2000, *Astrophys. J.* **543**, 373.
Fender, R.P., Garrington, S.T., McKay, D.J., *et al.*: 1999, *Mon. Not. R. Astron. Soc.* **304**, 865.
Giovannini, G. *et al.*: 2001, *ApJ* **552**, 508.
Mirabel I.F. and Rodríguez, L.F.: 1994, *Nature* **371**, 46.

THE ACCRETION-EJECTION INSTABILITY AND A 'MAGNETIC FLOOD' SCENARIO

M. TAGGER
Service d'Astrophysique (CNRS URA 2052), CEA Saclay, 91191 Gif sur Yvette, France

Abstract. I present a global view of recent results on the Accretion-Ejection Instability (AEI), described in more details in other contributions to this workshop. These results address essentially the characteristics of the AEI as a good candidate to explain the low-frequency QPO of X-ray binaries, in particular (at $\sim 1 - 10$ Hz) of micro-quasars. I then discuss how, if the AEI is considered as the source of the QPO, a possible scenario can be considered where the $\sim$ 30 mn. cycles of GRS 1915+105 are controlled by the evolution of magnetic flux in the disk.

1. Introduction

The Accretion-Ejection Instability was first described by Tagger and Pellat (1999), based on previous results of Tagger *et al.* (1991). It appears, in the inner part of disks threaded by a vertical (poloidal) magnetic field, essentially as a spiral wave, similar to the ones of galactic disks but driven by magnetic stresses rather than self-gravity. It becomes unstable by extracting energy and angular momentum from the disk, causing accretion, and storing them in a Rossby vortex it generates at its corotation radius (a few times the inner disk radius). The instability occurs if the field has a moderate amplitude (of the order of equipartition), and presents a sufficient radial gradient.

We proposed the AEI as a possible explanation for the low-frequency ('ubiquitous') QPO observed in many X-ray binaries, and in particular in micro-quasars. This is based on the frequency of the instability (a fraction of the rotation frequency at the inner disk radius, as observed), and other properties such as its effect on the disk corona or the fact that, as in galaxies, we expect it to form long-lived, quasi-stationary spiral features, leading to a quasi-periodic modulation of the disk emission. I will present here a global view of recent work, presented in more details in other contributions to these proceedings. I will then turn to the more prospective discussion of a 'magnetic flood' scenario for the $\sim$ 30 mn cycles of GRS 1915+105. Starting from the hypothesis that the QPO does correspond to the AEI, we consider various observed properties of the QPO and of the disk, and confront them with what we expect from the AEI. This leads us to this scenario, compatible with known facts, although it cannot be proven at this stage; considered

Astrophysics and Space Science is the original source of publication of this article. It is recommended that this article is cited as: *Astrophysics and Space Science* **276** (suppl.): 113–116, 2001.

as a possibility, and certainly not as the final word on these cycles, it can thus be used as guidelines for our future work.

2. New Results

2.1. Numerical simulations

We have performed 2D (r, ϕ) MHD simulations of a disk threaded by a vertical (poloidal) magnetic field (Caunt and Tagger, these Proceedings). The simulations globally confirm the linear analysis. The instability shows all the expected features (frequency, radial propagation, vortex at its corotation radius, etc.). Non-linearly it saturates at high amplitude, giving a long-lived, quasi-stationary spiral pattern in the disk. Accretion causes the magnetic flux to pile up in the central region, so that most simulations end up with an $m = 1$ (single-armed) spiral, as expected.

2.2. Emission of Alfvén waves

We have computed the flux of energy and angular momentum 'leaking' from the vortex to the corona as Alfvén waves (Varnière and Tagger, these Proceedings). A future step will be to determine the way these waves can be damped in the corona, ultimately powering a wind or a jet from the accretion energy.

2.3. QPO frequency

We expected the frequency of the $m = 1$ mode to be quite sensitive to the rotation profile in the disk. In particular, relativistic effects when the inner radius is close to the Last Stable Orbit allow the presence of an Inner Lindblad Resonance (familiar in galactic dynamics). We have studied this in a pseudo-newtonian potential (Varnière and Tagger, these Proceedings). We find that, when the inner radius $r_{int} \lesssim 1.4 r_{LSO}$, an ILR appears and reverses the correlation between r_{int} and the QPO frequency ν_{QPO}.

2.4. Comparison with observations

We have studied (Rodriguez *et al.*, these proceedings) the possibility that this could explain the reverse correlation between ν_{QPO} and the color radius r_{col}, noted by Sobczak *et al.* (2000) in GRO J1655-40 compared to other sources. Comparing with GRS 1915+105, we find that this is a definite possibility, although present results cannot prove it. However a detailed analysis of the data leads to interesting results:
1) In the 30 mn cycles of GRS 1915+105, the QPO appears *before* the dip in the X-ray curve: the QPO might be the cause, and not the consequence, of the transition to the low state. 2) Some data from GRS 1915+105 are compatible with the model

of Merloni *et al.*, where anomalously low measures of the color radius result from geometric effects in the disk and corona; but this is not the case in GRO J1655-40, where low radii correspond to high temperatures. It rather indicates that the emitting region is small and hot, i.e. that dissipation occurs in a spiral shock or hot point in the disk, which could be a consequence of the AEI.

3. Magnetic Flood

These results confirm that the AEI is a good candidate to explain the 'ubiquitous' QPO. In order to guide our investigations, we have attempted to go further by considering what this could tell us about the disk. This leads to a 'Magnetic Flood' scenario (Tagger, 2000) where the 30 mn cycles of GRS 1915+105 are controlled by the evolution of magnetic flux in the disk.

We start from the fact that a magnetized disk is subject to the Magneto-Rotational Instability (MRI), creating small-scale turbulence, when the field is weak ($\beta = 8\pi p/B^2 > 1$), and to the AEI when it is of the order of equipartition ($\beta \sim 1$). The transition from the high to the low state could thus correspond to a transition between a weakly and a more strongly magnetized configuration, as magnetic flux advected with the gas accumulates in the central region. In a weak field the MRI heats the disk by dissipation of the accretion energy; when the field approaches equipartition the MRI stops and the AEI sets in: this marks the appearance of the QPO, at the end of the high state. The heating of the disk decreases, because accretion energy is redirected to the corona, so that pressure drops and makes β even lower: this would explain why the transition from the high to the low state is so sudden (we note that in recent numerical simulations of the MRI Hawley and Krolik (2000) find that, after some time, the magnetic configuration becomes similar to the one we use here, and most of the accretion energy is carried away by waves. Technically these are fast magnetosonic waves, i.e. the same spiral waves involved in the AEI. Although the full 3D treatment limits the spatial resolution of these simulations, we believe that this might very well be a first manifestation of the magnetic flood).

Because of its decreased pressure the disk moves out, and then back in as accretion from outer regions is still going on (the MRI is then seen as the band-limited noise at lower frequencies, corresponding to larger radii; this would explain the correlation between the break and QPO frequencies). The end of the process is marked by a reconnection event (though other large-scale magnetic events could be considered) at the inner edge of the disk, leading to the ejection of a plasmoid and to the destruction of magnetic flux, allowing a return to the high state.

During the low state, some or most of the accretion energy from the inner region of the disk is redirected to the corona. Dissipation of this energy, by a process we still have to work out, would then heat the electrons, leading to the dominant comptonized tail seen in this low and hard state. This would also provide the link between disk accretion and MHD jet models.

This leaves the question of what exists in the region between the disk and the black hole. An ADAF would be possible, but we also consider another candidate: in the Blandford-Zjanek model, the black hole contains magnetic flux due to external currents, held in a force-free magnetic configuration between the black hole and the disk. Such a configuration, containing the magnetic flux advected to the central 'hole' in our scenario, would be very similar to an ADAF in many ways: it would form a fat torus of very low density plasma, where the heat exchange time between ions and electrons is much longer than the dynamical time. The inner radius of the disk would be fixed by the equilibrium with the magnetic pressure in the torus.

Although this scenario is still very schematic, it is compatible with the main observed properties of the 30 mn cycles of GRS 1915+105. It leads to a number of questions and prospects: the formation of the compact jet and of the ultra-luminous ejections, the plasma processes in the corona, and the longer term history of this source (which could be compared to other ones) are the most intriguing ones.

References

Hawley, J.F. and Krolik, J.H.: 2000, *Astrophys. J.*, in press (astro-ph/0006456).

Merloni, A., Fabian, A.C. and Ross, R.R.: 2000, *Mon. Not. R. Astron. Soc.* **313**, 193.

Sobczak, G.J., McClintock, J.E., Remillard, R.A., Cui, W., Levine, A.M., Morgan, E.H., Orosz, J.A. and Bailyn, C.D.: 2000, *Astrophys. J.* **531**, 537.

Tagger, M.: 2000, in: *Proc. of the 5th Compton Symposium*, AIP Conf. Proc., in press (astro-ph/9910365).

Tagger, M. and Pellat, R.: 1999, *Astron. Astrophys.* **349**, 1003.

THE 1997 HARD STATE OUTBURST OF THE X-RAY TRANSIENT GS 1354-64 / BW CIR

CATHERINE BROCKSOPP
Open University, U.K.

PETER JONKER, ROB FENDER and MICHIEL VAN DER KLIS
University of Amsterdam

PAUL GROOT
Harvard-Smithsonian Center for Astrophysics

STEVEN TINGAY
Australia Telescope National Facility

Abstract. We present observations of the 1997 outburst of the X-ray transient GS 1354-64 (BW Cir) at X-ray, optical and, for the first time, radio wavelengths. The X-ray outburst was unusual in that the source remained in the low/hard X-ray state throughout; the X-ray peak was also *preceded* by at least one optical outburst, suggesting that it was an 'outside-in' outburst – similar to those observed in dwarf novae systems, although possibly taking place on a viscous timescale in this case. It therefore indicates that the optical emission was *not* dominated by the reprocessing of X-rays but that instead we see the instability directly. While the radio source was too faint to detect any extended structure, spectral analysis of the radio data and a comparison with other similar systems suggests that mass ejections, probably in the form of a jet, took place and that the emitted synchrotron spectrum may have extended as far as infrared wavelengths. Finally, we compare this 1997 outburst of GS 1354-64 with possible previous outbursts and also with other hard state objects, both transient and persistent. It appears that a set of characteristics – such as a weak, flat spectrum radio jet, a mHz QPO increasing in frequency, a surprisingly high optical:X-ray luminosity ratio and the observed optical peak preceding the X-ray peak – may be common to all hard state X-ray transients.

1. Introduction

Soft X-ray transients are a class of low mass X-ray binaries in which instabilities in the accretion disc cause a sudden increase in mass accretion rate onto the compact object, resulting in an outburst (see e.g. Charles, 1998 and references within). These outbursts are generally accompanied by X-ray spectral softening and we would normally expect the sources to show characteristics of the soft state, as defined by Van der Klis (1994, 1995).

However, this is not necessarily the case and some outbursting systems have remained in the low/hard state throughout the outburst. GS 1354-64 is one such system which entered an outburst phase in 1997 November but did not leave the low/hard state. It is a recurrent transient source which has previously displayed

Astrophysics and Space Science is the original source of publication of this article. It is recommended that this article is cited as: *Astrophysics and Space Science* **276** (suppl.): 117–120, 2001.

TABLE I

Comparison between outbursts of Cen X-2, MX 1353-64 and GS 1354-64, which may all be the same source. X-ray peaks are quoted for the 1–10 keV range. See Tanaka and Lewin (1995) and references within

Source	Date	Soft X-ray Peak (mCrab)	X-ray State	Optical Peak (mags)
Cen X-2	1966–67	~ 8000	(Very?) high/soft	$V < 13.5$ (WX Cen)?
MX 1353-64	1972	75	low/hard	–
GS 1354-64	1987	≥ 300	high/soft	$V \leq 16.9$ (BW Cir)
GS 1354-64	1997	50	low/hard	$V \leq 17.3$ (BW Cir)

much softer and dramatic behaviour (Table I), if indeed Cen X-2 and GS 1354-64 are the same source (Kitamoto *et al.*, 1990).

2. Observations and Results

We have obtained radio data at four different frequencies (ATCA), BVR photometry (0.91 m Dutch telescope, ESO) and public $RXTE$/ASM, $RXTE$/PCA and $CGRO$/BATSE data for the 1997 outburst of GS 1354-64; the subsequent lightcurves can be found in Figure 1.

The X-ray lightcurves show well-correlated, triangular-shaped profiles and this observation, coupled with the hard power-law energy spectra, suggests that both hard and soft X-rays are emitted from a similar origin – i.e. the Comptonising corona surrounding the compact object. The energy spectra also show that the soft state was not reached; indeed, the power density spectrum computed from the PCA data shows typical low/hard state temporal properties including a mHz QPO which varies in frequency throughtout the outburst – this frequency variation is also shown in Figure 1. These X-ray properties suggest a black hole nature for the compact object, although optical observations were insufficient to produce a mass estimate.

The optical lightcurves are unusual for an X-ray transient in that they are not well-correlated with the X-rays and the first optical peak appears to take place ~ 10 days before that of the X-rays. Brocksopp *et al.* (2000) also find that the optical:X-ray luminosity ratio is surprisingly high – these properties suggest that reprocessing of X-rays is not the dominant source of optical light during this outburst and that instead we see the instability directly as it passes through the accretion disc.

Our radio observations show that the radio spectrum is flat (i.e. for $S_\nu \propto \nu^\alpha$, $\alpha \sim 0$ is the spectral index, ν is the frequency and S_ν is the flux density at ν in mJy) from 843 MHz, through the radio and into the near-infrared. If it is indeed the case that the radio jet of GS 1354-64 emits synchrotron radiation over such a

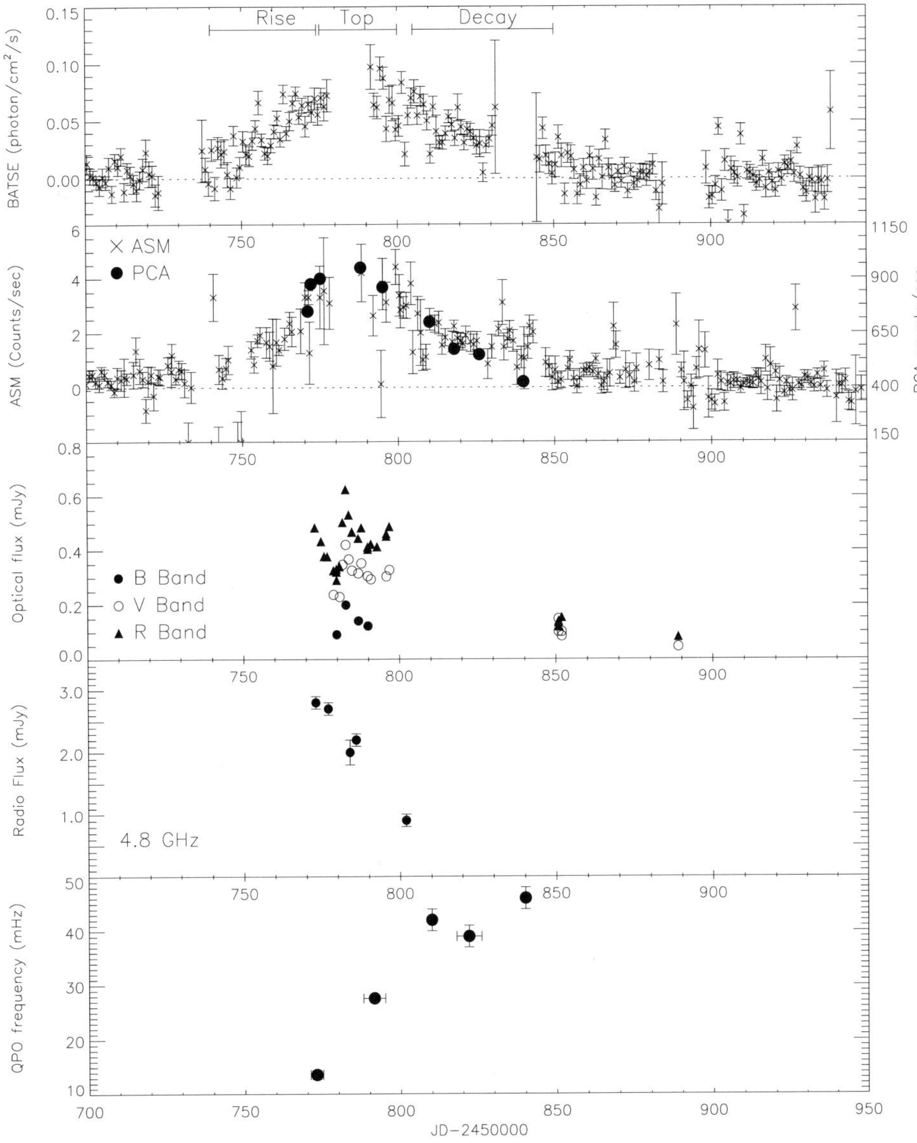

Figure 1. BATSE, ASM, PCA, optical and radio (4.8 GHz) lightcurves of the 1997 outburst for the period 8 October 1997–16 May 1998. Radio lightcurves at other frequencies are shown in Brocksopp *et al.* (2000). Error-bars on the optical points are smaller than the symbols. The variability of the QPO frequency is also shown in the bottom panel.

wide range of frequencies then it must use considerably more accretion power than current X-ray models allow.

3. Discussion

These multiwavelength observations of the 1997 outburst of GS 1354-64 show that we can obtain a wealth of information from outbursts remaining in the low/hard state. Large amounts of X-ray reprocessing dominate the optical outburst of a ‘normal’ soft X-ray transient and so we are not able to observe the instability pass through the disc – this appears to be possible in a hard state outburst. Likewise, during the soft state the radio jets are quenched and cannot be observed (although radio emission is often observed, probably originating in previously ejected material which then becomes decoupled from the system).

Low/hard state X-ray transients do not appear to be as unusual as once thought – V404 Cyg, GRO J0422+32, GRO J1719-24 and XTE J1118+480 have all entered an outburst phase and remained in the low/hard state throughout. We note that these systems share some of the properties of GS 1354-64 – not only do they show hard state behaviour but they also display a mHz QPO at a frequency which increases during the outburst and a flat spectrum radio jet. Some also show evidence for the ‘surprisingly high’ optical:X-ray luminosity ratio and preceding optical peak (see Brocksopp *et al.*, 2000 for references). It is probable that some or all of these properties will turn out to be reliable characteristics of low/hard state X-ray transients.

References

Brocksopp, C., Jonker, P.G., Fender, R.P., Groot, P.J., Van der Klis, M. and Tingay, S.J.: 2000, *Mon. Not. R. Astron. Soc.*, in press.

Charles, P.A.: 1998, in: M. Abramowicz, G. Bjornsson and J. Pringle (eds.), *Theory of Black Hole Accretion Disks*, Cambridge University Press.

Kitamoto, S., Tsunemi, H., Pedersen, H., Ilovaisky, S.A., Van der Klis, M.: 1990, *Astrophys. J.* **361**, 590.

Tanaka, Y. and Lewin, W.H.G.: 1995, in: W.H.G. Lewin, J. Van Paradijs and E. van den Heuvel (eds.), *X-ray Binaries*, Cambridge University Press, p. 126.

Van der Klis, M.: 1994, *Astron. Astrophys.* **283**, 469.

Van der Klis, M.: 1995, in: W.H.G. Lewin, J. Van Paradijs and E. van den Heuvel, E. (eds.), *X-Ray Binaries*, Cambridge University Press, p. 252.

VLBA OBSERVATIONS OF THE SUPERLUMINAL RADIO JET FROM CYGNUS X-1 IN THE LOW/HARD X-RAY STATE

CHRIS DE LA FORCE and RALPH SPENCER
Jodrell Bank Observatory, U.K.

A. STIRLING
University of Central Lancashire, U.K.

M. GARRETT
JIVE, The Netherlands

R. FENDER
University of Amsterdam, The Netherlands

Abstract. We present the detection of a resolved jet-like radio-emitting structure from the black-hole candidate and X-ray binary source Cygnus X-1 at both 8.4 GHz and 15 GHz. The structure in the milli-arcsecond resolution observation with the VLBA at 15 GHz shows an extended emission region to the north-west and possibly also perpendicular emitting regions, believed to be in the orbital plane. This X-ray state is believed to be associated with an advective accretion environment. These observations were spaced 2 days apart over the 5.6-day orbital period. The images confirm the existence of persistent emission from the source from both an unresolved core and a variable, and at least moderately ($> 0.6c$) relativistic jet, during the low/hard X-ray state. However, at this resolution and wavelength they do not show any sign of the emission in the orbital plane observed at a 5σ level at 15 GHz in the initial epoch.

1. VLBA and VLA Observation Details

Following the detections of resolved radio structure (Stirling *et al.*, 1998) further high resolution observations of Cygnus X-1 were taken in August 1998 with the VLBA. Three epochs of observations were made every 2 days to cover a large fraction of the period (5.6 days) of the binary system. The runs were made on Aug 10 (Epoch A), Aug 12 (Epoch B) and on Aug 14 (Epoch C) for six hours per epoch. Observations were made at 15.4 and 8.4 GHz, the frequencies being alternated each hour.

Since the source is relatively weak, the phased-up VLA was used as one element of the array, and the high data rate of 256 Mbits per second was used. The observations of Cygnus X-1 (using a position derived from proper motion measurements (Lestrade *et al.*, 1995)) were interleaved with scans on the nearby phase reference source B1955+335. Only the 8 GHz data from Epoch A are presented here; the 8 GHz images and the 15 GHz images and spectral index maps will be published in later papers (Stirling *et al.*, 2001 and de la Force *et al.*, 2001 respectively).

Astrophysics and Space Science is the original source of publication of this article. It is recommended that this article is cited as: *Astrophysics and Space Science* **276** (suppl.): 121–122, 2001.

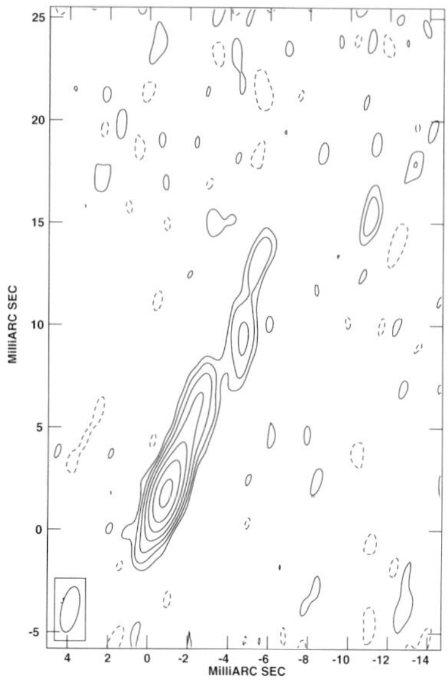

Figure 1. A high resolution image (Epoch A) of Cygnus X-1 at 8.4 GHz; lowest contour 0.15 mJy. Contouring represents a factor of two.

2. Conclusions

A relativistic jet with either a bending flow or a variable ejection angle exists in Cygnus X-1. We cannot confirm or deny the orbital plane emission seen previously at 15 GHz with the 8 GHz images. The unresolved core is likely to have an inverted spectrum, while the extended jets are likely to be optically thin, otherwise the 15 GHz observations should have detected them. It is likely that either self-absorption or free-free absorption occurs near the base of the jets, giving the inverted spectrum. However the optically thin jet contribution will tend to give a flatter integrated spectrum at radio wavelengths.

References

de la Force, C.J., *et al.*: 2001, in preparation.
Lestrade, J.-F., *et al.*: 1995, *Astron. Astrophys.* **304**, 182.
Stirling, A.M., *et al.*: 1998, *New Astron. Rev.* **42**, 657.
Stirling, A.M., *et al.*: 2001, *Mon. Not. R. Astron. Soc.*, in preparation.

VSOP OBSERVATIONS OF THE X-RAY BINARY LS I +61°303

M. PERACAULA and J.M. PAREDES
Departament d'Astronomia i Meteorologia, Universitat de Barcelona, Av. Diagonal 647, 08028 Barcelona, Spain

A.R. TAYLOR, S.M. DOUGHERTY and W.K. SCOTT
Department of Physics and Astronomy, University of Calgary, 2500 University Dr. NW, Calgary, T2N 1N4 Alberta, Canada

Abstract. We present preliminary results from the VSOP observations obtained of a radio outburst of the radio emitting X-ray binary system LS I +61°303. The image shows a central source surrounded by extended emission out to a radius of approximately 2 mas, and spanning along two preferred directions.

1. Introduction

The massive X-ray binary system LS I +61°303 shows regular nonthermal outbursts with a period of 26.5 days (Taylor and Gregory 1982, 1984). Each outburst lasts for about ten days, and seems to be formed by a superposition of different flares, with a rise to the peak flux density over a period of about two days (Taylor and Gregory, 1984; Peracaula *et al.*, 1997).

2. Observations and Analysis

In order to have a dynamical view of the evolution of the radio emitting region of LS I +61°303, the source was observed for 48 hours at 5 GHz, using HALCA and a ground array of 18 radio telescopes, including the VLA, the VLBA and the European VLBI network. In the left panel of Figure 1 the light curve of LS I +61°303, obtained with the Green Bank Interferometer at 8.3 GHz, from about 10 days before to 10 days after the observation, is represented. One cycle of the source periodic radio outburst is visible. As can be seen, the flux density around this peak was very irregular, with variations of the order of several tens of mJy.

In the right panel of Figure 1 we show the image of LS I +61°303 constructed from a 2 hour long segment of data where the flux density was relatively constant. The image shows a central source surrounded by extended emission out to a radius of approximately 2 mas. The extended components seem to span along two preferred directions: NE-SW and SE-NW. The strong variations in the GBI and

Astrophysics and Space Science is the original source of publication of this article. It is recommended that this article is cited as: *Astrophysics and Space Science* **276** (suppl.): 123–124, 2001.

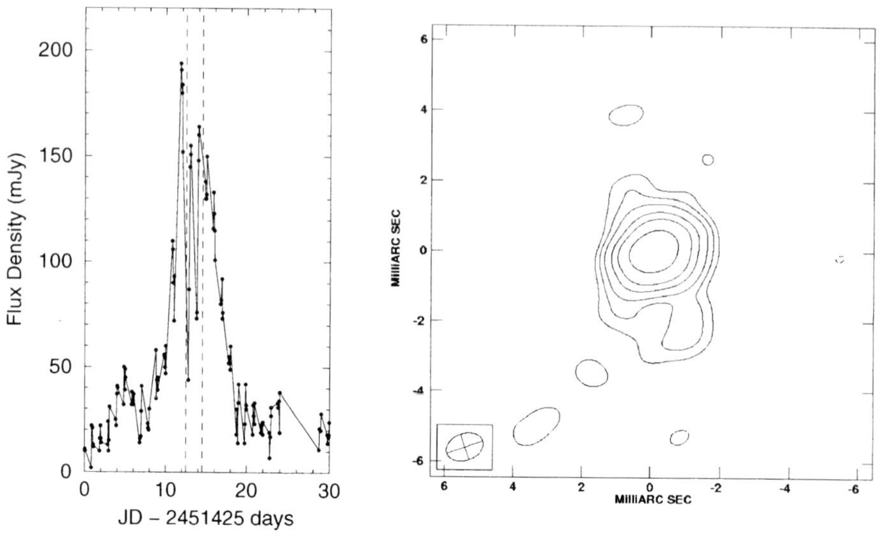

Figure 1. **Left:** Light curve of LS I +61°303 obtained with the Green Bank Interferometer at 8.3 GHz. The dashed lines indicate the interval of time when the VSOP observations took place. **Right:** Image of LS I +61°303 obtained from a 2 hour long subset of the data. Contours are represented at 1 mJy/beam × (−0.5, 0.5, 1, 2, 5, 10, 25). The map peak is 54.3 mJy/beam and the restoring beam FWHM is 1.1 × 0.76 mas at a PA = −71.2°.

VLA light curves suggest that we are seeing different flares that could be due to consecutive episodes of particle injection. The remaining data is currently being analyzed to check this hypothesis.

Acknowledgements

The GBI is a facility of the USA National Science Foundation operated by the NRAO in support of NASA High Energy Astrophysics programs. We gratefully acknowledge the VSOP Project, led by the Japanese Institute of Space and Astronautical Science in cooperation with many organizations and radio telescopes around the world. MP and JMP acknowledge partial support by DGESIC (PB97-0903). WKS, ART, and SMD acknowledge support from the Canadian Space Agency.

References

Peracaula, M., Martí, J. and Paredes, J.M.: 1997, *Astron. Astrophys.* **328**, 283.
Taylor, A.R. and Gregory, P.C.: 1982, *Astrophys. J.* **255**, 210.
Taylor, A.R. and Gregory, P.C.: 1984, *Astrophys. J.* **283**, 273.

ONE-SIDED ELONGATED FEATURE IN LS I +61°303

MARIA MASSI
Max Planck Institut für Radioastronomie, Auf dem Hügel 69, D-53121 Bonn, Germany

MARC RIBÓ, JOSEP M. PAREDES, MARTA PERACAULA and ROBERT ESTALELLA
Departament d'Astronomia i Meteorologia, Universitat de Barcelona, Av. Diagonal 647, E-08028 Barcelona, Spain

Abstract. An EVN map of LS I +61°303 presented in Massi *et al.* (2001) has shown the presence of a south-east extended structure. On the other hand, the measured expansion velocities for radio structures in LS I +61°303 seem small when compared to microquasars. We present here a possible explanation for both such a peculiar morphology and the low expansion velocity in the context of Doppler boosting.

1. LS I +61°303 in the Context of Doppler Boosting

LS I +61°303 is an X-ray binary system composed of a neutron star and a B0V star with a stable equatorial disk. It is controversial whether the system belongs to the so called class of microquasars, which have predictable luminosity in the X-ray range and relativistic radio jets.

LS I +61°303 is subluminous in the X-ray range (1–40 keV), having a luminosity of only 10^{35} erg s^{-1}. However, it is the most promising candidate to be the optical counterpart of the γ-ray source 3EG J0241+6103 (Hartman *et al.*, 1999), with a luminosity of 10^{37} erg s^{-1}. Therefore, LS I +61°303 could be another case of X-ray binary having the bulk of the energy shifted from X-ray to γ-ray wavelengths, like the recently discovered microquasar LS 5039 (Paredes *et al.*, 2000).

Several VLBI observations of LS I +61°303 have been undertaken in the past years, which have resolved the source on several occasions showing a structure of a few milliarcseconds (Massi *et al.*, 2000 and references therein). However, the measured expansion velocities apparently contradict the assumption that the outburst is related to super Eddington accretion events (Taylor *et al.*, 1992). The VLBI measurements give an expansion velocity in the range of (0.001–0.06)c. As reviewed by Mirabel and Rodríguez (1999), the expansion velocities for microquasars (as SS 433, Cyg X-3, GRS 1915+105, GRO J1655−40) range from 0.1c to 0.9c. The peculiarity of LS I +61°303 is evident. However, we note here that the measured velocity of (0.001–0.06)c is simply the apparent transverse velocity (v_a)

Astrophysics and Space Science is the original source of publication of this article. It is recommended that this article is cited as: *Astrophysics and Space Science* **276** (suppl.): 125–126, 2001.

and this might not be quite representative of the true velocity (v), especially for small values of the angle (θ) between the ejection and the line of sight.

Massi *et al.* (2001) performed an EVN observation 10 days after an outburst in 1994 June. The obtained map exhibits an extended structure elongated to the south-east. Assuming the source as a result of an expansion they derived a velocity of $0.004c$, in agreement with the above given range. The best fit of the visibility amplitudes and phases is obtained with a two-component model given by a point source and an elliptical gaussian component with position angle of 123 °. In some of the previous VLBI maps one may see extended structures at comparable position angle but not so pronounced. The asymmetry in the EVN map seems that of a one-sided jet typical to those due in the AGN to Doppler boosting. In this context we derive, from the peak value of our map and its noise level, $\beta \cos\theta > 0.3$, where $\beta = v/c$. On the basis of the relationship between apparent speed of the approaching component $\beta_a = 0.004$ and intrinsic speed β we derive: $\beta > 0.39$ and $\theta < 0.4°$. This lower limit of velocity of LS I +61°303 compares well with the other microquasars. We note that the combination of the values of θ and β we have found gives $\gamma = 1.1$ and a Doppler factor $\delta = 1.5$ for the approaching jet. Finally, the EVN one-sided jet seems to be aligned with a large jet-like structure, located 2' away, present in a deep VLA map. New observations with resolution intermediate between that of the VLA and VLBI are already in progress.

Acknowledgements

This research was partly supported by the DGESIC (PB97-0903), in Spain. MR acknowledges receipt of a fellowship from Generalitat de Catalunya (ref. 1999 FI 00199).

References

Hartman, R.C., Bertsch, D.L., Boom, S.D., *et al.*: 1999, *Astrophys. J. Suppl. Ser.* **123**, 79.

Massi, M., Ribó, M., Paredes, J.M., Peracaula, M. and Estalella, R.: 2001, *Astron. Astrophys.*, in press.

Mirabel, I.F. and Rodríguez, L.F.: 1999, *Annu Rev. Astron. Astrophys.* **37**, 409.

Paredes, J.M., Martí, J., Ribó, M. and Massi, M.: 2000, *Science* **288**, 2340.

Taylor, A.R., Kenny, H.T., Spencer, R.E. and Tzioumis, A.: 1992, *Astrophys. J.* **395**, 268.

THE X-RAY SPECTRUM OF THE JETS IN THE SS 433 SYSTEM USING THE CHANDRA HETGS

HERMAN L. MARSHALL, CLAUDE R. CANIZARES and NORBERT S. SCHULZ
MIT Center for Space Research

Abstract. Observations of SS 433 using the Chandra High Energy Transmission Grating Spectrometer are presented. Many lines of highly ionized elements are detected; all are detectably broadened by 2000–5000 km s^{-1} and the widths do not depend significantly on the characteristic emission temperature, suggesting that the emission occurs in a freely expanding region of constant collimation. Line ratios show that the base temperature of the jet is below 10 keV and may be closer to 2.5 keV. The X-ray luminosity during this observation was somewhat lower than average, which may be related to the lower jet temperature. The blue shifts of lines from low temperature gas are the same as those of high temperature gas, indicating that jet acceleration occurs closer to the compact object than $\sim 2 \times 10^{11}$ cm, where the emission lines originate. The image from zeroth order shows extended emission at a scale of $\sim 2''$, aligned in the direction of the radio jets.

1. Introduction

The Galactic X-ray binary SS 433 is the only known astrophysical system where red- and blue-shifted lines are observed from high Z elements. Lines from the H Balmer series were modelled kinematically (Margon *et al.*, 1981) as emission from opposing jets emerging from the vicinity of a compact object at $0.26c$ along a line inclined about 80° to the line of sight. There are radio jets which are oriented at a position angle of 100° (east of north) which show an oscillatory pattern that can be explained by helical motion of material flowing along ballistic trajectories (Hjellming and Johnston, 1981). The radio jets are about $2''$ long for a linear dimension of $\sim 3.5 \times 10^{19}$ cm, assuming a distance to SS 433 of 3 kpc.

EXOSAT X-ray observations showing that the 6.7 keV line appeared to shift with precession phase were interpreted as detecting Fe XXV from 10 keV gas (Watson *et al.*, 1986). Using Ginga, Brinkmann *et al.* (1991) concluded that kT was more like 30 keV while Kotani *et al.* (1996) obtained a value of 20 keV based on the ratio of Fe XXV and Fe XXVI line fluxes from ASCA data.

SS 433 was observed using the High Energy Grating Spectrometer (HETGS) while the compact object was not in eclipse and at a precession phase (0.6) that gives maximal separation between blue and red jets. An example of the high spectral resolution of the HETGS is shown by Canizares *et al.* (2000). An important feature of the HETGS is that spectra are obtained independently and simultaneously using the High Energy Gratings (HEG) and the Medium Energy Gratings

Astrophysics and Space Science is the original source of publication of this article. It is recommended that this article is cited as: *Astrophysics and Space Science* **276** (suppl.): 127–130, 2001.

(MEG). The HEG spectra have higher spectral resolution but lower effective area below 2 keV. Details of the spectroscopic analysis will be presented elsewhere.

2. Imaging

The zeroth order image is distinctly elongated in the east-west direction, which is the same orientation as the radio jet. Due to event pileup, we cannot determine if the image matches that of a point source within $1''$ of the core but profiles along the E-W and N-S directions are clearly different, as shown in Figure 1. There are 'shoulders' on the E-W profile that do not appear in the N-S profile. The total power in this extended component is less than 1% of that in the point source and is detectable out to $5''$ from the core or up to $\sim 3 \times 10^{17}$ cm.

3. Spectra

The X-ray spectra show strong emission lines of H- and He-like ions of Si, S, and Fe due to thermal gas at 3–6 keV and lines of Fe XXIV, Mg XII and Ne X from gas at $\sim$ 1 keV. There are two sets of lines: the blue system has a shift of $v/c = -0.078$ while the red system has a redshift of 0.153. The blue system lines are stronger than the corresponding red system lines, as expected from Doppler boosting. In most of what follows, the emission from the blue-shifted jet is modelled.

The lines are broadened by $\sim$ 2500 km s^{-1} (FWHM) as found in the (optical) H lines. The broadening is consistent with the range of line-of-sight velocities expected from a freely expanding, conical flow. From the FWHM, we infer an opening angle of $\sim 2°$. The redshifts of the He-like (resonance lines) and the H-like lines are identical, indicating uniform jet outflow because the bulk of the S XVI flux originates in hotter material closer to the jet base than the Si XIII flux – thus, we do not detect the jet acceleration zone. The line broadening is similar for all lines, indicating that the jet opening angle is the same for all line emission – thus, we are not detecting a 'nozzle' of expansion at the jet base. If the emitting plasma were entrained along the surface of a much faster jet, one might expect a 'hollow cone' geometry that would give rise to double-peaked emission lines which we do not observe. Thus, the cone is filled; i.e., the heavy elements are mixed into the bulk flow of the jet.

The flux ratios of the Fe, S and Si H-like to He-like lines gives temperature estimates near 2.5 keV. The Fe XXVI (H-like) line is too weak to allow temperatures in excess of 10 keV for emission measures comparable to that required for the Fe XXV line. The resonance to intercombination and forbidden flux ratio gives a similar temperature estimate for these ions.

A multi-temperature model of the combined spectra indicate that the Si XIII emission originates in gas with a temperature of order 1×10^7 K. The Si XIII

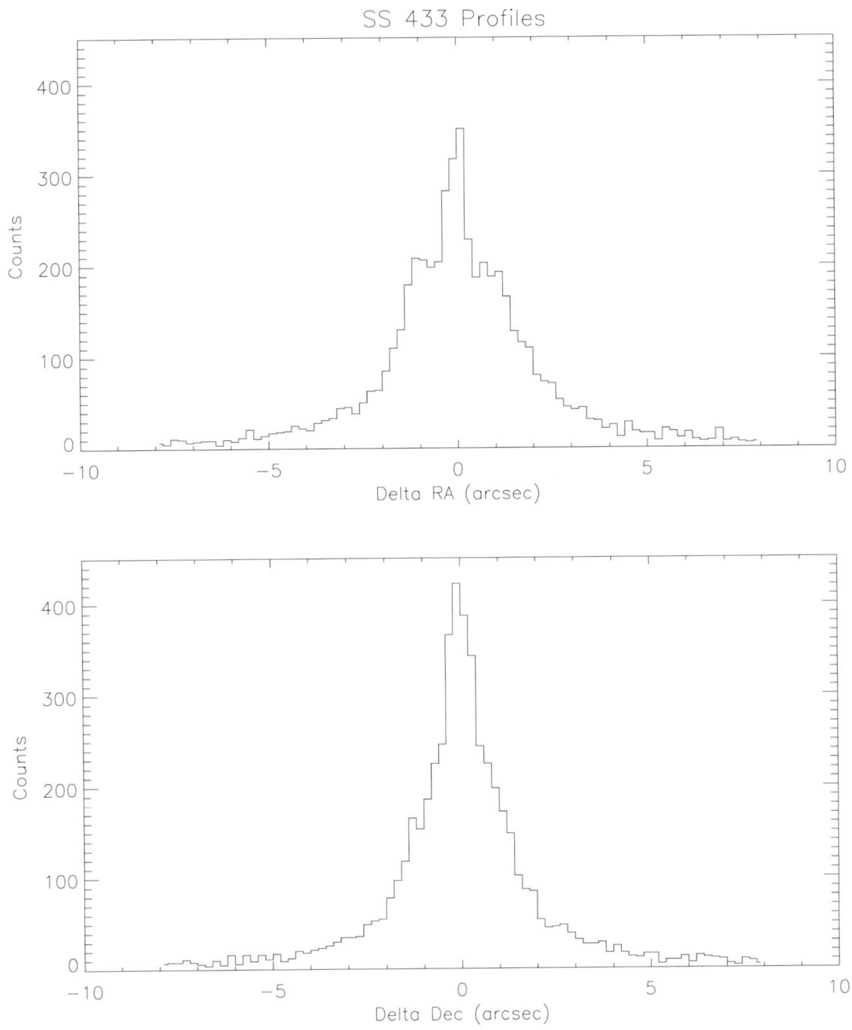

Figure 1. One dimensional image profiles of the zeroth order image of SS 433. *Bottom:* Profile along the north-south direction. The image is affected by pileup so the shape is not consistent with a Gaussian. *Top:* Profile along the east-west direction. The image is distinctly different than the north-south profile due to emission from a $1 - 2''$ scale jet.

forbidden lines are somewhat blended with the broadened resonance lines but are discernably weaker, indicating that the plasma density is high. Based on Gaussian fits to the lines in the triplet, we estimate that the electron density is $\sim 3 \times 10^{13}$ cm^{-3}. The total flux of the Si XIII lines and the estimated distance give us an estimate of the emission measure, assuming a filling factor f. For the Si XIII line, we find an emitting volume of $\sim 10^{34} f\ \mathrm{cm}^3$. Given the line widths and an assumed conical geometry, this volume corresponds to a distance from the core of $2 \times 10^{11 f - 1/3}$ cm, which is much smaller than the size scale of the radio and extended X-ray jets.

Although a simple absorbed power law gives a good fit to the continuum, it is dominated by bremsstrahlung from the same thermal plama that produces the emission lines. Thus, we find no evidence for nonthermal emission from the central source.

4. Conclusions

The X-ray spectrum of SS 433 was used to place constraints on the the acceleration of the material in the jet: it appears to arise in a center-filled cone with constant opening angle and constant flow velocity. For one temperature zone, the density is high enough to suppress the forbidden line and can be used to estimate the distance of the emission from the core.

Acknowledgements

This work has been supported in part under NASA contracts NAS8-38249 NAS8-39073 and SAO SV1-61010.

References

Brinkmann, W., Kawai, N., Matsuoka, M. and Fink, H.H.: 1991, *Astron. Astrophys.* **241**, 112.
Canizares *et al.*: 2000, *Astrophys. J. Lett.* **539**, L41.
Hjellming, R. and Johnston, S.: 1981, *Astrophys. J. Lett.* **246**, L141.
Kotani, T., Kawai, N., Matsuoka, M. and Brinkmann, W.: 1996, *Publ. Astron. Soc. Jpn.* **48**, 619.
Margon *et al.*: 1981, *Astrophys. J.* **246**, L141.
Watson, M.G., Stewart, G.C., Brinkmann, W., King, A.R., *et al.*: 1986, *Mon. Not. R. Astron. Soc.* **222**, 261.

RADIO PROPERTIES OF THE COMPACT JETS AND THE EQUATORIAL EMISSION REGION IN SS433

Z. PARAGI* and I. FEJES
FÖMI Satellite Geodetic Observatory, H–1373 Budapest, P.O. Box 546, Hungary
E-mail: paragi@sgo.fomi.hu

R.C. VERMEULEN
NFRA, P.O. Box 2, 7990 AA Dwingeloo, The Netherlands

R.T. SCHILIZZI
JIVE, P.O. Box 2, 7990 AA Dwingeloo, The Netherlands
Leiden Observatory, P.O. Box 9513, 2300 RA Leiden, The Netherlands

R.E. SPENCER
NRAL, Jodrell Bank, Macclesfield, Cheshire, SK11 9DL, United Kingdom

A.M. STIRLING
Univ. Central Lancashire, Preston, Lancashire, PR1 2HE, United Kingdom

Abstract. Results from multi-frequency VLBA experiments and a 1.6 GHz global VLBI observation of SS433 in 1998 are briefly summarized.

1. The Compact Jets of SS433 and RJXRBs

The innermost radio jets in SS433 are quasi-continuous, rather than knot-like (Vermeulen *et al.*, 1993). Within this compact jet region ('core-wings') there is a $\sim$ 25 AU central gap, which harbours the binary stellar system (Paragi *et al.*, 1999). The existence of this Central Radio Gap is confirmed in our VLBA monitoring experiments in 1998. Its size decreases slightly with observing frequency (7–8 mas at 5 GHz, 5–6 mas at 15–22 GHz), and depends on the precessional phase as well. There is also secular time variability unrelated to the changing orientation. To explain the gap, on top of synchrotron self-absorption we need free-free absorption by either a well bounded disk around the binary system with 25–30 AU diameter, or an ionization front. The variable absorbing environment could be explained by a variable equatorial outflow from the accreting environment. Existence of both slow outflow (Chakrabarti and Matsuda, 1992) and fast wind (King *et al.*, 2000) have been predicted in SS433.

* Financial support is acknowledged from the Hungarian Space Office (MŰI), the Netherlands Organization for Scientific Research (NWO) and the Hungarian Scientific Research Fund (OTKA, grant no. N31721 & T031723).

Astrophysics and Space Science is the original source of publication of this article. It is recommended that this article is cited as: *Astrophysics and Space Science* **276** (suppl.): 131–132, 2001.

The unusually flat spectra of radio-jet X-ray binaries (RJXRB) up to the mm regime have been interpreted by conical jet models (Fender *et al.*, 2000). The first Galactic radio-jet source, discovered to have a flat-spectrum compact jet region that can be described with conical jet models, was SS433 (Hjellming and Johnston, 1988; Vermeulen *et al.*, 1993; Paragi *et al.*, 1999). Recently Cyg X-1 (Stirling *et al.*, 1998), GRS 1915+105 (Dhawan *et al.*, 2000) and LS 5039 (Paredes *et al.*, 2000) were shown to have a faint compact jet region. However, a frequency-dependent core-shift (an indicator of self-absorbed compact jets) has not yet been observed in any of these sources.

2. The Equatorial Emission Region in SS433

An emission region spanning hundreds of AU, perpendicular to the radio beams, was discovered in SS433 (Paragi *et al.*, 1999). The presence of this Equatorial Emission Region is confirmed in our 1.6 GHz global VLBI experiment, but also in some 5 GHz VLBA observations in 1998. Its structure is variable with timescales much shorter than a year, but considerably longer than a day. The brightness temperature is too high for thermal emission, and it is probably due to optically thin synchrotron radiation originating in shocks in the ISM around the central engine. It was proposed by S. Fabrika (priv. comm.) that shocks develop due to the interaction of the slow outflow in the plane of the accretion disk and the faster disk wind, as the disk slowly turns during the precessional cycle. As similar 'equatorial' radio components seem to exist in Cyg X-1 (Stirling *et al.*, 1998), and there is evidence from IR spectral lines for an enchanced equatorial outflow in Cyg X-3 as well (Fender *et al.*, 1999), this phenomenon might be common in RJXRBs.

References

Chakrabarti, S.K. and Matsuda, T.: 1992, *Astrophys. J.* **390**, 639.
Dhawan, V., Mirabel, I.F. and Rodríguez, L.F.: 2000, *Astrophys. J.* **543**, 373.
Fender, R.P., Hanson, M.M. and Pooley, G.G.: 1999, *Mon. Not. R. Astron. Soc.* **308**, 473.
Fender, R.P., Pooley, G.G., Durouchoux, P., Tilanus, R.P.J. and Brocksopp, C.: 2000, *Mon. Not. R. Astron. Soc.* **312**, 853.
Hjellming, R.M. and Johnston, K.J.: 1988, *Astrophys. J.* **328**, 600.
King, A.R., Taam, R.E. and Begelman, M.C.: 2000, *Astrophys. J.* **530**, L25.
Paragi, Z., Fejes, I., Vermeulen, R.C., Schilizzi, R.T., Spencer, R.E. and Stirling, A.M.: 1999, *Astron. Astrophys.* **348**, 910.
Paredes, J.M., Martí, J., Ribó, M. and Massi, M.: 2000, *Science* **288**, 2340.
Stirling, A.M., Spencer, R.E. and Garrett, M.: 1998, *New Astron. Rev.* **42**, 657.
Vermeulen, R.C., *et al.*: 1993, *Astron. Astrophys.* **270**, 177.

THE GALACTIC MICROQUASAR W50 / SS 433 SYSTEM

SAMAR SAFI-HARB*
University of Manitoba & NSERC, Canada
E-mail: samar@physics.umanitoba.ca

PHILIPPE DUROUCHOUX
CEA, Saclay, France; E-mail: durvla@discovery.saclay.cea.fr

ROBERT PETRE
NASA/Goddard Space Flight Center; E-mail: rob@lheamail.gsfc.nasa.gov

Abstract. We review the X-ray and millimeter observations of the W50 lobes associated with the twin-jet source SS 433. This study is aimed at shedding light on the energetics from SS 433, understanding the asymmetries of the lobes, and testing for cosmic ray acceleration in a Galactic twin-jet source.

1. X-ray Observations of the W50 Lobes

W50 is the large supernova remnant (SNR) associated with the binary source SS433, famous for its twin-precessing jets. For a review of SS 433 see Kotani (1997) and references therein. *ROSAT* and *ASCA* observations of W50 are summarized in Safi-Harb and Ögelman (1997). The X-ray images show two knotty X-ray lobes filling the space between SS 433 and the radio ears. The X-ray spectrum softens away from SS 433. Both thermal and non-thermal models give adequate fits to the *ROSAT* plus *ASCA* spectra.

W50 was later observed with the Rossi X-ray Timing Explorer (RXTE) to pin down the nature of its X-ray spectrum (Safi-Harb and Petre, 1999). Fitting the PCA spectrum of the *eastern* lobe in the 10–20 keV range, a power law model gives a better fit than a thermal bremsstrahlung model (Figure 1, left). The parameters derived from fitting the *RXTE* data with a thermal model are inconsistent with those derived using the *ASCA* data (Figure 1, right). We conclude that the spectrum is non-thermal.

The *western* lobe was recently observed with *RXTE*. Our preliminary results indicate the presence of an Fe-K emission line, which is likely associated with the Galactic ridge. A power law model yields a photon index $\Gamma = 1.7 - 2.0$, while a thermal bremsstrahlung model yields a very high temperature $kT = 6.5 - 10.2$ keV (90% confidence level).

* NSERC fellow

Astrophysics and Space Science is the original source of publication of this article. It is recommended that this article is cited as: *Astrophysics and Space Science* **276** (suppl.): 133–134, 2001.

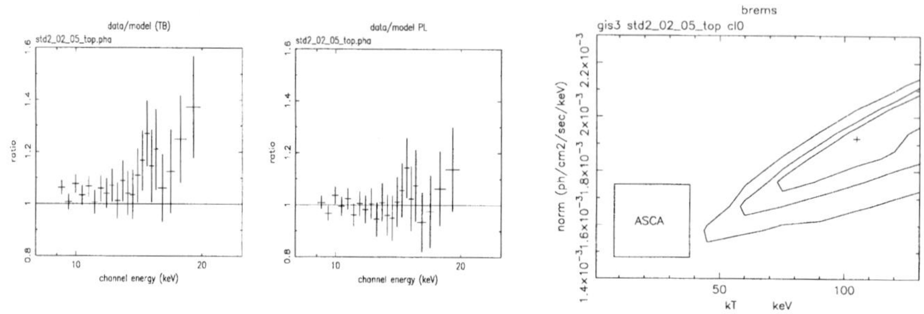

Figure 1. Left panel: The SS 433 *RXTE* PCA data to fitted model. Right panel: Confidence contours for fitting the *RXTE* data with a thermal model.

2. Millimeter Observations

W50 was observed with *NOBEYAMA*, *SEST*, & *MOPRA*. No emission was found from the eastern lobe. In the western lobe, a 50 km s^{-1} CO emission line was found along the jet axis and correlated with the X-ray spots detected with *ROSAT* (Durouchoux *et al.*, 1998). The implied distance is 3.24 ± 0.46 kpc, in agreement with Dubner *et al.* (1998).

3. Summary and Conclusions

The non-thermal X-ray emission from the eastern lobe is suggestive of synchrotron emission from the high-energy electrons accelerated at the jet-SNR shell interaction site, and hints that W50 could be an important site for cosmic ray acceleration. Millimeter observations of the western lobe show shock-excited regions arising from the western jet interacting with a molecular cloud. The total radiated luminosity from W50 is a very small fraction of the kinetic power of the SS 433 jets. Most of the power goes into bloating W50 in the east-west direction and energizing the surrounding medium.

References

Dubner, G., *et al.*: 1998, *Astron. Astrophys.* **116**, 1842.
Durouchoux, P., *et al.*: 1998, in: *Proc. of the 32nd COSPAR meeting*, Nagoya, Japan.
Kotani, T.: 1997, Ph.D. thesis, University of Tokyo, ISAS, RN 655.
Safi-Harb, S. and Ögelman, H.: 1997, *Astrophys. J.* **483**, 868.
Safi-Harb, S. and Petre, R.: 1999, *Astrophys. J.* **512**, 784.

MONITORING AND MODELING RADIO FLARES FROM MICROQUASARS

SERGEI TRUSHKIN, ELENA MAJOROVA and NIKOLAI BURSOV
Special Astrophysical Observatory of Russian AS

Abstract. We present results of long-term daily monitoring of a sample of Galactic radio-emitting X-ray binaries showing relativistic jets (RJXRB): SS433, Cyg X-3, and GRS 1915+105, with the RATAN-600 radio telescope in the 0.6–22 GHz range. We carried out the modeling calculations to understand the temporal (1–100 days) and spectral (1–22 GHz) dependence. We tested the finite jet segment models and we found that the geometry of the conical hollow jets is responsible for either a power law or an exponential decay of the flares. SS433 was monitored for 100 days in 1997 and 120 days in 1999. From the quiescent radio light curves, we obtained clear evidence of a 6.04-day 10–15% modulation. Three powerful flares (up to 13 Jy) from Cyg X-3 were detected in April 2000.

Results of the Multi-Frequency Daily Monitoring

Many (or all!) of the radio sources related with X-ray binaries have been resolved into relativistic jets, often with detectable proper motion of the radio-emitting blobs. We present results of long-term daily monitoring of a sample of these RJXRBs with the RATAN-600 radio telescope at 0.96, 2.3, 3.9, 7.7, 11.2 and 21.7 GHz. Many strong optically thin and thick flares were detected during the monitoring period (see Bursov and Trushkin, 1996; Trushkin, 1998). The basic model of the flares is the synchrotron emission evolution of two relativistic jets with conical geometry and an adiabatic expansion of the blobs, which are composed of relativistic electrons embedded in magnetic fields, moving away from the central source (Shklovskii, 1960; van der Laan, 1966; Hjellming and Johnston, 1988; Marti *et al.*, 1992). The flare decay law is determined by the geometry of the conical hollow jets. The jet velocity, thermal and relativistic electron densities, and magnetic field intensity within the blobs can be derived from the fits.

Our monitoring of radio variability of RJXRB shows that the decay of a flare follows a power law, $S_\nu = S_0 t^{-2p}$, as predicted by the Shklovskii and van der Laan models, or an exponential law $S_\nu = S_0 e^{-t/\tau}$ (Figure 1), which is due to the initial exponential increase of the jet cone radius. Often $\tau \sim \nu^\beta$, where β ranged from –0.8 to –0.4; these flares decay faster at higher frequencies (see current light curves and the spectra animation at `http://cats.sao.ru/~satr/XB/C`).

We computed the power spectra of the daily light curves of SS433 in the 2.3–22 GHz range during the long quiescent periods May–July 1997 and May–August

Astrophysics and Space Science is the original source of publication of this article. It is recommended that this article is cited as: *Astrophysics and Space Science* **276** (suppl.): 135–137, 2001.

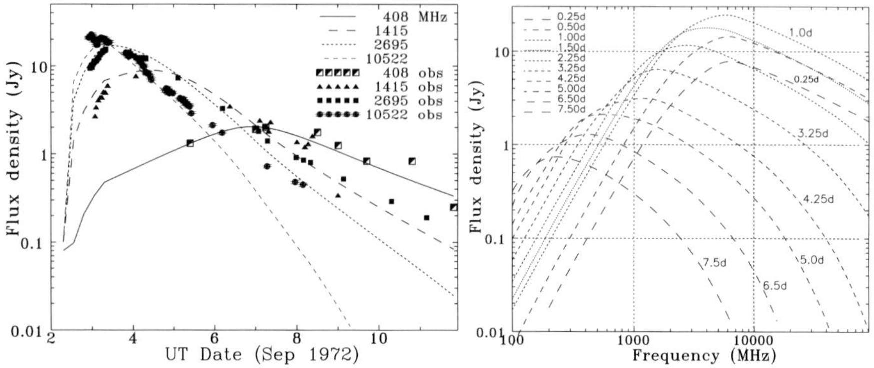

Figure 1. The model light curves at select frequencies (left) with the real data and the spectra of the famous Cyg X-3 flare at 2–11 September 1972 (right). The model parameters are from Marti *et al.* (1992).

1999. Both spectra show a mean harmonic at 0.165 ± 0.02 day^{-1}, which corresponds to a period of 6.04 ± 0.06 days.

We propose that Doppler boosting of the jet radiation could explain such modulation. The 5-elements kinematical model of the jets and ephemerides proposed by Vermeulen (1989) for the nodding motion of the jets and common formulas for boosting produce a modulation similar to the observed one. Thus we detect a 6-day modulation at the 10–15% level in the quiescent flux density, which is $S_\nu[\mathrm{Jy}] = 1.2\ \nu_{\mathrm{GHz}}^{-0.6}$.

Three powerful flaring events (up to 13 Jy) from Cyg X-3 were detected in April 2000. The daily spectra clearly show two components: i) a quasi-stable steep spectrum and ii) a very unstable flat spectrum.

In the wide range 0.6–22 GHz the optically thin spectrum of the detected strong flare of GRS 1915+10 (5 August, MJD 51751.85) is well fitted by the power law: $S_\nu[\mathrm{Jy}] = 0.690\ \nu_{\mathrm{GHz}}^{-0.43}$.

Acknowledgements

The authors are grateful to RFBR for supporting the project of monitoring X-ray binaries, grant N98-02-17577.

References

Bursov, N.N. and Trushkin, S.A.: 1995, *Pis'ma in Astron. Zh.* **21**, 163.
Hjellming, R. and Johnson, K.J.: 1988, *Astrophys. J.* **328**, 600.
Martí, J., Paredes, J.M. and Estalella, R.: 1992, *Astron. Astrophys.* **258**, 309.
Shklovskii, I.S.: 1960, *Soviet Astron.* **4**, 243.

Trushkin, S.A.: 1998, *Pis'ma in Astron. Zh.* **24**, 19.
van der Laan, H.: 1966, *Nature* **211**, 1131.
Vermeulen, R.C.: 1989, Ph. D. Thesis, University of Leiden, The Netherlands.

ELONGATED SNRS: POSSIBLE NEW JET SOURCES

PH. DUROUCHOUX[1], R. SOOD[2], S. SAFI-HARB[3], F. J. LU[4], P. O'NEILL[2], H. FLOHIC[1] and F. LEFEVRE[1]

[1] *CEA Saclay, DSM/DAPNIA/Service D'Astrophysique, Gif sur Yvette, France*
[2] *School of Physics, ADFA, University of NSW, Canberra, Australia*
[3] *GSFC, NASA, Greenbelt, Maryland, U.S.A.*
[4] *IHEP, Beijing, Republic of People of China*

Abstract. We summarise the properties of selected supernova remnants (SNRs) which have elongated envelopes that could house relativistic jets. We present MOPRA and SEST millimeter observations of these objects, showing correlation between known X-ray hot spots and molecular cloud density as signified by the detection of a CO(1-0) line flux.

1. Introduction

Some well known SNRs, e.g. Crab, SS 433, are known to be associated with relativistic jets and are thus natural laboratories for studying the jet interaction with the surrounding material (e.g. Gaensler, 1998). While examining SNR catalogues, we have found many more sources with elongated shapes which could be interpreted in terms of jet emission. Some of our observations are summarised below.

2. Observations of Sources and Results

SS 433: The X-ray binary SS 433, responsible for two oppositely directed relativistic jets, is located near the centre of the SNR W50. The jets are aligned with the radio wings of W50. We have observed a large molecular cloud in the direction of SS 433 / W50, and show that the CO(1-0) 50 km s^{-1} line flux is strongly correlated with the system's X-ray hot spots as detected by ROSAT and ASCA.

G309.2-00.6: G309.2-00.6 is a SNR analogous to SS 433, with an appearance indicating opposite jets or outflows from a central source. A jet like feature, and breaks in the SNR shell can both be seen along the proposed axis of the outflow. We detected a molecular cloud (–43 km s^{-1}) in the direction of G309.2-00.6. The CO(1-0) line flux showed a correlation with the radio map of the source. Moreover, there was a correlation between the maxima of the mm line flux and the ASCA X-ray intensity for this source, again indicating an interaction of the jet with the ambient medium.

Astrophysics and Space Science is the original source of publication of this article. It is recommended that this article is cited as: *Astrophysics and Space Science* **276** (suppl.): 139–140, 2001.

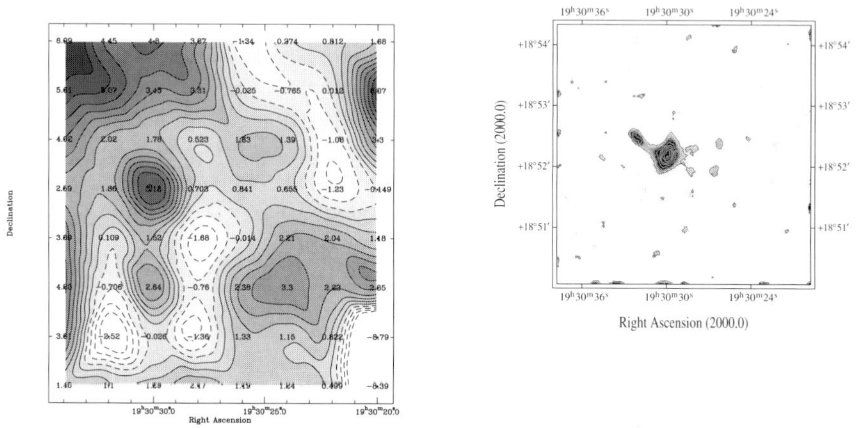

Figure 1. Millimeter (left) and X-ray (right) maps of G54.1+0.3.

G41.1-0.3: The 3C397 complex consists of two radio sources, the East source which is a SNR (G41.1-0.3), and the West source, which is an HII region. G41.1-0.3 has an elongated shape in the radio, with brighter regions to the east and west. We observed a strong correlation between the CO(1-0) flux and the X-ray hot spot. The evidence for the interaction of a jet with the surrounding medium was strengthened by the detection of a CO(2-1) line flux which traces out the jet axis towards the west.

G54.1+0.3: G54.1+0.3 was identified as a Crab-line SNR in both the radio and X-ray bands. In high resolution VLA maps, G54.1+0.3 has a filled-centre brightness distribution which extends to the northeast and north, making this object a jet source candidate. The source has recently been shown to contain a 40″ X-ray jet (Lu *et al.*, 2000, these Proceedings). Our mm observations show that the CO(1-0) maximum strongly coincides with the X-ray maximum (Figure 1).

3. Conclusions

Millimeter observations of galactic sources have shown that when a molecular cloud association is established with an elongated SNR, signatures of interactions with the ambient medium can be determined via CO(1-0) and CO(2-1) transitions. Such elongated SNRs may be further samples of galactic jet sources.

References

Gaensler, B.M.: 1998, *Astrophys. J.* **493**, 781.
Lu, F.J.: 2000, these proceedings.

DISCOVERY OF AN X-RAY JET ASSOCIATED WITH SUPERNOVA REMNANT G54.1+0.3 *

F.J. LU[1,2], B. ASCHENBACH[2], L.M. SONG[1] and PH. DUROUCHOUX[3]
[1] *Laboratory of Cosmic Ray and High Energy Astrophysics, IHEP, Beijing 100039, China*
[2] *Max-Planck-Institut für Extraterrestrische Physik, Postfach 1312, D-85741 Garching, Germany*
[3] *CE Saclay, DSM, DAPNIA, Service d'Astrophysique, 91191 Gif sur Yvette Cedex, France*

Abstract. We present in this paper the discovery of an X-ray jet associated with the Crab-like supernova remnant G54.1+0.3. It points to the northeast with a length about 40″ from the center of the nebula and has a flux $\sim$ 1/6 of the whole remnant. The X-ray jet is consistent with the radio extension to the northeast, suggesting that it is connected with the pulsar embedded in the remnant.

1. Introduction

The radio source G54.1+0.3 was identified as a Crab-like SNR with radio extensions to the northeast and north, which are reminiscent of radio jets (Figure 1, left panel) (Reich *et al.*, 1985; Velusamy and Becker, 1988). X-rays from G54.1+0.3 were firstly detected by the *Einstein* IPC but no extent to the X-ray emission was found (Seward, 1989). We present here its X-ray image and spectrum from *ROSAT* PSPC and *ASCA* GIS and SIS data.

We use the *ROSAT* PSPC data to derive the X-ray image, while the *ASCA* SIS and GIS data are used in the spectral and temporal analyses, respectively. In order to improve the spatial resolution, we have deconvolved the X-ray image obtained by the *ROSAT* PSPC with the Lucy formula (Lucy, 1974). In the iteration process the mean background is used as the lower limit constraint, in order to improve the quality of the deconvolved image, as proposed by Li and Wu (1994).

2. Results and Discussion

The deconvolved X-ray image (0.1–2.4 keV) is shown in Figure 1 (right panel), in which a jet-like feature (hereafter JLF) pointing to the northeast appears, in addition to the central bright nebula (hereafter CBN) coinciding with the brightest radio region. The angular distance from the head of the JLF to the center of the CBN is about 40″. The total photon flux of the CBN is about 2.07×10^{-2} counts s^{-1}, and that of the JLF is 3.9×10^{-3} counts s^{-1}.

* Send offprint requests to: F.J. Lu *lufj@astrosv1.ihep.ac.cn*

Astrophysics and Space Science is the original source of publication of this article. It is recommended that this article is cited as: *Astrophysics and Space Science* **276** (suppl.): 141–142, 2001.

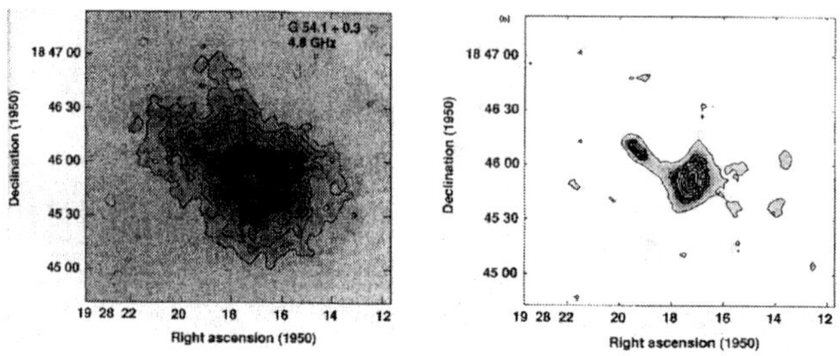

Figure 1. SNR G54.1+0.3. *(Left)* – The radio map obtained by VLA at 4.8 GHz (Velusamy and Becker, 1988). *(Right)* – The 0.1-2.5 keV X-ray intensity map obtained by *ROSAT* PSPC and image deconvolution. The contour intervals are linear with a step size of 4×10^{-5} cnt s^{-1} per $2'' \times 2''$ pixel. The lowest contour corresponds to 2×10^{-5} cnt s^{-1} per pixel.

We find that the power law spectral model gives the best fit of the data obtained by *ASCA* SIS. The derived parameters are: photon index α – $1.9^{+0.2}_{-0.2}$, column density N_H $17.9^{+2.8}_{-2.5} \times 10^{21}$ cm^{-2}, 0.7–10.0 keV energy flux 1.138×10^{-11} erg cm^{-2} s^{-1}, reduced χ^2 0.7. The N_H value indicates a distance comparable with the radius of the galaxy, $\sim$ 10 kpc. The X-ray luminosity of G54.1+0.3 is $1.4 \times 10^{35} d_{10}^2$ erg s^{-1}, where d_{10} is the distance to G54.1+0.3 in unit of 10 kpc.

We examined the *ASCA* GIS data for temporal variability and have not detected any pulsation with significance of more than 3σ between 0.01 and 2.0 seconds. The upper limit of the pulsed fraction is 31% with the confidence of 90%.

Figure 1 shows nice coincidence of the X-ray JLF and radio extensions. It suggests that (1) the JLF is indeed an X-ray jet produced by the nonthermal electrons and that (2) these electrons are probably connected with the putative pulsar.

References

Li, T.P. and Wu, M.: 1994, *Astrophys. Space Sci.* **215**, 213.
Lucy, L.: 1974, *Astron. J.* **79**, 745.
Reich, W., *et al.*: 1985, *Astron. Astrophys.* **151**, L10.
Seward, F.D.: 1989, *Astron. J.* **97**, 487.
Velusamy, T. and Becker, R.H.: 1988, *Astron. J.* **95**, 1162.

Chapter III
INNER DISK OSCILLATIONS

Conclusions

INNER DISK OSCILLATIONS

TOMASO BELLONI
Osservatorio Astronomico di Brera, Merate, Italy

Abstract. The RXTE observations of GRS 1915+105 have given a new impulse to the study of spectral and timing properties of the X-ray emission of black hole candidates. At variance with any other known source, GRS 1915+105 shows dramatic changes on time scales as short as a tenth of a second. These changes are associated to marked spectral changes which have been interpreted as changes of the observable inner radius of the accretion disk. I review the existing results and discuss the current evidence for such disk oscillations. Independently of the precise theoretical models, the detailed study of these fast variations can provide an extremely valuable insight on the accretion processes onto black holes. Making use of these results, I compare the properties of GRS 1915+105 with those of other black hole candidates.

1. Introduction: Black Hole Candidates

In the recent years, a classification scheme for the X-ray emission of black hole candidates has emerged. From the timing and spectral properties, four separate states (plus the quiescent state for transient systems) have been identified, probably in dependence of the accretion rate level (see van der Klis, 1995). Unfortunately, all known persistent sources, Cyg X-1, GX 339-4, LMC X-1 and LMC X-3, show state transitions very rarely, making the accumulation of extensive datasets on different states difficult. Transient systems are more promising, but their transient nature limits the number of states and state transitions that can be observed from one system, and the comparison between different systems is always problematic.

The spectral properties of black hole candidates are usually characterized in terms of a double-component model. The first component is thermal and (when present) contributes mostly to the flux below 10 keV. It is commonly interpreted as emission from an optically thick accretion disk (Mitsuda *et al.*, 1984). The second component is much harder and extends above 10 keV. In its simple form, it can be approximated as a power law, although more sophisticated models are often required (see e.g. Frontera *et al.*, 2000). An attractive feature of the optically thick disk model, the so-called *disk-blackbody*, is that one of its parameters is the value of the inner radius of the accretion disk. Therefore, the application of this model can in principle yield the measurement of this fundamental parameter, although a precise value depends on the distance and the inclination of the system. This model became rather popular when a number of black hole systems showed a rather constant inner disk radius, around 20–30 km, despite large changes in accretion rate (see Tanaka

Astrophysics and Space Science is the original source of publication of this article. It is recommended that this article is cited as: *Astrophysics and Space Science* **276** (suppl.): 145–152, 2001.

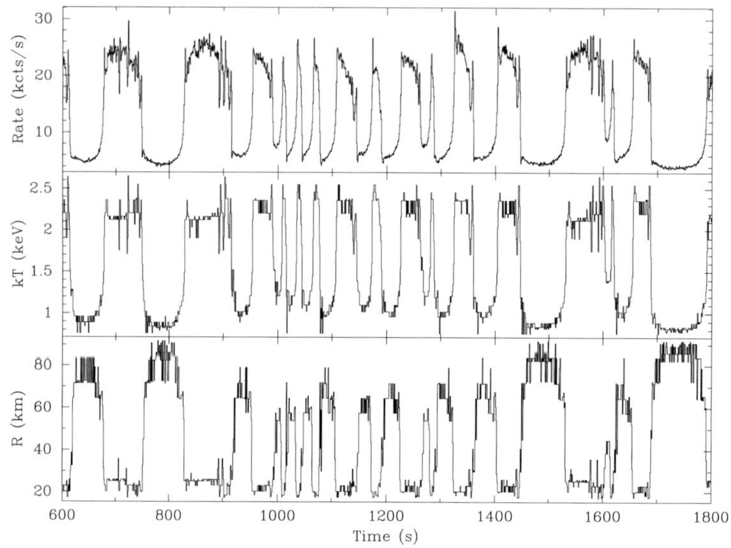

Figure 1. Count rate, inner disk temperature and radius as a function of time of the 1997 Jun 18th of GRS 1915+105 (from Belloni *et al.*, 1997b).

and Lewin, 1995). Although this model is simplified and leads to an underestimate of the values of the radii, it can detect large variations in the inner radius of the disk (see Merloni, Fabian and Ross, 2000).

2. GRS 1915+105: Inner Disk Oscillations

The galactic Microquasar GRS 1915+105 is known to exhibit dramatic variability in the soft (1–20 keV) energy band (see Greiner, Morgan and Remillard, 1996; Belloni *et al.*, 2000) and since its appearance in 1992 it has remained active up to the time of writing. This variability is accompanied by strong spectral changes (Greiner, Morgan and Remillard, 1996; Belloni *et al.*, 1997a,b; Muno, Morgan and Remillard, 1999), providing an ideal system to study spectral transitions in a black-hole candidates: the source is always observable and significant spectral variability is present on very short time scales. At the same time, a complex behavior is observed in the power density spectra of the source (see Rao, this volume).

Belloni *et al.* (1997a,b) proposed a model for the observed spectral variations, based on fits to RXTE/PCA data with the approximated spectral shape described above. They interpret the observations as the onset of a thermal-viscous instability, during which the innermost part of the accretion disk becomes unobservable and is slowly refilled from the outer parts. This interpretation is based on the measurement of a variable inner disk radius from energy spectra (see Figure 1) and is supported

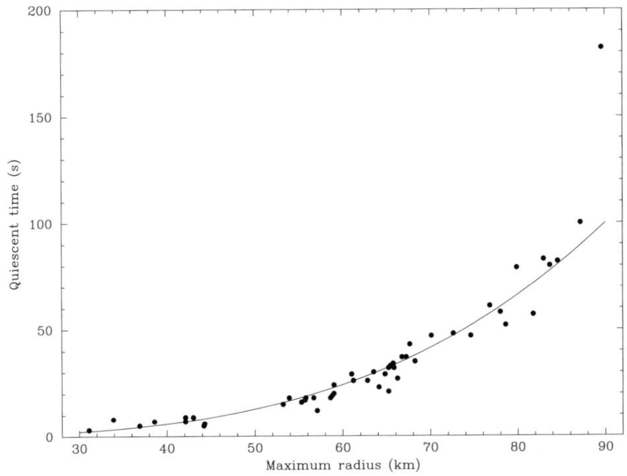

Figure 2. Relation between disk re-fill time and inner disk radius for the same observation shown in Figure 1 (from Belloni *et al.*, 1997b).

by the observation of the expected dependence between size of the unstable region and refill time (Figure 2). The observed radius variations are too large to be attributed to spurious effects due to the approximate form of the model. This is what I will refer to as 'inner disk oscillations'. The modeling by these authors was rather qualitative (with the exception of the estimate of the refill time scale), but other authors have developed more accurate models for this process (Szuszkiewicz and Miller, 1998; Nayakshin, Rappaport and Melia, 2000; Janiuk *et al.*, 2000).

3. Spectral States of GRS 1915+105

Belloni *et al.* (2000) analyzed a large number of RXTE/PCA observations of GRS 1915+105 and obtained a subdivision of the observed phenomenology into 12 separate classes. From this classification, they identified three basic states, the alternation of which cause all of the observed variability. The three classes, called A, B and C, are shown schematically in Figure 3, which represents a color-color diagram (both colors increase with increasing hardness of the spectrum).

These states correspond to the three states already identified by Markwardt, Swank and Taam (1999). Class B correspond to the 'normal' state of a black hole candidate at high accretion rate (the very high state), with an optically thick accretion disk extending down to the last stable orbit and a steep power-law component. State C corresponds to the instability periods: the accretion disk stops at a larger radius than in state B, while the power law component is harder. State A is a new state, not recognized earlier. It corresponds to an accretion disk like in state B, but with a lower temperature, and therefore lower local accretion rate. The relatively fast transition time between A and B is consistent with being the viscous time scale at the innermost stable orbit around the black hole. The observed variability

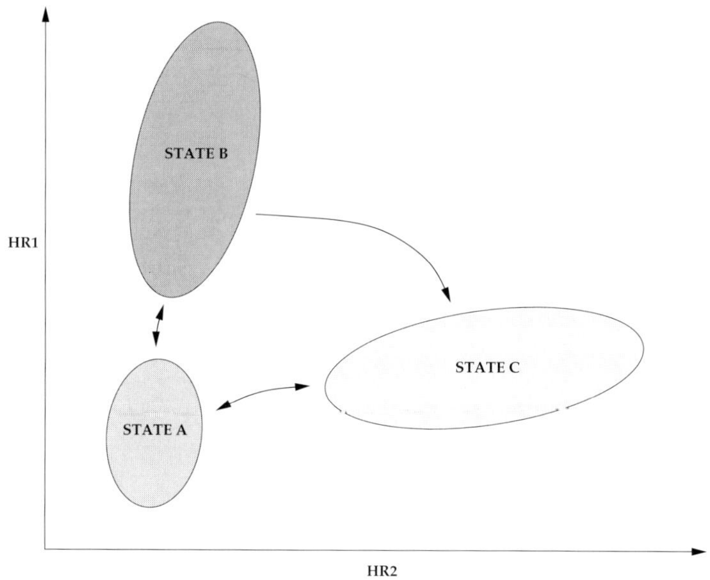

Figure 3. Schematic color-color diagram showing the basic A/B/C states and their observed transitions (from Belloni *et al.*, 2000).

consists of transitions between these three states: all possible transitions between two of the states are observed, with the exception of C to B transitions, of which no example has been found.

There are intervals of time, even as long as a month, when the source is found only in state C. These are the 'plateau' intervals observed in the radio band (see Fender, 1999). These are consistent with being long instability intervals, when a large missing inner disk is slowly refilled at a lower external accretion rate (Belloni *et al.*, 2000). On the other hand, there are very fast variations as well. During some observations, variability of a factor of five in 0.1 seconds has been observed, corresponding to fast A–B/B–A transitions (Belloni *et al.*, 2000).

4. Timing Properties

It is important to connect what is observed in the time domain and the state transitions described above. As a first step, I consider the three types of QPOs observed in GRS1915+105:

- 1–10 Hz QPO: (see Rao, this volume). This QPO is only observed during state C. Its central frequency varies systematically with time, count rate and hardness. Since state C is the only state when systematic changes in the inner disk radius are observed, it is natural to associate this oscillation to the inner radius, although often the missing part of the disk is so large that no disk

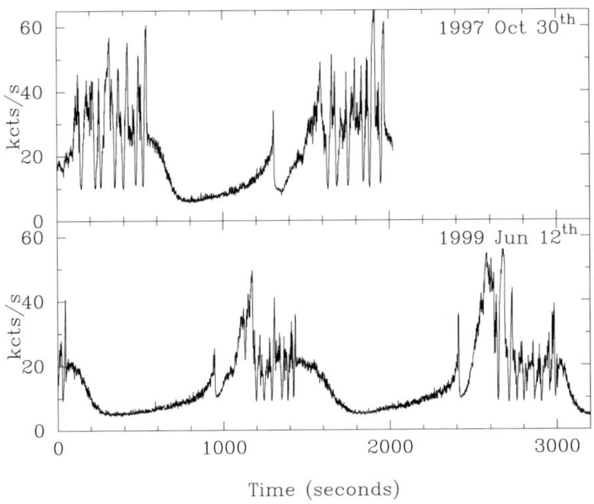

Figure 4. Two PCA light curves of GRS 1915+105 separated by almost two years. The similarity between the two panels, even in some small details, is evident.

component is observed directly, indicating that the QPO must be related to the power law component.

- 67 Hz QPO: This QPO appears only during state B. Its high and rather constant frequency and its association with a state with small (and possibly constant) inner disk radius also point to a connection to the inner disk radius. It is interesting to note that this QPO is not observed in all state-B intervals, indicating that there must be another parameter involved in its production.

- Low-ν QPO: These oscillations, in the range 10–100 seconds (see Morgan, Remillard and Greiner, 1997) *are* the regular transitions between states that are often observed. The instability model provides an interpretation for the variability, but does not cast light on why the light curves are so regular.

There is another important point about the timing properties of the source. Looking at the light curves (see Belloni *et al.*, 2000), one notices not only that their time structure is very complex, but also it often repeats in an almost undistinguishable way, so that all observations can be grouped in twelve classes. As mentioned above, there is no clear explanation for why the structure of the light curves is the observed one. More than this, sometimes even the finest structures of the light curves repeat at a distance of years (see Figure 4). The basic questions that need to be answered are: why do the light curves have those complex and repeatable shapes, and why do the light curves have *only* those complex shapes, with only a few possibilities to choose from? Answering these questions could provide crucial information about the accretion phenomenon.

5. A Unique Source?

Another important question is: are there other sources that show the same phenomenology seen here? To this date, the answer is: the C-state events (the instability) are observed only in GRS 1915+105, but A–B transitions have been observed in other systems, namely 4U 1630-47 and GRO J1655-40 (Trudolyubov, Borozdin and Priedhorsky, 2000, see Figure 5). Possibly, the dips and the so-called flip-flops seen earlier in GX 339-4 (Miyamoto *et al.*, 1991) are also of the same nature. It seems clear that this type of fast (and often very regular) temperature oscillations are more common among bright black hole candidates, although their origin is still unknown. The question of why the C-state instability is observed only in this source is still basically unanswered. Possibly, this type of instability appears only at very high levels of accretion rate, which would then be reached only by GRS 1915+105. The observation of the same phenomenon in another source would greatly help to understand this issue.

6. Relation to the Canonical Black Hole States

An important point is the comparison between the A/B/C states of GRS 1915+105 and the four canonical black hole states mentioned above. It is tempting to associate state C with the low/intermediate state (hard spectrum flat-top noise and QPO in the power spectrum), state B with the very high state (strong disk component, weaker noise level), and the A state with the high state (cooler disk component, low noise level). However, as mentioned above, the instability related to the C state is not observed in other sources. It is more likely that the similarities between the properties of GRS 1915+105 and the canonical black hole states are not indicative of them being the same, but rather of them looking the same. In other words, the onset of the instability in GRS 1915+105 influences the accretion structure in a way that makes it mimic the properties of the canonical states observed in 'normal' sources.

7. Inner Disk Oscillations and Radio Jet Ejection

The X-ray phenomenology described above has been positively linked to the variability observed in the radio band, and major events have been associated to the emergence of superluminal radio jets. This topic is treated by other authors in this volume. Here it is important to stress that the analysis of simultaneous X-ray/radio observations has shown that radio flares seem to be associated only to state C events, and therefore to the inner disk oscillations, while observations containing only states A and B do not correspond to significant radio detections of the source (see Klein-Wolt *et al.*, this volume). However, the presence of this instability has not yet been observed in the other galactic sources showing jet ejection in the radio.

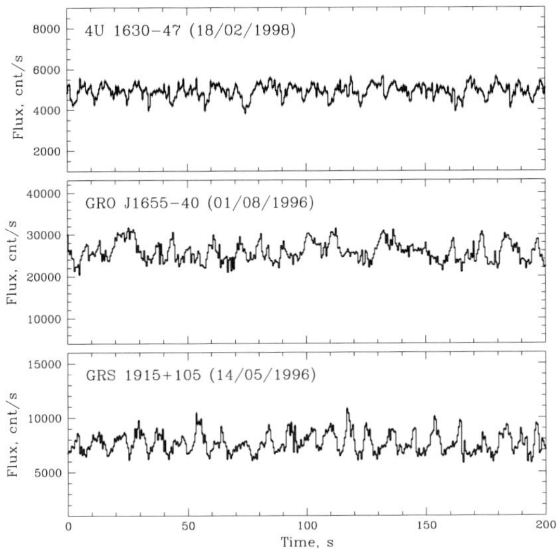

Figure 5. Comparison between selected RXTE/PCA light curves of 4U 1630-47 (upper panel, GRO J1655-40 (middle panel) and GRS 1915+105 (lower panel). From Trudolyubov, Borozdin and Priedhorsky (2000).

8. Conclusions

GRS 1915+105 is the only source up to now which can provide a large number of spectral transitions, enabling us to study different states of one source without having to wait for years. The most important of these transitions do not involve the canonical states of black hole candidates, but are associated to an instability of the innermost region of an optically thick accretion disk, which causes its inner parts to be evacuated and refilled on the local viscous time scale. Other variations observed in this source can be linked to those observed in other sources, strengthening the connection between this unique system and more conventional ones. On the X-ray side, it is now important to examine the spectral/timing behavior of GRS 1915+105 in more detail (see Migliari, Vignarca and Belloni, this volume), in order to provide a more complete phenomenological picture for further theoretical work. Modeling of the X-ray properties of GRS 1915+105 is at the moment the most promising way to make significant progress in the understanding of accretion onto black holes.

References

Belloni, T., *et al.*: 1997a, *Astrophys. J.* **479**, L145.
Belloni, T., *et al.*: 1997a, *Astrophys. J.* **488**, L109.
Belloni, T., *et al.*: 2000, *Astron. Astrophys.* **355**, 271.
Fender, R.P., *et al.*: 1999, *Mon. Not. R. Astron. Soc.* **304**, 865.

Frontera, F., *et al.*: 2000, *Astrophys. J.*, in press (astro-ph/0009160).
Greiner, J., Morgan, E.H. and Remillard, R.A.: 1996, *Astrophys. J.* **473**, L107.
Janiuk, A., *et al.*: 2000, *Astrophys. J.*, in press (astro-ph/0008354).
Markwardt, C.B., Swank, J.H. and Taam, R.E.: 1999, *Astrophys. J.* **513**, L37.
Morgan, E.H., Remillard, R.A. and Greiner, J.: 1997, *Astrophys. J.* **482**, 993.
Merloni, A., Fabian, A.C. and Ross, R.R.: 2000, *Mon. Not. R. Astron. Soc.* **313**, 193.
Muno, M.P., Morgan, E.H. and Remillard, R.A.: 1999, *Astrophys. J.* **527**, 321.
Mitsuda, K., *et al.*: 1984, *Publ. Astron. Soc. Jpn.* **36**, 741.
Miyamoto, S., *et al.*: 1991, *Astrophys. J.* **383**, 784.
Nayakshin, S., Rappaport, S. and Melia, F.: 2000, *Astrophys. J.* **535**, 798.
Szuszkiewicz, E. and Miller, J.C.: 1998, *Mon. Not. R. Astron. Soc.* **298**, 888.
Tanaka, Y. and Lewin, W.H.G.: 1995, in: *X-ray binaries*, Cambridge Univ. Press, p. 126.
Trudolyubov, S.P., Borozdin, K.N., Priedhorsky W.C., 2000, *Mon. Not. R. Astron. Soc.*, in press (astro-ph/9911345).
van der Klis, M.: 1995, in: *X-ray binaries*, Cambridge Univ. Press, p. 252.

ON THE NATURE OF THE MICROQUASAR V4641 SAGITTARII *

S. CHATY
Department of Physics and Astronomy, The Open University, United Kingdom
E-mail: s.chaty@open.ac.uk

I.F. MIRABEL
Service d'Astrophysique, CEA/Saclay, France & IAFE/Conicet, Argentina

J. MARTÍ
Departamento de Física, Escuela Politécnica Superior, Universidad de Jaén, Spain

L.F. RODRÍGUEZ
Instituto de Astronomía, Campus UNAM, Morelia, Michoacán 58190, México

Abstract. We present photometric and spectroscopic optical and near-infrared (NIR) observations taken during the outburst on September 1999 of the source V4641 Sgr = SAX J1819.3-2525 (in't Zand *et al.*, 2000) = XTE J1819-254 (Markwardt *et al.*, 1999). We show that our observations suggest a distance between 4 and 8 kpc, the spectral type of the companion star being constrained between B3 and A2 V, making the system a High Mass X-ray Binary System (HMXB). In view of the radio and optical/NIR observations, it is possible that the ejecta of the source interacted with the surrounding medium of the source.

1. Introduction

The source V4641 Sgr = SAX J1819.3-2525 = XTE J1819-254 attracted considerable attention after the detection of a giant optical outburst on 1999 September 15.7 UT, from the magnitudes 14.0 to 8.8 in the V-band (Stubbings, 1999). The X-ray source flared, from 1.6 to 12.2 Crab in the 2–12 keV X-rays on 1999, September 14th, as observed by XTE, through a brief but dramatic eruption, its position being coincident with the optical transient (Smith *et al.*, 1999). Less than 10 hours later, the source was fainter than 50 mCrab.

The VLA detected on Sep. 16.0 UT a strong radio source at 0.4 Jy at the position of the variable star. An elongation was extending $0.25'$ between 0.6–1.2 day after the X-ray flare, and at the same position on 17.9, 22 and 24 UT. This allows to classify the source as a new microquasar (Hjellming *et al.*, 2000). A detailed study of this source and details on the observations are reported in Chaty *et al.* (2001).

* Based on observations collected at the European Southern Observatory, Chile. S.C. thanks Rob Hynes for having alerted him to this new flaring source on September 1999, and Bob Hjellming for the communications he gave on the radio observations. S.C. is very grateful to the ESO/NTT team staff for their availability and skills. S.C. acknowledges support from grant F/00-180/A from the Leverhulme Trust. IFM acknowledges partial support from Conicet/Argentina.

Astrophysics and Space Science is the original source of publication of this article. It is recommended that this article is cited as: *Astrophysics and Space Science* **276** (suppl.): 153–156, 2001.

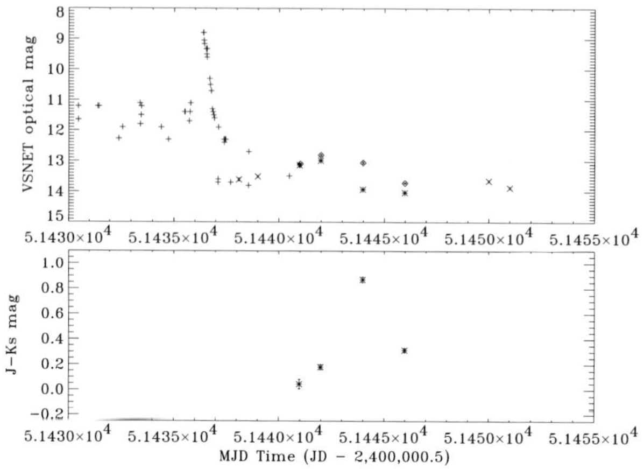

Figure 1. Top: +:VSNET, x:V, *:J, diamond:Ks magnitudes. Bottom: J-Ks color.

2. Observations and Results

The optical observations were performed with the NTT telescope and the instrument EMMI RILD. We imaged the source in V, R, I and Z filters, and took some spectra with the grism #1. The infrared observations were performed with the NTT and the instrument SOFI through the filters J, H and Ks, with an exposure time of 15 min. The optical and infrared lightcurves and the J-Ks color during the outburst are reported in Figure 1. In the optical, after the big outburst there was still some flaring activity through 0.5 mag in V, R, I with no significant change in the colors. There was also some flaring activity in NIR through 1 mag in J and K, with a significant change in the J-Ks color during the post-outburst stage (between 2 and 5 days after the giant burst). This suggests an increased K-contribution compared to J, either due to the emission of a jet, the appearance of heated dust, or even by the interaction with the interstellar medium.

We report the normalized optical spectra offset in intensity to get an easier reading in Figure 2. On a timescale of one day, the lines were changing from emission to absorption. All the Balmer serie is visible: Hα, β, γ, δ, ϵ, ζ ... The Hα emission line is extraordinarily strong: one day after the outburst, its equivalent width was $\sim$ 100 Å, with a FWZI of $\sim$ 6700 km s^{-1} and a blue wing. There was also a strong He I 5876 Å. The Na-D gives E(B-V) = 0.25 implying $N(\mathrm{H}) = 0.13 \times 10^{22}$ cm^{-2}. The strong variability of the lines and the blue continuum suggest the emission from an accretion disk, or of a corona, with accretion of matter onto a compact object with a high-velocity wind component ($\sim$ 6000 km s^{-1}), also perhaps the presence of a cocoon or a jet.

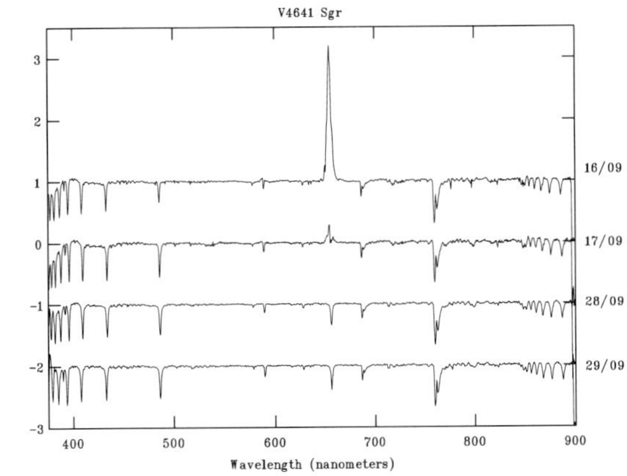

Figure 2. Normalized and offset optical spectra.

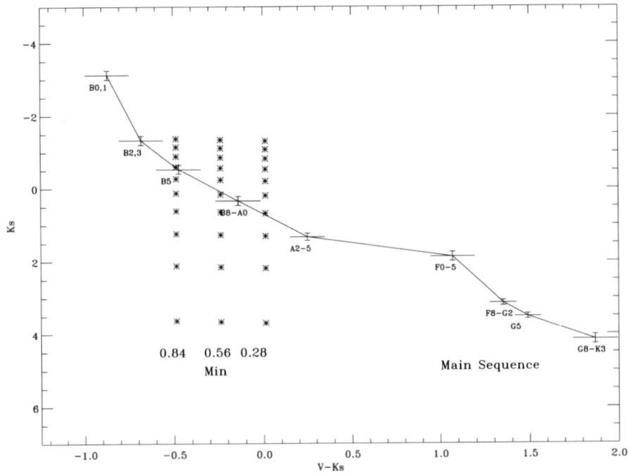

Figure 3. Color-magnitude [V-Ks,Ks] diagram. *: Min magnitudes of V4641 Sgr.

3. Discussion

Taking the magnitudes at the faintest stage of the source in the optical and NIR wavelengths, we plotted them on a color-magnitude diagram (Figure 3). We also plotted three different absorptions, corresponding respectively to 0.05, 0.1 and 0.15×10^{22} cm^{-2}, and different distances of the source, from 1 to 10 kpc. We can see that the distance is constrained to $4 < d < 8$ kpc in order to have a spectral stellar type consistent with the stars of our Galaxy. This location on the color-magnitude diagram suggests that the spectral type of the companion star of the binary system is only consistent with an early type main sequence star B3–A2 V. Therefore the mass is constrained between $2 < M < 10\ M_\odot$, suggesting that it is a high-mass X-ray binary. This is consistent with optical spectra in quiescence, taken by Orosz *et al.* (2000), who derived a stellar spectral type of A2 V at a distance of 6.1 kpc.

If the elongation seen in the radio is interpreted as a moving component, the proper motion is between $224 < \mu < 788$ mas/d depending on the exact time of the ejection. In the following we will assume that this is the approaching (brighter) condensation with $\mu_a = 500$ mas/d. Since $D \leq \frac{c}{\sqrt{\mu_a \mu_r}}$ and from our results $D \geq 4$ kpc, we conclude that the apparent velocity in the plane of the sky would be strongly superluminal, v_a of the order of 12. However, no movement of this elongation was detected between Sept. 16.02 and 24.1 UT, suggesting an interaction with surroundings at $0.25'' \geq 1.5 \times 10^3$ AU at the distance of 6 kpc. This is possible if the ejections began to take place 10 days before the radio detection e.g. on September, 8th, and we can see from Figure 1 that the source was already active at this date. It seems therefore that the activity of this source was not as sporadic as we could have thought at the beginning. Indeed, *RXTE* could detect this source during 270 days before the outburst (in't Zand; Markwardt, private communications).

References

Chaty, S., Mirabel, I.F., Martí, J. and Rodríguez, L.F.: 2001, *Proceedings of the 4th INTEGRAL workshop*, ESA-SP Conf. Proc., in press.

Hjellming, R.M., Rupen, M.P., Mioduszewski, A.J., *et al.*: 2000, *Astron. Astrophys. Suppl.* **195**, 13403.

in't Zand, J.J.M., Kuulkers, E., Bazzano, A., Cornelisse, R., Cocchi, M., Heise, J., Muller, J.M., Natalucci, L., Smith, M.J.S. and Ubertini, P.: 2000, *Astron. Astrophys.* **357**, 520.

Markwardt, C.B., Swank, J.H. and Marshall, F.E.: 1999, *IAU Circ.* **7120**.

Orosz, J.A., Kuulkers, E., van der Klis, M., McClintock, J.E., Jain, R.K., Bailyn, C.D. and Remillard, R.A.: 2000, *IAU Circ.* **7440**.

Smith, D.A., Levine, A.M. and Morgan E.H.: 1999, *IAU Circ.* **7253**.

Stubbings, R.: 1999, *IAU Circ.* **7253**.

Address for Offprints: Department of Physics and Astronomy, The Open University, Walton Hall, MK7 6AA, Milton Keynes, United Kingdom

NUMERICAL SIMULATIONS OF THE ACCRETION-EJECTION INSTABILITY

S.E. CAUNT and M. TAGGER

DSM/DAPNIA/Service d'Astrophysique (CNRS URA 2052), C.E.A. Saclay, 91191 Gif sur Yvette, France

Abstract. The Accretion-Ejection Instability (AEI) is explored numerically using a global 2d model of the inner region of a magnetised accretion disk. The disk is initially currentless but threaded by an external vertical magnetic field created by external currents, and frozen in the flow. In agreement with the theory a spiral instability, similar in many ways to those observed in self-gravitating disks but driven by magnetic stresses, develops when the magnetic field is close to equipartition with the disk thermal pressure. The present non-linear simulations give good evidence that such an instability can occur in the inner region of accretion disks.

1. Introduction

The accretion-ejection instability (AEI) (Tagger and Pellat, 1999; see also Varnière *et al.*, Rodriguez *et al.* and Tagger *et al.*, this volume) can occur close to the inner edge of a magnetised accretion disk when the plasma beta (the ratio of gas to magnetic pressures) is around unity. The instability appears, if the magnetic field decreases outwards sufficiently fast, as a spiral wave of low azimuthal wavenumber, made unstable through the interaction with a Rossby vortex (associated with the gradient of vorticity) which it generates at its corotation radius.

We have undertaken numerical simulations, with a setup optimised according to our previous knowledge of the AEI. We consider an infinitely thin disk threaded by a moderate vertical magnetic field: this is justified by the properties of the AEI, which are essentially constant vertically across the disk (unlike the magneto-rotational instability (Balbus and Hawley, 1991) for which vertical modes are required). The disk is embedded in vacuum and, in order to separate different physical effects, we consider here only configurations where initially the equilibrium magnetic field is due to external currents. Starting with small scale initial random fluctuations in the fluid velocity, the simulations show large-scale spiral perturbations evolving in the disk, and corresponding to the expected properties of the AEI.

The following sections present the essentials of the model, basic results and conclusions. More detailed discussions of these are covered elsewhere (Caunt and Tagger, 2000).

Astrophysics and Space Science is the original source of publication of this article. It is recommended that this article is cited as: *Astrophysics and Space Science* **276** (suppl.): 157–160, 2001.

2. The Model

Under the assumption, valid for the physics of the AEI, that perturbations are essentially independent of z in the disk, we integrate vertically the standard MHD equations (momentum, continuity, induction) describing the fluid motion in a disk around a central object. We thus solve for the 2d (r, ϕ) evolution of the fluid properties (surface density, velocity, magnetic field) averaged over z. A conservative scheme of the type described by Stone and Norman (1992) is used to ensure that mass, angular momentum and magnetic flux are conserved to numerical precision throughout the simulation.

We assume that the disk is threaded by a poloidal magnetic field which is symmetric about the midplane, hence purely vertical at $z = 0$. In the vacuum surrounding the disk the magnetic field can be described by a magnetic potential which, in turn, results from perturbed currents in the disk. It can thus be calculated from a Poisson equation, similar to the one describing the gravitational potential of a self-gravitating disk, but whose source is the vertical component of the magnetic field at the disk surface (rather than density in the gravitational case). Currents within the disk can be derived from the jump in the horizontal component of the field across the disk. Hence magnetic stresses can be included in an otherwise purely hydrodynamic model.

We assume that the central object has a mass of $M = 10\ M_{\odot}$ and the disk extends from between $r = 1,000$ km to $r = 50,000$ km with a constant aspect ratio of $\epsilon = 0.1$. We use a logarithmic radial grid which allows us to get a better resolution in the inner region of the grid, where it is necessary, and to model a very large radial extent of the disk. It also has the added advantage that unwanted effects of the boundary condition at the outer radius of the simulation are avoided (waves are damped more effectively as they approach the outer radius).

3. Results

The results presented here are aimed at illustrating certain properties of the instability as determined by linear theory, specifically the growth rate/magnitude of instability for different field strengths and development of the spiral wave and Rossby vortex.

Figure 1 shows the evolution of the rms velocity within the disk for values of $\beta = 0.1, 0.5, 1.0, 5.0$ and 10.0. As is shown, the perturbed velocity grows rapidly for $\beta = 0.5$ and $\beta = 1.0$, exceeding the initial random noise after approximately 10 orbits. The growth is approximately exponential during this time. For the other values of β we see very little action in comparison. The $\beta = 0.1$ and $\beta = 10.0$ cases are virtually stable with the $\beta = 5.0$ being somewhat more active but failing to achieve the same amplitude as the most unstable cases. This run is still inter-

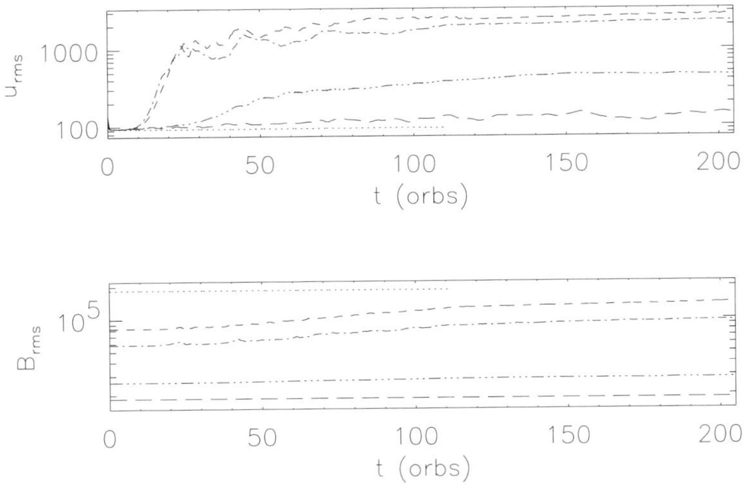

Figure 1. Comparison of the evolution of the instability for different values of the initial β. Values of $\beta = 0.1, 0.5, 1.0, 5.0$ and 10.0 are shown as dotted, dot-dashed, dashed, tripple-dot-dashed and long-dashed lines respectively. Time is given in units of the orbital time at the inner radius. The plots show the evolution of (top) the rms velocity, and (bottom) the rms magnetic field.

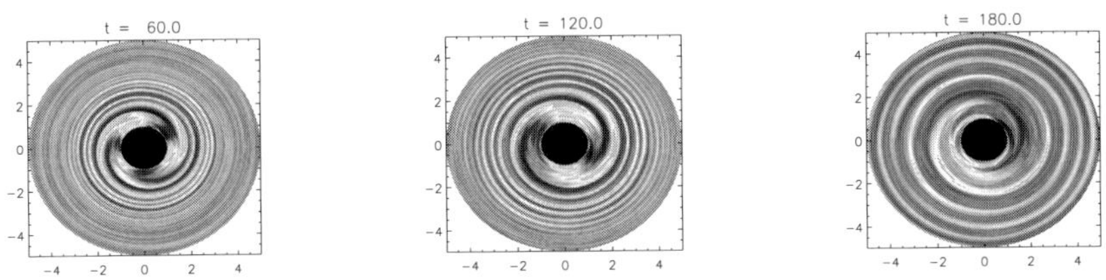

Figure 2. Plots of the spiral structure in the inner region of the disk at times $t = 60$, 120 and 180 orbits. Starting from initially random fluctuations at $t = 0$, the disk has clearly developed 3-armed followed by 2-armed and finally 1-armed spirals.

esting to observe as the growth of the instability is slower and the evolution of the spiral wave is somewhat cleaner.

The evolution of the instability into a well-developed spiral wave is best seen for the $\beta = 5.0$ case. Figure 2 shows the different number of spiral arms that occur over time (plotted is the radial velocity): initial random perturbations develop to $m = 3$, $m = 2$ and $m = 1$ spiral structures, as the magnetic flux is advected toward the central region. The transition is generally through a mixture of these for all values of unstable β. For the more active values ($\beta = 0.5$ and $\beta = 1.0$) the transition is much quicker and always ends up with $m = 1$.

The AEI relies on the existence of a Rossby vortex generated at co-rotation. Figure 3a shows the radial variation of vorticity in the disk. Clearly shown is a discontinuity close to the inner radius. The velocity field, shown in Figure 3b

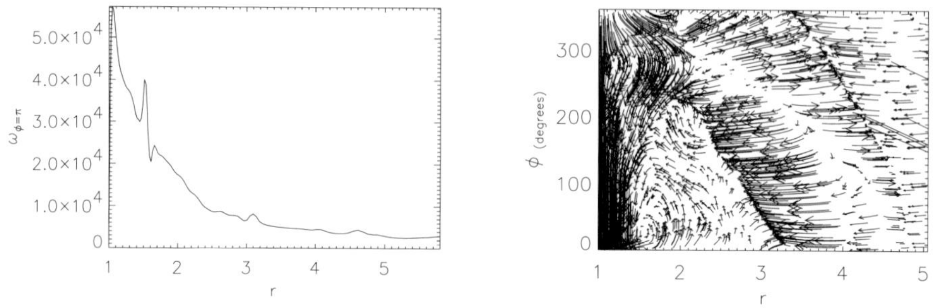

Figure 3. (left hand): Vorticity in the inner region of the accretion disk. (right hand): The velocity field within the disk in Cartesian projection. The vortex is clearly visible around $r \sim 1.5$.

clearly shows that the fluid flow contains a vortex at the same radius. This is a good indication that a Rossby vortex has indeed been generated at corotation.

4. Conclusions

Using this model we have shown the essential features predicted by linear theory of the AEI. We have seen that the instability is clearly strongest for field strengths of the order of equipartition, and leads to the development of low azimuthal wavenumber spiral waves and a Rossby vortex.

Future work will link in the third type of wave (namely the vertical emission of Alfvén waves) from which the formation of winds and jets is believed to be possible.

References

Balbus, S.A. and Hawley, J.F.: 1991, *Astrophys. J.* **376**, 214.
Caunt, S.E. and Tagger, M.: 2000, submitted to *Astron. Astrophys.*
Stone, J.M. and Norman, M.L.: 1992, *Astrophys. J. Suppl.* **80**, 753.
Tagger, M. and Pellat, R.L.: 1999, *Astron. Astrophys.* **349**, 1003.

TRAPPED DISK OSCILLATIONS AND STABLE QPOs IN MICROQUASARS

XUE-BING WU[1,2]
[1]*Max-Planck-Institut für Astrophysik, Karl-Schwarzschild Str. 1, D-85741 Garching, Germany*
[2]*Department of Astronomy, Peking University and CAS-PKU Joint Beijing Astrophysical Center, Beijing 100871, China*

Abstract. The observed quasi-periodic oscillations (QPOs) in low-mass X-ray binaries have been suggested to be probably related to the characteristic oscillation modes in the inner part of a relativistic accretion disk. In this paper we study the trapped feature of these oscillation modes by considering the radial perturbations and viscosity effects. We find that the trapped oscillations are amplified significantly when the viscosity parameter increases and the trapped size varies when the radial wavenumber and Mach number change. Our results support that the stable hundred-hertz QPOs and their low fractional rms amplitudes found in two Galactic microquasars may be explained by some trapped non-damping oscillation modes in the inner part of accretion disks around rotating black holes.

1. Introduction

In recent years, hundred-hertz QPOs have been observed in several black hole X-ray binaries with *RXTE* satellite. These high-frequency QPOs are probably similar to the kilohertz QPOs observed frequently in a dozen neutron star low-mass X-ray binaries. However, the high frequency QPOs in two Galactic microquasars, GRO J1655-40 and GRS 1915+105, are very stable in frequency at 300 Hz and 67 Hz respectively (Morgan, Remillard and Greiner, 1997; Remillard *et al.*, 1999). These stable QPOs have also very low fractional rms amplitude ($\sim 1\%$). Theoretical models involving frame-dragging (Cui, Zhang and Chen, 1998), diskoseismic g-mode (Nowak *et al.*, 1997), and periastro precession (Stella, Vietri and Morsink, 1999) have been proposed to explain these stable QPOs. But the origin of them is still uncertain.

Recently Psaltis and Norman (2000; hereafter PN) suggested that the observed QPOs in X-ray binaries may be related to the strong resonances peaked at some characteristic oscillation frequencies in the inner part of accretion disks. However, they introduced a simplified term for angular momentum transfer and did not the damping and growth rates of oscillation modes. It is also important to check if some of these modes can be trapped in the inner disk and if the low fractional rms amplitude of QPOs observed in two microquasars can be explained by the trapped feature of oscillation modes.

Astrophysics and Space Science is the original source of publication of this article. It is recommended that this article is cited as: *Astrophysics and Space Science* **276** (suppl.): 161–164, 2001.

2. Non-Damping Oscillation Modes in Accretion Disk

The basic equations of a geometrically thin accretion disk are the same as those in PN except that the viscosity term in the angular momentum equation is now expressed as: $N_\phi = (1/\rho r)\partial(\nu\rho r^3 d\Omega/dr)/\partial r$. The viscosity is assumed to be described by the usual α prescription, namely, $\nu = \alpha c_s H = \alpha c_s^2/\Omega$. Like in PN, we consider the general relativistic effects in an approach to specify the the gravitational potential ψ through the dependences of angular velocity (Ω), radial and vertical epicyclic frequencies (κ and $\Omega_\perp$) on the mass, radius and the spin parameter of the central object. We consider small perturbations to all physical quantities in the form of $Q = Q_0 + \delta Q e^{i(\omega t - kr - m\phi)}$, where k and m are radial and azimuthal wavenumbers. We also assume that the perturbations do not alter the local sound speed and the disk is isothermal in the vertical direction. Under the local approximation $kr \gg 1$, we can obtain four perturbed equations and derive the dispersion relation as

$$\bar{\sigma}^4 + \alpha\epsilon^2\bar{\sigma}^3 + (\bar{\kappa}^2 + \epsilon^2 + n\bar{\Omega}_\perp^2)\bar{\sigma}^2 - \epsilon[m\frac{H}{r}(\frac{\bar{\kappa}^2}{2} - 2) + 2\alpha\epsilon(\frac{\bar{\kappa}^2}{2} - 2)$$

$$-\alpha\epsilon^3 - n\epsilon\bar{\Omega}_\perp^2]\bar{\sigma} + n\bar{\kappa}^2\bar{\Omega}_\perp^2 = 0. \qquad (1)$$

Here we have defined $\bar{\sigma} = i(\omega - ku_r - m\Omega)/\Omega$, $\bar{\kappa} = \kappa/\Omega$, $\bar{\Omega}_\perp = \Omega_\perp/\Omega$, and $\epsilon = kH$. The value of n represents the number of nodes of the oscillations in the vertical direction.

By solving the dispersion relation (1), we found that the characteristic frequencies of oscillation modes depend on the radial wavenumber and radius significantly but are nearly independent of the viscosity. However, the growth (or damping) rates of these modes depend strongly on the viscosity. According to their non-damping feature, we found that the most possible oscillation modes in the disk are the acoustic modes with approximate frequencies of $ku_r \pm \sqrt{\kappa^2 + k^2c_s^2}$ (in the case of $m = 0$ and $n = 0$) and $ku_r + \Omega \pm \sqrt{\kappa^2 + k^2c_s^2}$ (in the case of $m = 1$ and $n = 0$), neutral modes with frequency of ku_r (in the case of $m = 0$ and $n = 0$) and $ku_r + \Omega$ (in the case of $m = 1$ and $n = 0$), and the corrugation modes with approximate frequencies of $ku_r + \Omega \pm \Omega_\perp$ (in the case of $m = 1$ and $n = 1$). In Figure 1 we show the dependences of oscillation frequencies of all non-damping modes on the disk radius and black hole spin parameters. Note that the lower-frequency acoustic modes and the lower-frequency corrugation mode are trapped in the inner part of accretion disks since $\omega^2 > \kappa^2$. The trapped sizes of these modes depend significantly on the radial wavenumber, Mach number and black hole spin parameter. The relative luminosity variations caused by the oscillations in the trapped area may be estimated using the expression of local flux emitted from a relativistic accretion disk. These relative variations may be comparable to the observed fractional rms magnitudes if the X-ray luminosity variations are caused

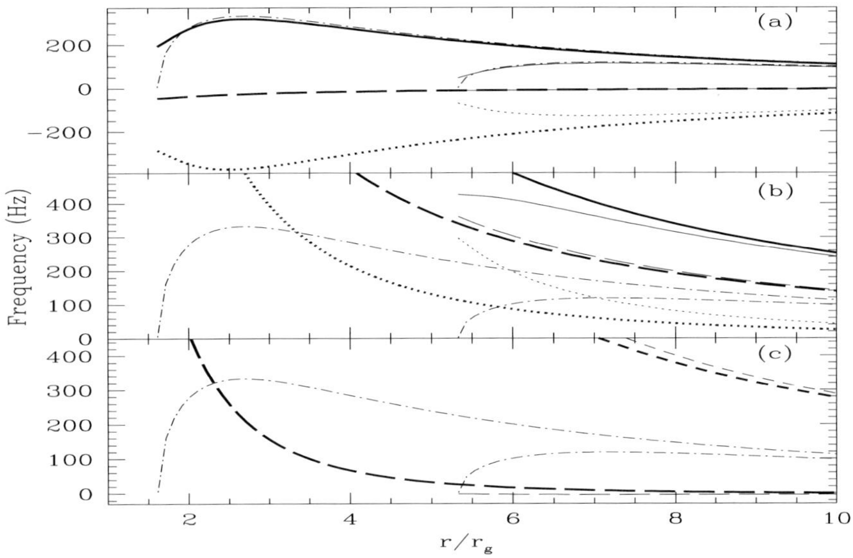

Figure 1. Oscillation frequencies of non-damping modes when $kH = 0.1$ and $|u_r|/c_s = 0.3$. The weighted and light lines represent the cases with black hole spin parameter $a_* = 0.98$ and 0.2. The panel (a), (b) and (c) correspond to $(m, n) = (0, 0)$, $(1, 0)$ and $(1, 1)$. In (a) and (b) the solid, dotted and dashed lines represent two acoustic modes and the neutral mode. In (c) the long and short dashed lines represent two corrugation modes. The dot-dashed lines show the radial epicyclic frequencies. The black hole mass was taken to be $7M_{\odot}$and r_g was defined as GM/c^2.

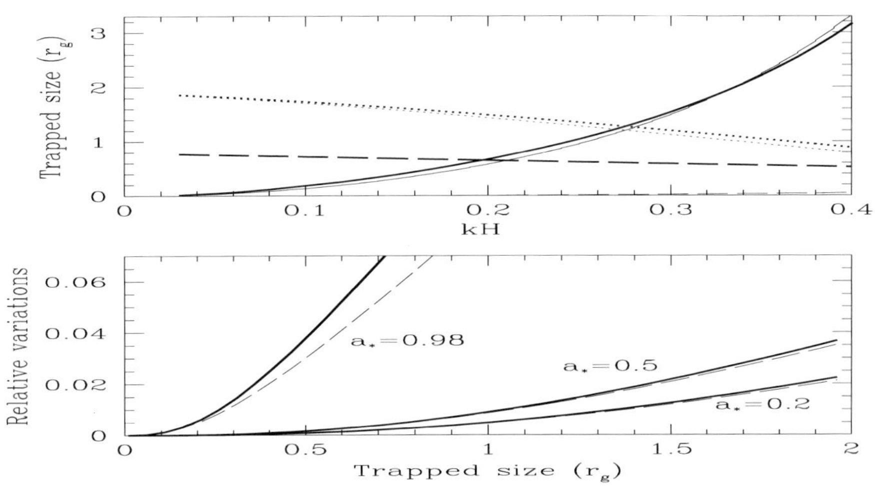

Figure 2. Upper panel: Trapped sizes of three oscillation modes. The solid, dotted and dashed lines represent the lower-frequency acoustic modes in the cases of $(m, n) = (0, 0)$ and (1,0), and the lower-frequency corrugation mode in the case of $(m, n) = (1, 1)$. The weighted and light lines correspond to the cases with $a_* = 0.98$ and 0.2. Lower panel: Relative variations caused by the trapped oscillations. The solid and dashed lines correspond to the cases with accretion rate of 10^{17} g s^{-1} and 10^{18} g s^{-1}.

by the disk oscillations in the trapped area. In Figure 2 we show the trapped sizes of the oscillation modes and the relative luminosity variations.

3. Conclusion

RXTE observations on GRO J1655-40 have shown 300 Hz QPOs with $\sim 1\%$ rms magnitude. If this stable QPO originates from the trapped disk oscillations, from Figures 1 and 2 we see that it may be explained by either the lower-frequency acoustic mode in the case of $m = 1$ and $n = 0$ or the lower-frequency corrugation mode in the case of $m = n = 1$ and $a_* \sim 0.9$. However, the rapidly spinning black hole is more favored to explain thc high QPO frequency. The unambiguous determinations of the inner disk radius and the black hole spin are still needed to confirm this explanation. Further studies on the origin of low-frequency QPOs and their correlation with high-frequency QPOs, as well as the correlation between spectral and temporal properties, will help us to understand more clearly the origin of stable QPOs in microquasars.

References

Cui, W., Zhang, S.N. and Chen, W.: 1998, *Astrophys. J.* **492**, L53.
Morgan, E.H., Remillard, R.A. and Greiner, J.: 1997, *Astrophys. J.* **482**, 993.
Nowak, M.A., Wagoner, R.V., Begelman, M.C. and Lehr, D.E.: 1997, *Astrophys. J.* **477**, L91.
Psaltis, D. and Norman, C.: 2000, *Astrophys. J.*, in press (astro-ph/0001391).
Remillard, R.A., Morgan, E.H., McClintock, J.E., Bailyn, C.D. and Orosz, J.A.: 1999, *Astrophys. J.* **522**, 397.
Stella, L., Vietri, M. and Morskink, S.M.: 1999, *Astrophys. J.* **524**, L63.

SPECTRA OF THERMALLY UNSTABLE SLIM DISCS

EWA SZUSZKIEWICZ
Institute of Physics, University of Szczecin, Szczecin, Poland
Toruń Centre for Astronomy, Nicolaus Copernicus University, Toruń, Poland
Dipartimento di Fisica 'Galileo Galilei', Università degli Studi di Padova, Italy
International School for Advanced Studies, SISSA, Italy

ROBERTO TUROLLA and LUCA ZAMPIERI
Dipartimento di Fisica 'Galileo Galilei', Università degli Studi di Padova, Italy

Abstract. Thermal instability driven by radiation pressure might be relevant for intrinsically bright accreting sources. The most promising candidate where this instability seems to be at work is one of the two known galactic superluminal sources, GRS 1915+105 (Belloni *et al.* 1997). In spite of being of relevance, this scenario has not yet been confirmed by proper time-dependent modelling. Non-linear time-dependent calculations performed by Szuszkiewicz and Miller (1998) show that thermally unstable discs undergo limit-cycle behaviour with successive evacuation and refilling of the central parts of the disc. This evolution is very similar to the one proposed by Belloni *et al.* (1997) in their phenomenological model. Further investigations are needed to confirm the thermal instability being operational in this source. First of all the spectra emitted from the disc during its evolution should be calculated and compared with observations. Here such spectra are computed assuming local blackbody emission from the best studied transonic disc model.

1. Limit-Cycle Behaviour of Accretion Flows

The simplest disc model which properly describes physical conditions around black holes is a transonic (slim) accretion disc. It is an axisymmetric, vertically integrated, non-self-gravitating, differentially rotating fluid configuration. The assumption about self-gravity being negligible defines the outer edge of the disc and the requirement that the radial velocity equals the local sound speed somewhere in the disc, and then exceeds it, determines its radial structure. Once the mass, M, of the central black hole, accretion rate, $\dot{M}$, and viscosity coefficient α are specified the structure of the stationary disc is uniquely determined. However, such a disc might be subject to secular, thermal and pulsational instabilities. Here we focus on the thermal instability in which a temperature perturbation out of the equilibrium is enhanced by thermal processes. Moreover, we consider the thermal instability which is triggered by one particular mechanism, namely by the radiation pressure. The possibility that a global instability might develop in this model was indicated by the results of a local stability analysis. In the temperature-surface density diagram the sequence of equilibrium models forms a characteristic S-shape curve. Such curves reflect the possibility of a transition between low and high accretion rate states,

Astrophysics and Space Science is the original source of publication of this article. It is recommended that this article is cited as: *Astrophysics and Space Science* **276** (suppl.): 165–168, 2001.

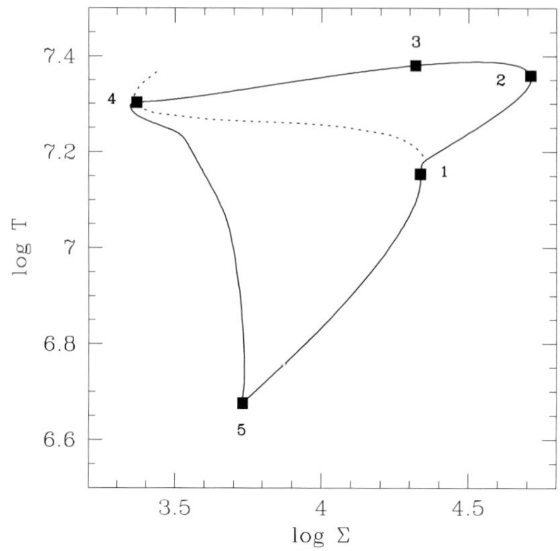

Figure 1. The phase portrait in the $\log T - \log \Sigma$ plane for $r = 10r_G$ (here r_G is the Schwarzschild radius). The dotted line represents the equilibrium states, the so called S-curve. The lower branch of the S-curve overlaps with the trajectory of the evolving model. The continuous line shows the trajectory of the model during the evolution. The points marked 1 to 5 indicate those stages for which spectra are calculated. The points 1 to 4 correspond to the phase of disc evacuation, the point 5 is at the beginning of the refilling phase.

associated with an increase of the temperature. The first complete calculation of the global evolution of thermal instability in transonic discs was performed by Szuszkiewicz and Miller (1998). They have shown that at least for some values of M, $\dot{M}$ and α this evolution has a cyclic character and consists of successive evacuation and refilling phases. An example of the limit-cycle for the best studied thermally unstable model with $M = 10M_{\odot}$, $\dot{M} = 0.06$ (in critical units) and $\alpha = 0.1$ is shown in Figure 1. This particular non-stationary disc behaviour might have interesting observational consequences.

2. Spectra in the Blackbody Approximation

The global dynamical calculations, described in Section 1, determine the overall properties of the transonic disc evolution. However, in order to get detailed theoretical predictions for the directly observed quantities, we must compute the spectra emitted from thermally unstable discs during the evacuation and refilling of their central parts.

In the case of a geometrically thin accretion disc, it is possible to consider the vertical and radial disc structures separately. It is also reasonable to assume

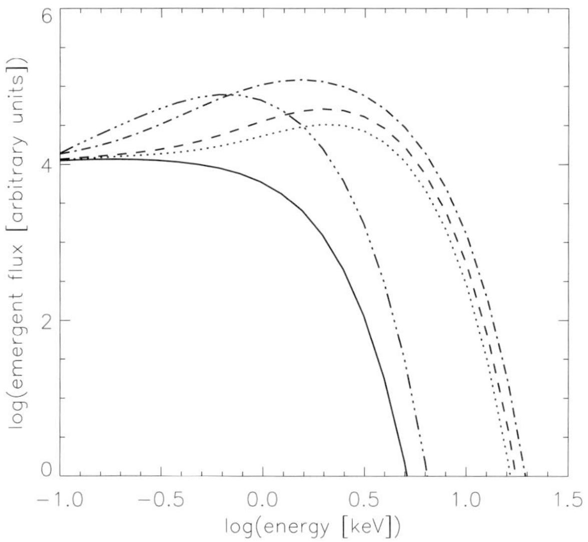

Figure 2. The disc spectra for the evolutionary stages shown in Figure 1: stage 1 (solid line), stage 2 (dotted line), stage 3 (dashed line), stage 4 (dash-dotted line), and stage 5 (triple dashed-dotted line).

local heat balance. In order to calculate the spectrum, it is sufficient to divide the disc into a number of concentric annuli, calculate the spectrum emitted by each annulus and then sum them all together. The simplest assumption is that each annulus radiates like a blackbody at the local effective temperature. We stress that this assumption does not necessarily apply to all relevant physical situations. Most of the previous studies were concentrated on the modelling of the vertical structure and radiative transfer through each annulus in Shakura-Sunyaev discs (e.g. Shimura and Takahara, 1995). However, the thin disc hypothesis breaks down at high accretion luminosities and it may not be a good approximation at very low accretion rates either. Slim disc models are more appropriate for astrophysical applications (Szuszkiewicz *et al.*, 1996; Wang *et al.*, 1999). Here as a first approximation we have calculated spectra assuming that the disc radiates locally as a blackbody (Figure 2). During the disc evacuation phase the spectra become significantly harder. Detailed spectra calculations, which account properly for radiative transfer in the disc atmospheric layers are under way and will be presented in the forthcoming paper (Szuszkiewicz, Turolla and Zampieri, in preparation).

3. Concluding Remarks

Transonic discs with accretion rates larger than a critical value which depends on the black hole mass and the viscosity parameter, show time dependent, cyclic behaviour due to the onset of thermal instabilities driven by radiation pressure. This behaviour can be described in terms of the cycle period, the duration of the bursting and quiescent phases, the amplitude and shape of the burst and finally by the spectral variability. For the different parameters of the system $(M, \alpha, \dot{M})$ we expect different variability properties and emitted spectra. Detailed theoretical predictions are definitely needed for modelling high-resolution timing and spectral data presently available for black hole systems.

Acknowledgements

E.S. gratefully acknowledges financial support from the Polish State Committee for Scientific Research (grant KBN 2P03D01817), the Italian MURST and the University of Padova. Work partially supported through MURST grant COFIN9802154100.

References

Belloni, T., Mendez, M., King, A.R., van der Klis, M. and van Paradijs, J.: 1997, *Astrophys. J.* **479**, L145.

Shimura, T. and Takahara., F.: 1995, *Astrophys. J.* **445**, 780.

Szuszkiewicz, E., Malkan, M.A. and Abramowicz, M.A.: 1996, *Astrophys. J.* **458**, 474.

Szuszkiewicz, E. and Miller., J.C.: 1998, *Mon. Not. R. Astron. Soc.* **298**, 888.

Wang, J.-M., Szuszkiewicz, E., Lu, F.-J. and Zhou, Y.-Y.: 1999, *Astrophys. J.* **522**, 839.

DISK-CORONA INTERACTIONS IN SOFT SPECTRAL STATES OF BLACK HOLE BINARIES

PIOTR T. ŻYCKI
N. Copernicus Astronomical Center, Warsaw, Poland
E-mail: ptz@camk.edu.pl

Abstract. Re-analysis of some archival X-ray data of black hole binaries in soft spectral states is presented. Results of detailed spectral fitting, including proper physical description of the X-ray reprocessed component, indicate that in most cases the soft, thermal disk emission cannot be described by a simple multi-color blackbody model. Additional Comptonization of those seed photons provides an acceptable description of the soft component, implying a presence of a warm ($kT_e \sim 1$ keV), mildly optically thick ($\tau \sim 10$) phase of accreting plasma.

Combined spectral/temporal analysis of GRS 1124-68 during 'flip-flop' transitions in very high state suggests that the optically thick accretion flow can proceed in (at least) two modes giving the same time-integrated behaviour (i.e. energy spectra) but very different timing behaviour.

1. The Comptonized Soft Component

Accreting black holes, both stellar (including the Galactic μ-quasars) and supermassive, show strong, soft, thermal components in their EUV-X-ray spectra, when accreting above certain fraction of Eddington rate. This component is thought to come from an optically thick accretion disk, and its spectrum is usually modeled as a simple, multi-color blackbody. For re-examining a number of such spectra, I will use the spectral models described in Życki *et al.* (1998).

1.1. THE INTERMEDIATE STATE

An example of IS is the May 18th observation of GS 1124-68 (Nova Muscae 1991). Using first the frequently employed but unphysical model, DISKBB + SMEDGE (POWERLAW), gives an unacceptable fit, $\chi^2_\nu = 38/25$. Adding a Gaussian line at 6.4 keV does *not* improve the fit, unless the line width is $\sim$ 10 keV ($\chi^2_\nu = 28/23$). Allowing the line energy to be free gives $\chi^2_\nu = 21/23$ for $E = 4 \pm 0.5$ keV. These values have little to do with Fe spectral features, but rather indicate that it is the continuum shape that is not properly modeled. Using now the DISKBB + THCOMP + REL-REPR model gives very bad fit, $\chi^2_\nu = 308/24$. Replacing the DISKBB component with a Comptonized blackbody (COMPTT) results in a dramatic improvement, $\chi^2_\nu = 13/22$. Thus the best fit is obtained for the proper physical description of X-ray reprocessing and the *Comptonized blackbody* model

Astrophysics and Space Science is the original source of publication of this article. It is recommended that this article is cited as: *Astrophysics and Space Science* **276** (suppl.): 169–172, 2001.

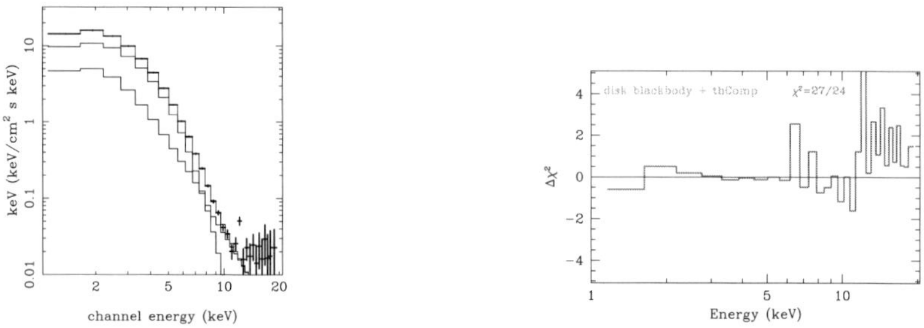

Figure 1. Spectrum of Nova Muscae 1991 in high state, showing a soft thermal component broader than a disk blackbody.

for the soft component. The same effect is observed in GRO J1655-40, GS 2000+25 and XTE J1550-564 (Życki *et al.*, in preparation).

1.2. The high state

Nova Muscae 1991 was in the High State when observed on Mar 18th. The spectrum shows a strong soft component, and the hard component is negligible (Ebisawa *et al.*, 1994). Modeling the spectrum with the DISKBB + THCOMP model gives $\chi^2_\nu = 27/24$ (see Figure 1). The Comptonized component is steep and makes a significant contribution to the soft component. Again, the soft component has to be described by a model *broader* than the multi-color disk blackbody.

1.3. The Very High State

On Jan 11th, while still on the rise to the peak of the outburst, Nova Muscae 1991, showed a spectrum consisting of two Comptonized components: the softer one again *cannot* be described by the DISKBB model. It can be modeled as Comptonization of $\approx$ 0.3 keV blackbody by a warm, $kT_e \approx 3$ keV, optically thick, $\tau \sim 7$, plasma.

1.4. Physical origin of the Comptonization

The additional Comptonization seems to be a common feature of disk-dominated spectral states of accreting black holes. There are two possible scenarios to explain it. The first one is Comptonization in a "hot skin" on top of the accretion disk. Such a hot layer could not be heated by X-ray irradiation (as postulated for the low/hard state), but it would require enhanced viscous energy dissipation compared to the bulk of the disk. The second possibility is that the electron energy distribution in the Comptonizing plasma is a hybrid, thermal-non-thermal one, rather than a pure thermal (Maxwellian) distribution. In fact, the COMPPS model (J. Poutanen, private communication) using the hybrid distribution gives very good fits to the

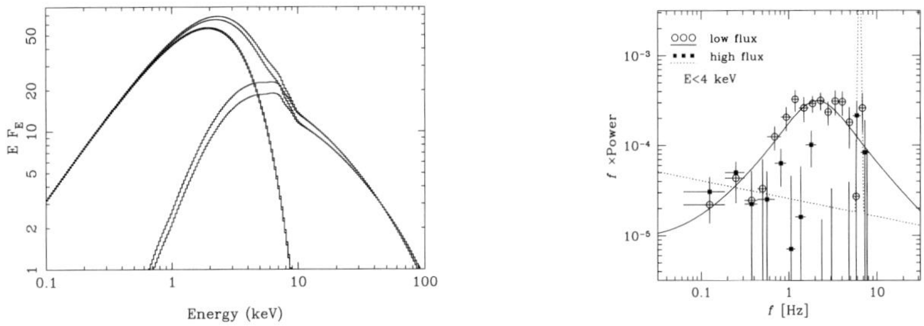

Figure 2. Energy and power spectra of Nova Muscae 1991 on Jan 14th (VHS), when the source showed 'flip-flop' transitions, probably related to re-structuring of the accretion disk. Two spectra in each panel are for two flux states.

high state (HS) and intermediate state (IS) data sets: $\chi^2_\nu = 17/24$ for IS, $\chi^2_\nu = 16/23$ for HS, but it fails for the very high state (VHS).

2. Inner disk oscillations in GRS 1124-68 in very high state?

Shortly before reaching the peak of outburst Nova Muscae 1991 showed 'flip-flop' transitions (Takizawa *et al.*, 1997): the flux in the 2–20 keV band changed by $\approx 20\%$. Spectral and timing analysis during this period suggests that the sources were undergoing re-structuring of the inner accretion disk.

2.1. Spectral analysis

Energy spectra both in the high- and low-flux state can be adequately described by the DISKBB + THCOMP + REL-REPR models (no additional Comptonization of the soft component is required). The best-fit model spectra differ most at energies 5–10 keV: the differences are small both at the lowest and highest energies (Figure 2). The difference spectrum can be described as Comptonization of single blackbody emission, but it cannot be described as Comptonized *disk* blackbody.

2.2. Timing analysis

X-ray power density spectra (PDS) show both energy and flux dependence. The soft X-rays ($E < 4$ keV) PDS is very different in the two flux states (Figure 2). In the high flux state the PDS resembles that of typical HS, i.e. a power law $f^{-\alpha}$ with $\alpha \approx 1.2$, although there is also a narrow QPO at ≈ 8 Hz. However, in the low flux state the PSD has a Lorenzian shape peaking (in the $f \times Power$ representation) at ≈ 2 Hz.

The PDS of the hard X-ray component ($E > 8.5$ keV) is similar in both flux states except for the QPO energy which is higher in the low flux state.

3. Discussion

Interactions between accretion disk and corona in the soft spectral states of accreting black holes are clearly complex. One outcome of those interactions is a presence of warm, mildly optically thick plasma modifying the disk thermal emission. Proper modeling of the resulting spectra is important e.g. for correct determination of system parameters.

The 'flip-flop' transitions observed in VHS of GRS 1124-68 close to the peak of its outburst suggest that the optically thick accretion flow can proceed in two modes giving the same time-integrated characteristics. One mode, usually observed in HS, shows the f^{-1}-like PDS, i.e. a scale-invariant behaviour, perhaps realized in magnetic turbulence (Kawaguchi *et al.*, 2000). The other mode clearly possesses certain spatial/temporal scale, since the PDS is described by a Lorenzian, i.e. it corresponds to exponential shots of a single time scale.

Acknowledgement

This work was supported in part by KBN grants no. 2P03D01816 and 2P03D01718.

References

Ebisawa, K., *et al.*: 1994, *Publ. Astron. Soc. Jpn.* **46**, 375.
Kawaguchi, T., *et al.*: 2000, *Publ. Astron. Soc. Jpn.* **52**, 1.
Takizawa, M., *et al.*: 1997, *Astrophys. J.* **489**, 272.
Życki, P.T., Done, C. and Smith, D.A.: 1998, *Astrophys. J.* **496**, L25.

INNER DISK OSCILLATIONS AND QPOS IN RELATIVISTIC JET SOURCES

A.R. RAO
Tata Institute of Fundamental Research, Mumbai, INDIA

Abstract. Recent results on the inner disk oscillations found in GRS 1915+105 are reviewed. QPOs during the low state are used as a marker for such oscillations and the physical picture emerging from a combined X-ray spectral and timing analysis is examined. The relationship between inner disk oscillations and synchrotron radio emission is critically evaluated.

Keywords: accretion, accretion disk, X-ray:stars, GRS 1915+105

1. Introduction

One of the remarkable features of the X-ray variability of the Galactic relativistic jet source GRS 1915+105 is the rapid intensity variations associated with repeatable and distinct spectral characteristics. Belloni *et al.* (1997) found that the source shows distinct and different inner disk radius during the "quiescence" and outburst which can be interpreted as the vanishing of the bright inner disk during the quiescence. Detection of rapid radio oscillations near the onset of superluminally moving ejecta prompted Fender *et al.* (1999) to suggest that such X-ray inner disk oscillations could be the underlying reason for the ejection of the synchrotron cloud responsible for the radio emission.

Ever since its discovery (Castro-Tirado *et al.* 1992) GRS 1915+105 has been exhibiting a rich variety of X-ray variability characteristics. Greiner *et al.* (1996) presented the RXTE observations of the source which included repeated patterns of brightness sputters, large amplitude oscillations, fast oscillations and several incidences of prolonged lulls. Belloni *et al.* (1997) classified some of the oscillatory X-ray variations as "outbursts" and from a time resolved spectral analysis found evidence for the disappearance of the inner disk during these oscillations. They also discovered a strong correlation between the quiescent phase and burst duration. Taam, Chen, and Swank (1997) detected a wide range of transient activity including regular bursts with a recurrence time of about one minute and irregular bursts. Paul *et al.* (1998) detected several types of such bursts using the IXAE data and found evidence for matter disappearing into the event horizon of the black hole. Yadav *et al.* (1999) made a systematic analysis of these bursts and classified them based on the recurrence time.

Belloni *et al.* (2000a) made a detailed evaluation of the X-ray variability characteristics of GRS 1915+105 and classified the variations in 12 different classes. To

Astrophysics and Space Science is the original source of publication of this article. It is recommended that this article is cited as: *Astrophysics and Space Science* **276** (suppl.): 175–181, 2001.

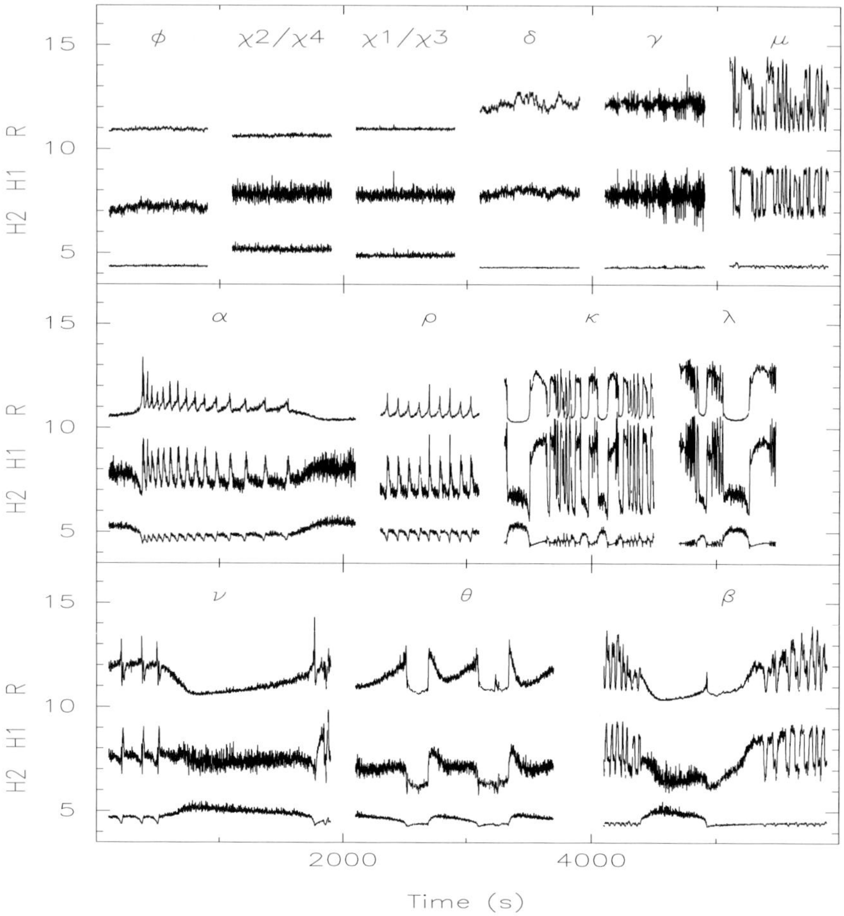

Figure 1. X-ray count rates (R) and the hardness ratios for the various classes of variability seen in GRS 1915+105.

understand the basic nature of these classes, in Figure 1 we give the representative light curves and hardness ratios obtained from the *RXTE* archives. The total count rate, R, is in the units of 10^4 counts s^{-1} and it is vertically shifted by 5 units. The hardness ratio H1 (5 – 13 keV to 2 – 5 keV ratio) is scaled by 5 and shifted by 2 units and the hardness ratio H2 (> 13 keV to 2 – 13 keV ratio) is scaled up by 10 and shifted by 4 units. The hardness ratio H1 is a measure of the disk blackbody temperature and H2 is a measure of the intensity of the hard power-law, presumably arising from a thermal-Compton cloud.

The X-ray variability classes which show steady behavior during one orbit of RXTE are shown in the top panel of the figure. Class ϕ is characterized by low count rate, low H1 and very low H2. It is also devoid of any strong variability. Class

χ is characterized by strong 0.5 – 10 Hz QPO and it resembles the intermediate state of other black hole sources (Trudolyubov *et al.* 1999b). Classes δ, γ and μ are high-soft states of the source and occasionally the 67 Hz QPOs are seen in these states (Morgan *et al.* 1997).

The various X-ray variability classes which exhibit inner disk oscillations are shown in the middle panel of Figure 1. The distinguishing feature is the sharp ($<$ 10 s) change in the intensity state. During the high state of these oscillations the source exhibits all the temporal and spectral characteristics of the high-soft state of the source and during the low state it exhibits all characteristics of the low-hard state of the source (Rao *et al.* 2000b). The drastic change in H2, signifying a change in the Compton cloud, is a common feature of this class. The behavior can be regular (class ρ) and it can remain in this class for periods up to several tens of days. In class α the source reverts back and forth between class χ and class ρ. During the quiescent periods of all these classes the 0.5 – 10 Hz QPO is always present.

In contrast to such inner disk oscillations, classes ν, θ, and β show a more gradual change from a high-soft to low-hard state and they are characterized by a sharp soft dip when both H1 and H2 are low and the variability too drops to a low level. Radio oscillations are associated with these classes (Eikenberry *et al.* 1998; Mirabel *et al.* 1998).

2. Inner Disk Oscillations and QPOs

One of the remarkable features of the inner disk oscillations is the presence of the ubiquitous 0.5 – 10 Hz QPO during the quiescent state of the oscillations. This type of QPO is found during the low-hard state of the source (and also in the low-hard state of other jet sources as well), and it is used as a marker for the low-hard state of the source (Muno *et al.* 1999; Trudolyubov *et al.* 1999b).

Reig *et al.* (2000) made a detailed study of the QPO phenomena in the $\chi 1$ and $\chi 3$ classes and found a correlation between the QPO frequency and the observed phase lag in the Fourier cross-spectrum in two energy channels. Nobili *et al.* (2000) have interpreted this correlation as due to arising from the geometry of the region producing the Compton spectrum, assumed to be arising from a corona formed by the puffed up inner regions of the accretion disk. Rao *et al.* (2000a) made a detailed spectral study during the $\chi 3$ state of the source and identified three spectral components. They found that the QPOs are due to the Compton component and they invoked an accretion disk shock to explain the QPO phenomena.

The large rms amplitude (about 10%), narrow width and the relative stability over considerably long durations makes it difficult to explain the 0.5 – 10 Hz QPO on the basis of any disk oscillation model. Chakrabarti & Manickam (2000) interpreted the QPOs as due to the oscillation of the region responsible for the hard radiation, i.e., the post-shock region. They also detected a correlation between

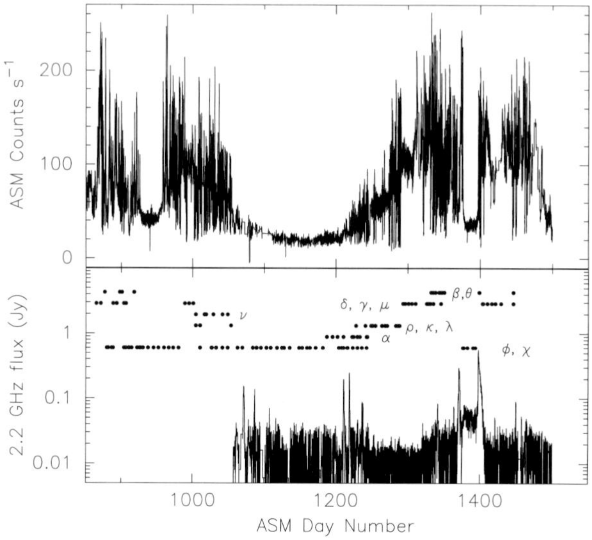

Figure 2. The ASM count rates for GRS 1915+105 is shown in the top panel and the 2.2 GHz GBI data is shown in the bottom panel. The time of occurrence of different intensity classes as established from the pointed RXTE PCA observations (Belloni *et al.* 2000a) are also shown in the bottom panel of the figure.

the average frequency of the QPO and the duration of the 'off' states. A similar correlation with centroid frequencies were also found by Trudolyubov, Churazov, & Gilfanov (1999a).

There have been attempts to explain the rapid variability seen in GRS 1915+105 using disk instability models. Belloni *et al.* (1997) have tried to explain the repeated patterns as due to the appearance and disappearance of the inner accretion disk. Nayakshin *et al.* (1999) have investigated the different accretion models and viscosity prescriptions and attempted to explain the temporal behavior of GRS 1915+105. In particular, they have shown that the accretion instability in a slim disk is not likely to adequately account for the behavior of GRS 1915+105 (because of the difficulty in maintaining the high state for long) and the appearance of the inner accretion disk, as postulated by Belloni *et al.* (1997), will take a time scale comparable to the burst time scale. Though Nayakshin *et al.* (1999) were able to reproduce many characteristics in the X-ray variability for GRS 1915+105, they were unable to explain the rise/fall times or the < 10 s oscillations.

Yadav *et al.* (1999) presented a comprehensive picture for the origin of these bursts. They suggested that the peculiar bursts are characteristic of the change of state of the source. The source can switch back and forth between the low-hard state and the high-soft state near critical accretion rates in a very short time scale, giving rise to the irregular and quasi-regular bursts. It was pointed out that changes in total accretion rate cannot manifest itself in short time scales and hence the fast spectral state changes in GRS 1915+105 cannot be explained by changes in total accretion rates. Chakrabarti & Manickam (2000) provided an explanation for this remarkable observation by invoking mass loss. Unlike the canonical low-hard state where the hard power-law is due to Comptonisation, in the hard state observed during the bursts there should be an additional feature due to winds. Chakrabarti *et al.* (2000) have observed subtle differences in the spectra during the bursts compared to the spectra during the long duration low-hard and high-soft states. This change is above $\sim$ 10 keV where the hard component of the low state softens. This is the effect predicted in Chakrabarti (1998) where the addition of about 10% mass loss in wind has an effect of pushing the spectral cross-over point from 15-17 keV (for a 10 $M_\odot$ black hole) to higher energies.

3. Inner Disk Oscillations and Synchrotron Radio Emission

There have been attempts to connect the X-ray variability characteristics to the onset of jet emission in GRS 1915+105. Fender *et al.* (1999) made a detailed analysis of the super-luminal jet ejection events observed in GRS 1915+105 in 1997 October/November and detected continuous short period (20–40 minute) radio oscillations during the start of the jet emission. They proposed that these are indications of repeated ejection of the inner accretion disk, quite similar to the events seen in X-rays by Belloni *et al.* (1997). A causal connection between disk and jet was thus attempted. Naik *et al.* (2000) have reported the detection of a series of X-ray dips during a huge radio flare of strength 0.48 Jy (at 2.25 GHz). They argue that a large number of such X-ray dips can account for the radio flare emission.

To put such disk-jet connections in perspective, we have shown in Figure 2 (top panel) the observed ASM count rates during 1996/1997. The concurrently obtained RXTE-PCA data have been classified by Belloni *et al.* (2000a) and these are shown in the bottom panel of the figure, along with the 2.2 GHz radio data obtained from GBI public domain. The 1997 slow state transition is particularly interesting (see also Rao *et al.* 2000b).

It can be seen from the figure that the source was in a stable low-hard state up to 1997 April 25 (ASM day number 1210). The radio flux during this state is 20 – 30 mJy. This is the radio-quiet low-hard state of the source (Muno *et al.* 1999). The source started a steady increase in its X-ray emission with an average increase in the ASM count rate of 0.65 s^{-1} day^{-1} and reaching a count rate of 76 s^{-1} in the middle of July (day number 1290). During the beginning of the transition, the source showed several episodes of α states, accompanied by brief radio flares. For more than a month the source showed several episodes of the inner disk oscillations (classes ρ, κ and λ) (Yadav *et al.* 1999; Belloni *et al.* 1997). The source reached a steady high-soft state (classes δ, γ and μ) when the radio emission is very low ($\leq$

20 mJy). The ringing flares started again in the beginning of 1997 August (Yadav *et al.* 1999) and towards the end of this state the peculiar outbursts (class β) accompanied by infrared flares were observed on 1997 August 14-15 (Eikenberry *et al.* 1998). On September 9 the peculiar "outburst" was seen again (Markwardt *et al.* 1999). In October 1997 the source reached the radio-loud hard-steady state when both the X-ray and radio fluxes were high without the evidence of any oscillations. Fender *et al.* (1999) described this class as the "plateau" state.

From these we can conclude that the inner disk oscillations involving rapid transitions (classes ρ, κ and λ) do not lead to any appreciable radio emission. Apart from the radio-loud hard state, classes θ and β are associated with enhanced radio emission. Class β are associated with the synchrotron flares in radio (Mirabel *et al.* 1998; Fender & Pooley 1998) and infrared (Eikenberry *et al.* 1998). From simultaneous X-ray and infrared observations, Eikenberry *et al.* (1998) made a strong argument that the onset of radio/infrared flare is associated with the soft dip rather than the gradual change to the low-hard state. Radio oscillations on a time scale of 20 – 30 minutes are seen to be accompanied by a series of soft X-ray dips (Fig. 10, Dhawan *et al.* 2000), similar to the oscillations observed by Fender *et al.* (1999). Since the soft dip events are associated with the jet emission, Naik *et al.* (2000) proposed that the huge radio flares are produced by a series of such soft X-ray dips.

Naik and Rao (2000) made a systematic study of the radio and X-ray emissions from GRS 1915+105 and found an one-to-one association between radio flares and the β and θ classes. If the onset of a huge radio flare (signifying the emission of a superluminal ejecta) is associated with an X-ray emission characteristic observed so far, it has to be necessarily with the β or θ classes. Belloni *et al.* (2000b), however, associate the change of state with the radio emission rather than the peculiar dips seen in these two classes. A continuous X-ray monitoring during a radio flare should clarify this question.

Acknowledgements

I thank the members of the RXTE and NSF-NRAO-NASA GBI team for making the data publicly available. This research has made use of data obtained through the High Energy Astrophysics Science Archive Research Center Online Service, provided by the NASA/Goddard Space Flight Center. I also thank S. Naik and S. V. Vadawale for helpful discussions and assistance in preparing this manuscript.

References

Belloni, T., Klein-Wolt, M., Mendez, M., van der Klis, M. and van Paradijs, J., 2000a, *A&A*, 355, 271

Belloni, T., Mendez, M., King, A. R., van der Klis, M. and van Paradijs, J., 1997, *ApJ*, 479, L145

Belloni, T., Migliari, S. and Fender, R. P., 2000b, *A&A*, 358, L29

Castro-Tirado, A.J., Brandt, S. and Lund, N., 1992, *IAU Circ.*, 5590

Chakrabarti, S. K., 1998, *Indian Journal of Physics*, 72B(6), 565

Chakrabarti, S. K. and Manickam, S. G., 2000, *ApJ*, 531, L41

Chakrabarti, S. K., Manickam, S. G., Nandi, A. K. and Rao, A. R., 2000, *A&A*, submitted

Dhawan, V., Mirabel, I. F. and L. F. Rodriguez, L. F., 2000, *ApJ*, 543, 373

Eikenberry, S. S., Matthews, K., Morgan, E. H., Remillard, R. A. and Nelson, R. W., 1998, *ApJ*, 494, L61

Fender, R. P., Garrington, S. T., McKay, D. J., Muxlow, T. W. B., Spencer, R. E., Sterling, A. M. and E. B. Waltman, E. B., 1999, *MNRAS*, 304, 865

Fender, R. P. and Pooley, G. G., 1998, *MNRAS*, 300, 573

Greiner, J., Morgan, E. H. and Remillard, R. A., 1996, *ApJ*, 473, L107

Markwardt, C.B., Swank, J. H., Taam, R. E, 1999, *ApJ*, 513, L37

Mirabel, I. F., Dhawan, V., Chaty, S., Rodriguez, L. F., Marti, J., Robinson, C. R., Swank, J. and Geballe, T., 1998, *A&A*, 330, L9

Morgan, E. H., Remillard, R. A. and Greiner, J., 1997, *ApJ*, 482, 993

Muno, M. P., Morgan, E. H. and R. A. Remillard, R. A., 1999, *ApJ*, 527, 321

Naik, S., Agrawal, P.C., Rao, A. R., Paul, B., Seetha, S. and Kasturirangan, K., 2000, *ApJ*, in press (astro-ph/0008414)

Naik, S. and Rao, A. R., 2000, Â&A, in press (astro-ph/0008433)

Nayakshin, S., Rappaport, S. and Melia, F., 2000, *ApJ*, 535, 798

Nobili, L., Turolla, R., Zampieri, L. and Belloni, T. 2000, *ApJ*, 538, L137

Paul, B., Agrawal, P. C., Rao, A. R., Vahia, M. N., Yadav, J. S. Seetha, S. and Kasturirangan, K. 1998, *ApJ*, 492, L63

Rao, A. R., Naik, S., Vadawale, S. V. and Chakrabarti, S. K., 2000a, *A&A*, 360, L25

Rao, A. R., Yadav, J. S. and Paul, B. 2000b, *ApJ*, in press (astro-ph/0007250)

Reig, P., Belloni, T., van der Klis, M., Mendez, M., Kylafis, N. D. and Ford, E. C., 2000, *ApJ*, 541, 883

Taam, R. E., Chen, X. and Swank, J. H., 1997, *ApJ*, 485, L83

Trudolyubov, S., Churazov, S. E. and Gilfanov, M., 1999a, *A&A*, 351, L15

Trudolyubov, S., Churazov, S. E. and Gilfanov, M., 1999b, *Ast. L.*, 25, 718

Yadav, J. S., Rao, A. R., Agrawal, P. C., Paul, B., Seetha, S. and Kasturirangan, K., 1999, *ApJ*, 517, 935

CHASING THE INSTABILITY: SPECTRAL AND TIMING ANALYSIS OF GRS 1915+105

S. MIGLIARI[1,2], F. VIGNARCA[1,3] and T. BELLONI[1]
[1] *Osservatorio Astronomico di Brera, Merate (LC), Italy*
[2] *Dipartimento di Fisica, Università degli studi di Milano, Milano, Italy*
[3] *Dipartimento di Scienze, Università dell'Insubria, Como, Italy*

Abstract. The superluminal galactic microquasar GRS 1915+105 shows complex variability in the X-ray band. This has been interpreted as an instability of the inner region of an optically-thick accretion disk. Making use of RXTE/PCA data, we inspected in detail the instability intervals of the source. We concentrated upon the light curves for which the disappearance of the inner region lasts sufficiently long to trace the behaviour of single parameters. Fits of time-resolved power spectra yielded the parameters of the 1-10 Hz QPO, while time-resolved energy spectra allowed us to find the disk inner radius and temperature, and the power-law photon index during the evolution of the instability. We present the analysis of the parameters and their correlations and discuss them in terms of current accretion disk models.

1. Introduction

The BHC GRS 1915+105 is an extremely variable X-ray transient which shows superluminal jets in the radio. The large variability in the X-ray band is compatible with the oscillations of the inner region of the accretion disk around the black hole (see Belloni *et al.*, 2000). Our aim is to follow in more detail the evolution of spectral and timing parameters during the instability intervals (see Belloni, this volume).

We analyzed RXTE/PCA data of selected GRS 1915+105 observations, corresponding to classes α, β, ν, κ and λ as in the classification by Belloni *et al.* (2000), in which the instability intervals last sufficiently long enough to trace the evolution of spectral and timing parameters. We isolated long low-flux intervals corresponding to states C and A (Belloni *et al.*, 2000). For each interval, we produced energy spectra with an integration time of 16 s and fitted each spectrum using the 'standard' model for BHCs, consisting of the superposition of a multicolor disk-blackbody and a power-law. From the same data, we extracted power spectra of selected intervals of 4 and 16 seconds. We fitted these power spectra, using Lorentzian functions, in order to find the characteristic frequency of the QPO in the 1–10 Hz frequency range.

Astrophysics and Space Science is the original source of publication of this article. It is recommended that this article is cited as: *Astrophysics and Space Science* **276** (suppl.): 183–184, 2001.

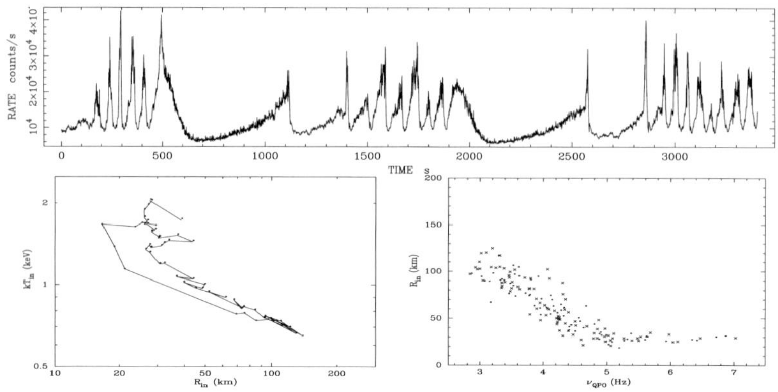

Figure 1. Top: PCA light curve of GRS 1915+105 corresponding to class β. Left: Correlation between inner disk radius and temperature during the instability of the accretion disk. Right: Correlation between QPO frequency and inner radius for many instability intervals.

2. Results

After obtaining the spectral and timing parameters, we investigated their behaviour by first following the time evolution of single parameters and then linking them to each other. Correlations between spectral parameters ($kT_{in} - R_{in}$, $kT_{in} - \Gamma$, $R_{in} - \Gamma$) yield constraints for more detailed instability models. Figure 1 (left panel) shows an example of a $kT_{in} - R_{in}$ evolution plot during a β-class instability. An important aspect of accretion disk physics is the connection between disk accretion and jet ejection. We can investigate in more detail this connection, comparing disk parameters and jet features from mutiwavelength observations in Radio, IR and X-ray band (see Klein-Wolt *et al.*, this volume). Other results are found when correlating QPO frequencies, obtained from timing analysis, and inner radii, obtained from spectral analysis. It is evident (Figure 1, right panel) that the 1-10 Hz QPO is strictly linked to the status of the accretion disk. Moreover, a correlation is also found between QPO frequency and power-law index Γ, showing that the power-law component is important for the origin of QPO. Our analysis shows that only by taking into account both spectral and timing parameters we have a complete picture of the complex phenomenology of GRS 1915+105.

References

Belloni, T., *et al.*: 2000, *Astron. Astrophys.* **355**, 271.

THERMAL COMPTONIZATION IN GRS 1915+105

OSMI VILHU and PETTER NIKULA
Observatory, Box 14, FIN-00014 University of Helsinki, Finland

JURI POUTANEN
Stockholm Observatory, SE-133 36 Saltsjöbaden, Sweden

JUKKA NEVALAINEN
ESA/ESTEC Astrophysics Division

Abstract. The Rossi X-ray Timing Explorer (*RXTE*) data of GRS 1915+105 from several observing periods were modelled with a thermal Comptonization model. In the model (Figure 1), seed soft photons from an optically thick cool disk are Comptonized in a hot spherical corona surrounding the inner disk. Although the coronal temperature and optical depth can not be determined simultaneously, a qualitative trend was observed: at high soft states the inner disk just touches the outer edge of the central corona, while during low hard states the corona is larger and denser and surrounds the inner disk.

1. Introduction and Results

The X-ray transient GRS 1915+105 was discovered by Castro-Tirado *et al.* (1992) using WATCH all-sky monitor on the GRANAT satellite. Since then it has been an active X-ray source with superluminal jets found by Mirabel and Rodriquez (1994).

The *Rossi X-ray Timing Explorer* (*RXTE*) has been monitoring it frequently and a rich pattern of variability has emerged from these data (see Belloni *et al.*, 2000; Vilhu and Nevalainen, 1998; Muno *et al.*, 1999). We collected several Proportional Counter Array (PCA) and High-Energy X-ray Timing Experiment (HEXTE) observations of GRS 1915+105, performed during 1996–97, from the TOO-archive of *RXTE*, and modelled the spectra with a thermal Comptonization model in a specific sombrero-geometry.

In the model (Figure 1), the disk penetrates into the central spherical coronal region with electron temperature kT_e, Thompson optical depth τ and coronal radius R_{cor}. The optically thick multi-temperature disk, with inner disk radius R_{in}, radiates the observed soft spectrum and serves as the source of seed photons for Comptonization in the corona. Inside the corona the disk temperature was assumed constant. The radiative transfer in the corona was handled by the iterative scattering method developed by Poutanen and Svensson (1996) and incorporated into the XSPEC spectral fitting software.

The results are shown in Figure 2 where the coronal temperature was fixed to 70 keV (T_e and τ can not be determined separately, only their product, the Kom-

Astrophysics and Space Science is the original source of publication of this article. It is recommended that this article is cited as: *Astrophysics and Space Science* **276** (suppl.): 185–186, 2001.

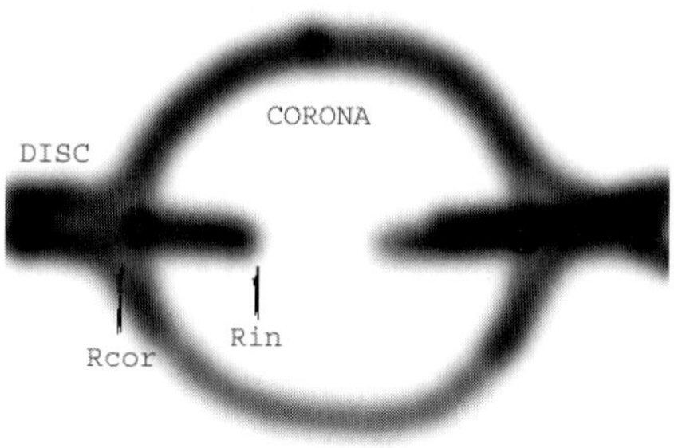

Figure 1. The Sombrero-model geometry

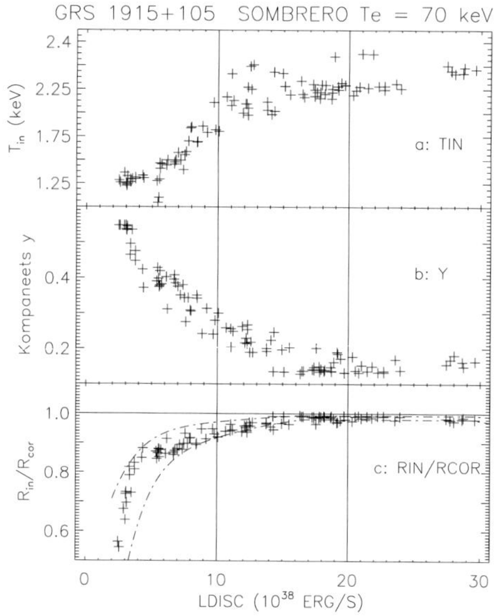

Figure 2. Model parameters as a function of the disk luminosity

paneets y-parameter, is well defined by the observations). However, the following trend is at least qualitatively independent on this assumption: *At high soft states the inner disk just touches the outer edge of the central corona, while in the low hard states the corona is larger and denser and totally surrounds the inner disk.*

References

Belloni, T., Klein-Volt, M., Mendez, M., *et al.*: 2000, *Astron. Astrophys.* **355**, 271.

Castro-Tirado, A.J., Brandt, S. and Lund, N.: 1992, *IAU Circ.* **5590**.

Mirabel, I.F. and Rodriquez, L.F.: 1994, *Nature* **371**, 46.

Muno, M.P., Morgan, E.H. and Remillard, R.A.: 1999, *Astrophys. J.* **527**, 321.

Poutanen, J. and Svensson, R.: 1996, *Astrophys. J.* **470**, 249.

Vilhu, O. and Nevalainen, J.: 1998, *Astrophys. J.* **508**, L89.

NONLINEAR TIME-SERIES ANALYSIS OF THE X-RAY FLUX OF COMPACT OBJECTS

M. THIEL, M. ROMANO, U. SCHWARZ and J. KURTHS
Universität Potsdam

G. HASINGER
AIP Potsdam

T. BELLONI
Astronomical Institute 'Anton Pannekoek', University of Amsterdam

Abstract. We analyse the X-ray light curves of compact objects using linear and nonlinear time series analysis methods. A Power Density Spectrum (PDS) describes the overall second order properties of the observed data well. To look beyond we propose the nonlinear Q-statistic to detect an asymmetry of the time series. This allows us to find relevant time scales. This method even grants a subclassification of the known states of X-ray sources.

1. Q-Statistic

We consider the nonlinear quantity 'asymmetry in time' $Q(m)$ of the set of data x_k, corresponding to an irreversibility in time, to characterize the observation beyond linearity (Timmer *et al.*, 2000). The significance $S(m)$ of this quantity $Q(m)$ corresponding to the null hypothesis of a linear Gaussian process is computed by the method of surrogate data (Kantz and Schreiber, 2000). The general idea is to simulate time series whose statistical properties are constrained to the null hypothesis one wants to test for. In testing for a linear Gaussian process (colored noise) we randomize the phases of the Fourier Transform of the data and transform the modified series back to the time domain.

2. Results for Cyg X-1 / Conclusions

We analyze the X-ray flux of Cyg X-1 in each low-, high- and intermediate state. The data are recorded on various days with the PCA on board the RXTE. The energy range is 2.0–14.1 keV. The sampling frequency is 256 Hz. The data sets consist of 100 000 points. Applying the Q-statistic to the observational the data, we get a graphical representation of both $Q(m)$ (left panel) and $S(m)$ (right panel) of a low-, an intermediate- and a high-state:

Astrophysics and Space Science is the original source of publication of this article. It is recommended that this article is cited as: *Astrophysics and Space Science* **276** (suppl.): 187–188, 2001.

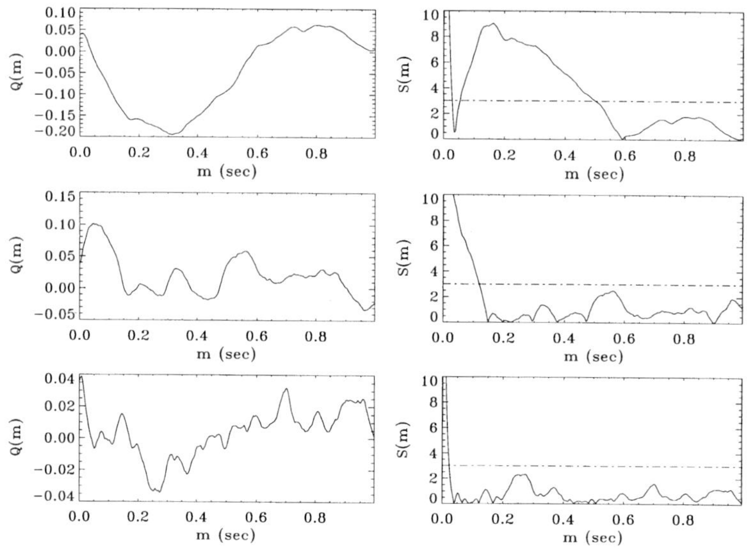

Figure 1. $Q(m)$ and $S(m)$ of low-, intermediate- and high-state (top to bottom).

Even though we have not yet a phyiscal interpretation of the structures in $S(\tau)$ and $Q(\tau)$, the results have some striking points: i) Changes of both $Q(m)$ and $S(m)$ in the low states occur on larger time scales than in the other states. ii) The decay of the first peak of $S(m)$ in an intermediate state is about double time as in the other two states. iii) The peaks of $Q(m)$ and $S(m)$ are anticorrelated in the low- and high- but correlated in the intermediate state. iv) The fine structure allows a subclassification of the different states of Cyg X-1. v) As we will show there is a close relationship between Fourier coefficients and the Q-statistics (Thiel *et al.*, 2001). vi) The Q-statistic provides information about the time-series that a PDS cannot discover.

Our next step is to try to connect physical processes, e.g. turbulence, to the structures of $S(\tau)$ and $Q(\tau)$ (Thiel *et al.*, 2001).

References

Kantz, H. and Schreiber, T.: 1997, in: *Nonlinear time series analysis*, Cambridge University Press.

Thiel, M., Romano, M., Schwarz, U. and Kurths, J.: 2001, in press.

Timmer, J., Schwarz, U., Voss, H.U., Wardinski, I., Belloni, T., Hasinger, G., van der Klis, M. and Kurths, J.: 2000, *Phys. Rev. E* **61**, 1342.

Chapter IV
WINDS AND DISK-CORONA OSCILLATIONS

ADVECTIVE FLOW PARADIGM AND THE MICROQUASAR GRS 1915+105

SANDIP K. CHAKRABARTI
SNBNCBS, JD Block, Salt Lake, Calcutta 700098

Abstract. Advective accretion disks and winds are the most self-consistent solutions today. We describe this paradigm briefly and show how it attempts to explain some of the interesting observations of the galactic microquasar GRS 1915+105.

1. Introduction

The standard accretion disk model of Shakura and Sunyaev (1973) was modified immediately after it was proposed. The spectrum of the black hole candidate Cyg X-1 showed (Sunyaev and Trümper, 1979) that apart from the modified cool black body spectrum, it also exhibits a power-law emission at much higher energies, indicating the presence of electrons much hotter than a standard disk. Subsequently, Sunyaev and Titarchuk (1980) showed that the power-law emission is the result of the Comptonization of soft photons from a standard disk by hot electrons whose source was thought to be *outside* the standard disk, in the form of floating 'Compton clouds' or 'magnetic corona'.

Meanwhile, transonic flow models were developed in the early eighties (Muchotrzeb and Paczyński, 1982; Matsumoto *et al.*, 1984) to modify the inner edge of the standard disk to ensure that it passes through a sonic point. These solutions ensure that the specific energy of the flow remains negative as in a Keplerian disk. However, there is another class of transonic disk solutions which are the generalizations of the classical Bondi (1952) solutions when angular momentum, viscosity, heating, cooling, etc. were taken into account. These are known as advective flow solutions (Chakrabarti, 1990, 1996a, 1998ab, 2000a). Particularly interesting is that these solutions did away with the external Compton clouds as the inner sub-Keplerian region itself is found to behave like one for all practical purposes.

In this review, I briefly present these solutions for viscous and non-viscous flows. The solutions can be suitably combined to obtain a paradigm of the black hole astrophysics. Truly speaking, a cool, Keplerian disk with a negative energy cannot smoothly join to most of these *steady* solutions which require positive energy (at least in quasi-adiabatic flows). For time-dependent flows such restrictions do not apply. In fact, steady transition from a Keplerian disk to an advective disk may require systematic heating of a Keplerian disk or the assumption that all disks

Astrophysics and Space Science is the original source of publication of this article. It is recommended that this article is cited as: *Astrophysics and Space Science* **276** (suppl.): 191–198, 2001.

are fundamentally non-steady, only the time scale varies. We also try to explain some of the observed features of the galactic micro-quasar GRS 1915+105. We find that it always helps to keep a paradigm in the back of one's mind while attempting to explain any observations. Otherwise, each observational feature would demand a separate 'model' which may or may not fit with the global solutions of accretion and winds.

2. Advective Flow Paradigm

2.1. Hydrodynamic and Magnetohydrodynamic Flows

Though most of the literature concentrates on disks and jets separately, it is advisable (and *economical*) to study them simultaneously. This is because in global studies, always both types of solutions appear together. Examples are Bondi (1952) and generalized Bondi solutions (Chakrabarti, 1990, 1996b,c). Numerical simulations have also shown outflows in much of the region of the parameter space (Molteni, Lanzafame and Chakrabarti, 1994 [MLC94]; Molteni, Ryu and Chakrabarti, 1996 [MRC96]; Ryu, Chakrabarti and Molteni, 1997 [RCM97]). Figure 1 shows the broad classification of all possible steady, inviscid solutions of advective disks (adapted from Chakrabarti, 2000a). Solutions shown are obtained for inviscid adiabatic flows (Chakrabarti 1989, 1996b) but they remain valid close to the black hole even when these conditions are relaxed. One important aspect which requires emphasizing is the presence of a centrifugal pressure supported boundary layer (CENBOL) in some of these steady solutions and in most of the non-steady solutions. Rapid inflow keeps the angular momentum to an almost constant value. Thus, centrifugal force becomes a dominant force. It slows downs matter in this barrier, heats them up so that hard X-rays could come out and drives hot matter perpendicular to the disk to form outflows. The barrier even oscillates with a large amplitude if conditions are right (Molteni, Sponholz and Chakrabarti, 1996; RCM97) thereby giving rise to quasi-periodic oscillations of X-rays. Beside each solution in Figure 1 a schematic picture is given of accretion and outflows. Inward and outward pointing arrows indicate accretion wind solutions respectively. In (A) and (a), low energy and low angular momentum flow behaves like a conical Bondi inflow and Parker type winds well known in solar physics. In (B) and (b), a steady shock does not form, the inflow becomes hotter due to slowing down at the centrifugal barrier. Numerical simulation shows that an oscillating shock forms (RCM97) which is accompanied by a non-steady outflow. In (C) and (c), the steady inflow solution has a standing shock. This has been tested by numerical simulations (Chakrabarti and Molteni, 1993 [CM93]; MLC94; MRC96). Positive energy outflow is also found to be steady. In (D) and (d), the inflow passes through an inner sonic point, but the outflow forms a steady shock (CM93). In (E) and (e), no shock is possible in the steady solution though there are three sonic points and we feel that *non-steady* shocks would form. In (F) and (f), both the inflow and the outflow have no shocks, and energetic steady solutions are present as both pass through the inner sonic point. In (G) and (H), solutions are incomplete since they are not from infinity to the horizon. Thus, no steady solution is possible. Simulations (Ryu and Chakrabarti, 1996) indicated that the flow is highly unsteady in these cases which are schematically shown in panels (g) and (h) respectively. Especially interesting is the solution of O*, where disks could be 'thick' not because of gas or radiation pressure, but because of turbulent pressure. Solutions from (g) are with negative energy.

When viscosity is present, closed topologies open up to join with Keplerian disks and when heating (such as magnetic, viscous) flow from a Keplerian disk can pass through shocks and sonic points (Chakrabarti, 1990, 1996, 2000a). For instance, solutions in (C) become similar to those of (I) and (i) while those in I* become similar to those of (J) and (j). These solutions are the backbones of the advective disk paradigm and are useful in obtaining complete solutions of the disk/jet systems. When magnetic field is added, the flow passes through the fast and the slow magnetosonic point and the Alfv́en point. Possible outflowing solutions with or without shocks have been presented in Chakrabarti (1990). It is likely that in some regions of the parameter space, these magnetohydrodynamic shocks should also show oscillations.

2.2. Effects of Radiative Transfer

Fully self-consistent two temperature flows in the advective disks were studied very recently. Chakrabarti and Titarchuk (1995) computed temperatures of electrons and ions for various accretion rate parameters in the CENBOL region and found that for high Keplerian rate CENBOL cools down completely and shock disappears. Soft photons intercepted by CENBOL are reprocessed by hot electrons to produce hard X-rays which are observed as a power-law tail in the hard states. In soft states, the hard tail is formed due to bulk motion Comptonization.

Hot CENBOL drives outflows from the disk along the vertical axis. When the outflow rate is very high, the sonic sphere (i.e., region of the wind till its first sonic surface) becomes dense enough to cool down due to Comptonization by the soft photons from the Keplerian disk. Estimation of the rate of wind generation (Chakrabarti, 1998a,b; Chakrabarti, 1999a) shows that in purely soft states no outflow is possible and in purely hard states, outflow is very small. Thus above mentioned cooling of the sonic sphere is not possible. When the compression ratio of the shock is intermediate, the outflow rate is significant.

After the sonic sphere is cooled, the sonic surface comes closer to the black hole and matter below it returns back to the disk, while matter above it separates as blobs. Thus blobby jets are expected in this intermediate states (this may also be called flare/quiescence or On/Off transition states). The return flow acts as a feedback on the already accreting flow and the count rate undergoes very interesting behaviour. Figure 2 (Chakrabarti, 2000b) shows how the ratio ($R_{\dot{m}}$) of the outflow

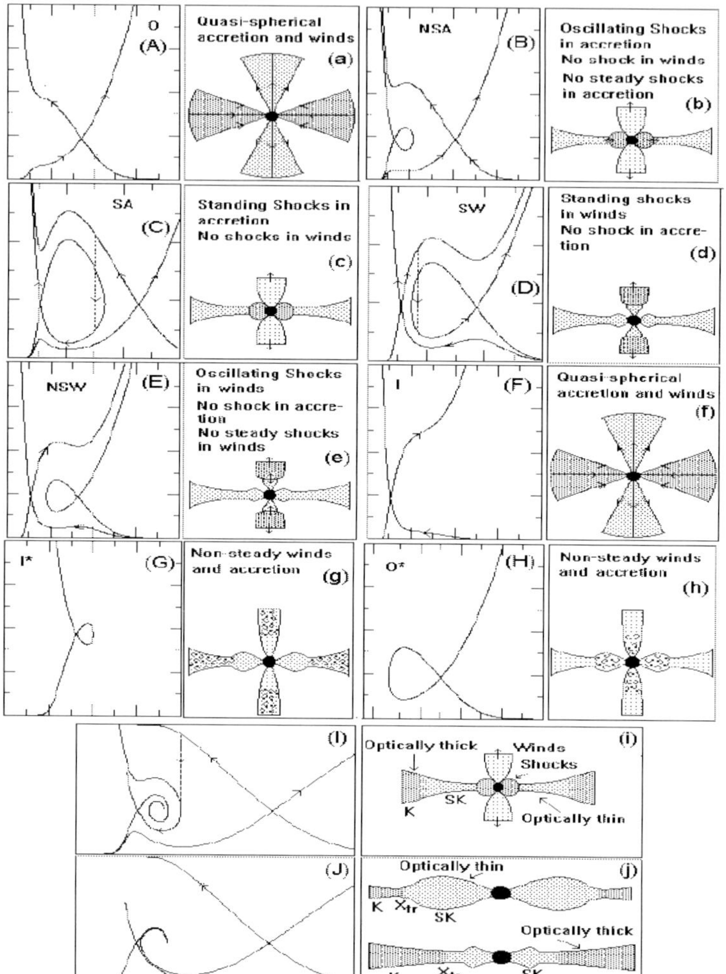

Figure 1. Nature of the solutions (Mach number along Y-direction and logarithmic radial distance along X-direction) of inviscid (A-H) and viscous (I-J) advective disks. Schematic flow behaviour of these solutions in meridional plane are shown in (a-j). Shock locations (C, D) and incomplete solutions (G, H) are indicated by puffed up (c, d) and turbulent regions (g, h) respectively. Solutions with three sonic points having no steady shock are shown oscillating with double arrows (b, e).

rate $\dot{M}_{out}$ and the inflow rate ($\dot{M}_{in}$ assumed to be proportional to $1/R^2$; $R \rightarrow 1 \rightarrow$ soft state; $R \rightarrow 7 \rightarrow$ hard state) depend on the compression ratio R of the gas at the shock. This clearly shows that there is a distinct relation between the spectral states and the outflow rate. Solid angles of inflow and outflow are assumed to be identical.

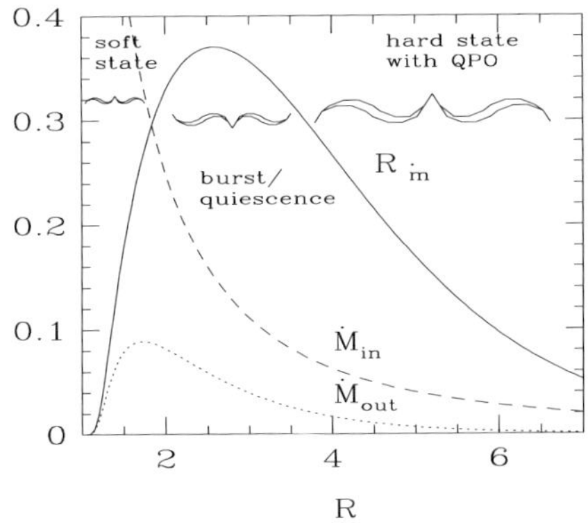

Figure 2. Variation of ratio of outflow and inflow rates as a function of the compression ratio at the accretion shock. Also shown are the expected outflow rate (short-dashed curve) for a given inflow rate (long-dashed curve).

An important characteristics the spectrum should show is that spectral slops of the power-law component should be harder in soft states and softer in the hard states. This is because, in the hard states, outflows from CENBOL reduced electron density but the soft-photon intensity from the pre-CENBOL flow remains the same. This softens the spectra in hard states (Chakrabarti, 1998a). Similarly return flow hardens the spectrum. A corollary of these effects is that the pivotal point of the power-law tail shifts to a larger energy in presence of winds/return flows.

3. Explanations of Observational Aspects of GRS 1915+105

GRS 1915+105 exhibits very intriguing light curves. Belloni *et al.* (2000, hereafter B00) divided them into twelve types. They draw HR1 vs HR2 diagrams where HR1=B/A and HR2=C/A (A: 2–5 keV, B: 5–13 keV, C: 13–60 keV). Color-Color diagrams showed very intricate structures (shapes of atoll, banana, etc.). If the pre-shock flow is indeed the source of the soft photons, photons originating in the 0–3 keV range should be roughly proportional to the Keplerian accretion rate. Thus, photon number may show time variation (due to periodic changes in the ‘accretion rate’ due to return flow mentioned above). However, no QPO should be seen from these photons. Chakrabarti and Manickam (2000, hereafter CM00) demonstrated this (see also, Rao *et al.*, 2000). The harder photons ($E > 3$ keV) would usually come from the post-shock flow. Since the spectra intersect at around 17 keV, and for $E > 17$ keV, photon number is not large, we make our choice of A, B and C to be in the ranges of 0–3 keV, 3–17 keV and 17–60 keV respectively (B and C would be related to sub-Keplerian rate). According to our paradigm, roughly speaking, A, B and C should be proportional to each other (since B and C are produced by interception of soft photons measuring A. Of course, soft X-ray absorption makes matter more complex) and whenever hardness or softness ratios are plotted basically straight lines are expected, instead of Atoll, Banana and Z shapes. The results are presented in Nandi *et al.* (2000).

One important conclusion of B00 was that the disk apparently has three states: A (low rate and low HR1, HR2), B (high rate, high HR1) and C (low rate, low HR1, variable HR2 depending on length of the event). It seems that the State C exhibits QPO. State A and state B do not exhibit QPO. More interestingly, except for the $C \rightarrow B$ transition, all other transitions of states are allowed. Nandi *et al.* (2001) found evidence of QPO in some of the State A light curve.

As discussed in Chakrabarti (2000a,b) there are practically four distinct ways that the solutions in Figure 1 could be combined for a realistic accretion-wind system. If the accretion rate is very low and shock does not develop, QPO may not be seen. In GRS 1915+105 we need not be concerned with this type of flow. The remaining three are shown in Figure 3 (Chakrabarti *et al.*, 2000a). If the shock develops, QPO would be seen but wind may be negligible or absent depending on the compression ratio of the shock. The object would be in State C of B00 or hard class of Nandi *et al.* (2000). If the accretion rate is generally increased, shock is weakened (compression ratio goes down), and in the intermediate state, burst/quiescence or On/Off may be seen. There could be two types of 'On' states (State A and State B of B00). After the winds of State C fills in the sonic sphere and cools it down by Comptonization, CENBOL and the region till the sonic sphere collapse. This is the State A of B00. Now there are two possibilities (Chakrabarti, 1999b): either the flow separates completely as a blob and returns to State C, or the flow mostly returns back to the accretion disk and enhances the accretion. This would be the State B of B00. This may in turn increases the outflow (see Figure 2). But shock itself is getting weakened because of post-shock cooling, hence the outflow is very mild, but may be in a threshold so that a bit more outflow can cause the sonic sphere to collapse again. Thus, occasional trips to State A from State B are possible. Once the enhanced matter is drained out and the shock bounces back to roughly the original location (compatible with its specific energy and angular momentum) the State C forms again. Since State C produces fewer soft photons, State B is not directly possible from State C without first producing return flow and enhanced accretion. This may explain why a transition from C to B is difficult.

Among other confirmations of our paradigm, we note that Nandi *et al.* (2001) demonstrated that the spectral slopes of GRS 1915+105 are hardened by enhanced accretion and softened by return flows. CM00 found that the duration of the Off states strongly depend on QPO frequency. Dhawan *et al.* (2000) observationally demonstrates that jets originate from what we term CENBOL region.

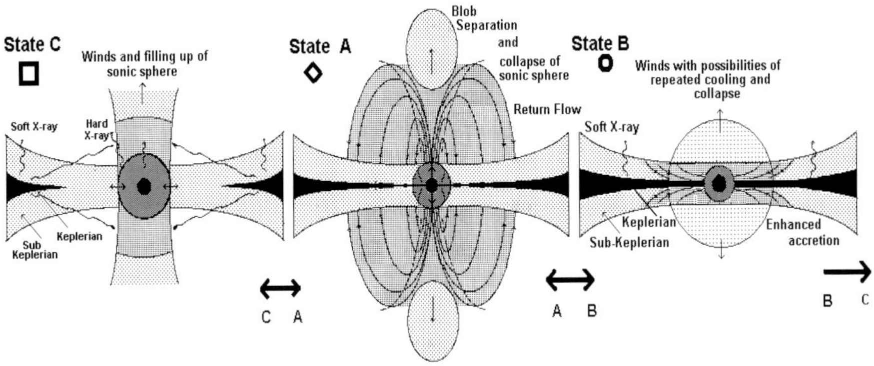

Figure 3. Suggested accretion/wind configurations in States A, B and C of B00.

4. Concluding Remarks

The advective disk paradigm being made up of the most complete accretion disk/wind solutions, it is not surprising that most of the observations could be explained by this paradigm. However, our explanations of the light curves of GRS 1915+105 have been over simplified since we ignored the effects of magnetic fields altogether. Apart from a new spectral component due to synchrotron radiation and perhaps increasing the outflow velocity to a much larger value, no other effect is expected as far as our State description is concerned.

Acknowledgement

This work is partly supported by a grant from ISRO for the project ‘Quasi-Periodic Oscillations of Black Holes’.

References

Belloni, T., Wolt, M.-K., Méndez, M., van Der Klis, M. and van Paradijs, J.: 2000, *Astron. Astrophys.* **355**, 271.

Bondi, H.: 1952, *Mon. Not. R. Astron. Soc.* **112**, 195.

Chakrabarti, S.K.: 1989, *Astrophys. J.* **347**, 365.

Chakrabarti, S.K.: 1990, in: *Theory of Transonic Astrophysical Flows*, World Scientific, Singapore.

Chakrabarti, S.K.: 1996a, *Phys. Rep.* **266**, 229.

Chakrabarti, S.K.: 1996b, *Mon. Not. R. Astron. Soc.* **283**, 325.

Chakrabarti, S.K.: 1996c, *Astrophys. J.* **464**, 664.

Chakrabarti, S.K.: 1998a, *Ind. J. Phys. (Rev.)* **72B**, 565.

Chakrabarti, S.K.: 1998b, in: *Observational Evidence for Black Holes in the Universe*, Kluwer Academic Publishers, Holland, p. 19.

Chakrabarti, S.K.: 1999a, *Astron. Astrophys.* **351**, 185.

Chakrabarti, S.K.: 1999b, *Ind. J. Phys.* **73B**, 931.
Chakrabarti, S.K.: 2000a, *Nuovo Cimento* **115**, 897.
Chakrabarti, S.K.: 2000b, in: F.A. Aharonian and H. Völk (eds.), *Proceedings of the Heidelberg Workshop on High Energy Gamma-Ray Astronomy*.
Chakrabarti, S.K. and Manickam, S.G.: 2000, *Astrophys. J.* **531**, L41 [CM00].
Chakrabarti, S.K. and Molteni, D.: 1993, *Astrophys. J.* **417**, 671 [CM93].
Chakrabarti, S.K. and Nandi, A.: 2000a, *Ind. J. Phys.* **75B**, 1.
Chakrabarti, S.K. and Titarchuk, L.G.: 1985, *Astrophys. J.* **455**, 623.
Dhawan, V., Mirabel, I.F. and Rodríguez, L.F.: 2000, *Astrophys. J.* **543**, 373.
Matsumoto, R., Kato, S., Fukue, J. and Okazaki, A.T.: 1984, *Publ. Astron. Soc. Pacific* **36**, 71.
Molteni, D., Lanzafame, G. and Chakrabarti, S.K.: 1994, *Astrophys. J.* **425**, 161 [MLC94].
Molteni, D., Ryu, D. and Chakrabarti, S.K.: 1996, *Astrophys. J.* **470**, 460 [MRC96].
Molteni, D., Sponholz, H. and Chakrabarti, S.K.: 1996, *Astrophys. J.* **457,** 805.
Muchotrzeb, B. and Paczyński, B.: 1982, *Acta Astron.* **32**, 1.
Nandi, A., Manickam, L.G. and Chakrabarti, S.K.: 2000a, *Ind. J. Phys.* **74B**, 331.
Nandi, A., Manickam, L.G. and Chakrabarti, S.K.: 2000b, *Mon. Not. R. Astron. Soc.* **324**, 267.
Rao, A.R., Naik, S., Vadawale, S.V. and Chakrabarti, S.K.: 2000, *Astron. Astrophys.* **360**, L25.
Ryu, D., Chakrabarti, S.K. and Molteni, D.: 1997, *Astrophys. J.* **474**, 378 [RCM97].
Ryu, D. and Chakrabarti, S.K.: 1996, unpublished.
Shakura, N.I. and Sunyaev, R.A.: 1973, *Astron. Astrophys.* **24**, 337.
Sunyaev, R.A. and Titarchuk, L.G.: 1980, *Astron. Astrophys.* **86**, 121.
Sunyaev, R.A. and Trümper, J.: 1979, *Nature* **279**, 506.

POWER LAW EMISSION
NO CUTOFF
f = 67 Hz
NEUTRON
STAR ?

A REVIEW OF DISK-CORONA OSCILLATIONS

JEAN SWANK
NASA/Goddard Space Flight Center

Abstract. Low frequency ($\approx 0.1 - 35$ Hz) quasi-periodic oscillations of the X-ray flux characterize many of the black hole candidates, in particular those which have radio evidence for jets. These QPO have amplitudes of up to 20% in the states of black hole novae which are called very high and intermediate and are correlated with the break frequency of the band-limited low frequency white noise. With transition to the state called the soft high state, both the QPO and the noise disappear. While the noise is strong in the low hard states like that of Cyg X-1, the QPO, when present, are weak and broad. In their strong manifestations, these QPO have the curious property of appearing to have the spectrum of the power-law component which dominates in the low state, while correlations between their frequency and the disk component in the spectrum imply control by the disk. The correlations, the harmonic structure of the QPO, and the phase lags have complex behavior in the same source (GRS 1915+105, XTE J1550-564). The phenomena point to interaction between the disk and the corona, for which there are several interesting ideas.

1. Introduction

Disk-corona oscillations, as discussed in this paper, refer to the oscillations in X-ray flux, appearing as quasi-periodic oscillations (QPO) in the power density spectra, which are in the non-thermal spectra possibly originating in a corona, but which appear to be related to optically-thick emission associated with the accretion disk. The QPO can have an amplitude of 10–20%. In the power density spectra of black hole candidates (BHCs), a subset of which are known to be microquasars, several low frequency features have appeared. For a given source several QPOs can appear simultaneously, in which case it is clear that there are more than one phenomenon manifested. The characteristics of the features seen in the range 1–15 Hz are best known, but they have been seen below 1 Hz and above 15 Hz.

These oscillations were seen with *Ginga* in GX 339–4 (Miyamoto *et al.*, 1991) and Nova Muscae (Takizawa *et al.*, 1997). The frequencies increased with the source intensity, not always in the same way. *RXTE* discovered GRS 1915+105 to be active in 1996. The oscillations were soon found to be prominent during long episodes of hard flux and relative quiescence, and the oscillation frequencies were found to be strongly correlated with the source intensity (Chen *et al.*, 1997). GRO J1655–40 became active and showed the same kind of features (Remillard *et al.*, 1999). By the end of 1999, *RXTE* had observed recurrences of the black hole candidate transient 4U 1630–47, and 4 new transients, all of which exhib-

Astrophysics and Space Science is the original source of publication of this article. It is recommended that this article is cited as: *Astrophysics and Space Science* **276** (suppl.): 201–208, 2001.

ited at some stages very high amplitude low frequency QPOs: GRS 1739–278, XTE J1748–288, XTE J1550–564, and XTE J1859+226.

Cyg X–1, 1E 1740.7–2942, GRS 1758–258, and several transient sources which did not get very bright, have been known for band limited white noise. Sometimes there have been peaks and even a harmonic in the frequency range where the spectra roll over (Cui *et al.*, 1997; Smith *et al.*, 1997). These have lower amplitudes, 5%, and sometimes are hard to distinguish from a sharp transition in the power density spectrum. In Cyg X–1 the amplitude was highest during the transitions between low and high states, appropriately called the 'intermediate state'.

2. Characteristics of the QPO

Usually the QPO feature has at least one harmonic. Sometimes a peak at what would be the subharmonic is present. For simplicity we refer to the dominant peak as representing the phenomenon, whatever may be the real fundamental of the oscillation.

Almost always it is present over a range of luminosities and the frequency varies with luminosity. In most cases the frequency has increased with count rate and indeed with luminosity, when the spectrum has been taken into account. GRO J1655–40 was a notable exception, with the frequency generally decreasing with the total luminosity. However, Sobczak *et al.* (2000) found that frequency was positively correlated with the flux of the soft spectral component. There may be some counterexamples. In 4U 1630–47, a square wave oscillation of flux at a lower frequency was observed (Dieters *et al.*, 2000) and two low frequency QPOs were observed in the 1–10 Hz range, one moving with the flux, the other stationary. These frequencies are at least a factor of 10 too low to be related to the inner most stable orbit of the accretion disk, for which the frequency would be $\geq$ 2200 Hz $(M/M_{\odot})^{-1}$.

The energy spectra of the sources can usually be well or almost described by an optically thick component and a non-thermal component. A physical model would be a disk spectrum Comptonized in a hot corona, but it can be approximated by the a multicolored disk and a power-law. The QPO usually appears when the flux is dominated by a relatively hard power-law-like component.

The energy spectrum of the QPO fundamental is that of the power-law component. Cui *et al.* (1999) plot (for XTE J1550–564) the root mean square (rms) amplitude of the QPO fundamental and harmonic with energy. The fundamental tracks the energy spectrum. Curiously the harmonic reverses and declines with energy, in comparison.

It was shown (Wijnands and van der Klis, 1999b) that for a number of observations of BHCs, the break frequency of the band limited white noise component was about a fifth of the QPO frequency. However Morgan *et al.* (1997) exhibit power density spectra with the QPO frequency clearly corresponding to the break frequency. The exact connection between the two requires further study.

For GRS 1915+105 there is a great deal of data from different states and correlations between characteristics have been studied with several approaches. Markwardt *et al.* (1999) looked at many observations in which on time scales of minutes, the flux varied strongly (factors of 2–10). They averaged power spectra for given flux levels and found that the average exhibited a strong correlation of frequency of the QPO with the bolometric flux of the disk component in a spectral fit (4 s spectra), above the frequency of about 4 Hz. Muno *et al.* (1999) computed spectral parameters and power spectra for 16 s intervals of a larger set of observations, including those in which variations were slow. Frequency appears correlated with disk inner temperature for many of these. Plots with respect to disk black body flux, temperature, and inner radius all show runs of correlated data points, but also subsets clearly violating the correlation or having a very different correlation, indicating the behavior is not unique.

It is tempting to think of the frequency as related to the Kepler frequency at some radius, although the factor and physical origin of the relation may not be clear. Trudolyubov *et al.* (2000) leapfrog from the dependences of f_K and the viscous time scale at the radius to the relation between the duration of the hard state and the minimum frequency observed. The duration and frequency are easy to measure. The inner radius of the disk is difficult to measure, especially when the disk inner temperature appears to be low and the radius large, so that the *RXTE* PCA can only 'see' the Wien tail of the energy spectrum. This gives $t \propto f_K^{-7/3}$ with a proportionality depending on the rate of mass flow through the shell. For the sets of observations with hard dips, this correlation appears to be a good fit. For the observations with soft dips, the dependence is significantly steeper.

For a given mass accretion rate in the disk, the temperature has a constrained relation to the radius, of course. Belloni *et al.* (2000) examined whether this relation holds and allows identification of the mass flow in the disk. Their aim was to compare to the mass loss rate in the radio emitting plasma that is ejected. These questions are just the sort that can give physical insight into the correct description of the disk accretion and the generation of winds and jets, but the X-ray calibration to date limits the conclusions.

The confusion of opposite dependence between frequency and total count rate in GRO J1655–40 and GRS 1915+105 appeared clarified by their having similar correlation with the disk flux alone. Furthermore, Sobczak et al. (2000) found that for XTE J1550–564, while the dependence of the QPO frequency with either temperature or radius was complex, the frequency did increase with disk flux. But from the point of view of the Accretion Ejection Instability (Tagger and Pellat, 1999; Rodriguez *et al.*, 2000) the complex dependence of the QPO frequency on radius is just what would be expected if the inner disk of GRO J1655–40 is nearer to the black hole, where relativistic effects can cause a turnover of the curve.

To interpret these correlations, physical models are needed for which the data can determine the parameters reliably. In deducing radii of the disk, even with the caveat that it is the optically thick disk, we are using zero-order approximations, when we know that we should consider the scattering atmosphere of the disk together with advective and wind flows (Merloni *et al.*, 2000). However, although the meaning of the slopes may need modification, the fits to the data show that correlated changes occur, with values of the parameters often in realistic ranges.

3. Relation of Disk-Corona Oscillations to other QPO

Although much of the discussion has been in terms of the inner radius of an optically thick accretion disk that is being seen in the X-ray spectrum, it should be remembered that the QPO has appeared to occur simultaneously with the QPO that appears at much higher frequencies. It has seemed reasonable that the latter could reflect either the Kepler orbital frequency near the Innermost Stable Circular Orbit about black holes or the eigenfrequency of the fundamental modes of oscillation of the inner disk. High frequencies have been reported for 5 BHCs. They do not invariably occur together with the low frequency QPO, but often have. Obviously if the higher frequency reflects emission from near the inner edge of the disk, the lower frequency cannot be the same thing, although it could be a different time scale at the same place. Alternatively, the inner disk may not be otherwise seen directly, while the lower frequency could reflect a transition region in the disk rather than the inner radius.

For GX 339–4 and GRS 1739–278 the low frequencies were only seen around 6 Hz and 5 Hz respectively. For Nova Muscae, GRS 1915+105, XTE J1550–564, 4U 1630–472, XTE J1859-277, values sampled include similar ranges within 0.08–18 Hz, while the high frequencies of the last 4 of these were 67 Hz, 161–237 Hz, 185 Hz, and 150–200 Hz. For two of the sources, GRO J1655–40 and XTE J1748–288, the low frequencies were higher, 14–35 Hz. GRO J1655–40's high frequency was also higher, 300 Hz. Reported high/low frequency pairs are shown in Figure 1. Recently a higher frequency, 450 Hz, has been found for GRO J1655-40 (Strohmayer, 2001). Dynamical masses are so far known only for GRO J1655–40 and Nova Muscae. How the frequencies scale with black hole mass is not yet clear.

QPOs at frequencies an order of magnitude below these disk-corona oscillations also occur. One type of variation which generates very low frequency oscillations in the power spectra are square-wave modulations of the flux originally called 'flip-flops' by Miyamoto *et al.* (1991), and also dips. GRS 1915+105 has a 15 mHz feature whose folded light curve looks almost like an occultation (Morgan *et al.*, 1997).

The disk-corona oscillation is in a frequency range just below the Horizontal Branch Oscillations (HBO) of Low Mass Binary Neutron Stars and they can look similar in harmonic structure. Usually the amplitude is lower. The HBO and Normal Branch frequencies of the Atoll and Z sources have been found to be correlated to the lower of two kilohertz frequencies. For the BHCs, Psaltis *et al.* (1999) see features, sometimes very broad, that they consider to be counterparts of the neut-

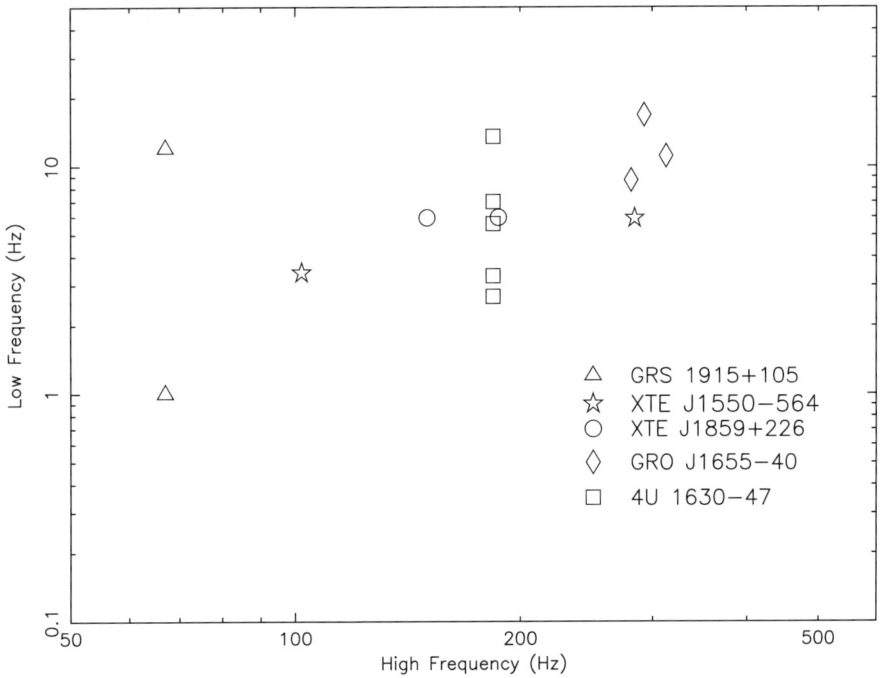

Figure 1. Low frequency QPO for BHCs with high frequency QPO. Selected published values are shown for GRS 1915+105 (Morgan *et al.*, 1997), XTE J1550–564 (Remillard *et al.*, 1999), XTE J1859+226 (Cui *et al.*, 2000), 4U 1630–47 (Dieters *et al.*, 2000), and GRO J1655–40 (Remillard *et al.*, 1999).

ron star lower kilohertz frequency. They find the disk-corona oscillation and this oscillation to be related in the same way as are the HBO and the lower kilohertz oscillation. The ratio of the average lower frequency in the BHCs and of the HBO in the neutron star sources could be consistent with being inversely proportional to the mass of the compact object (Chen *et al.*, 1997).

4. Relation to Black Hole Accretion States

Most black hole candidate transients have outbursts which last several months, but not as long as has GRS 1915+105. They usually have a well defined onset, rise, peak, and decay, although there are rather frequently one or more brightenings during the decay. GRS 1915+105 has for several years now evolved from one recurrence to another. GRO J1655–40 and XTE J1550–560 recurred after a year, but they had clearly gone into quiescence first. In a typical case there is a rising phase, frequently a very high state (VHS), a high soft state (HS) and decay through an intermediate state (IS) to a low hard state (LS) and finally a quiescent state.

RXTE has had a number of successful observations of rising phases (XTE J1550–564, Cui *et al.*, 1999; 4U 1630–472, Dieters *et al.*, 2000; XTE J1859+226, Markwardt *et al.*, 1999). In these cases the spectra started out harder than achieved subsequently in the outburst. The spectrum softens as the flux rises. The disk-corona oscillation starts below 1 Hz with high amplitude and rises in frequency, increasing in width and decreasing in amplitude somewhat. But the time scale for these changes is days rather than the minutes required for the GRS 1915+105 hard dip recoveries.

At the end of the rising phase, the transient may be in the VHS with a strong power law and a strong disk flux. In both the VHS and IS the QPO is present. In the HS the soft component is very bright (several tenths of the Eddington limit). A relatively steep power-law component extending to high energy may be of a completely different origin than the Compton up scattering in a hot corona that is favored for explaining the VHS. There are NO QPOs in the HS.

In the case of GRS1915+105, we discuss a scenario of mass flow in the disk, removal of the disk into a corona and ejection into jets. Does this scenario apply to the transient outbursts? In the VHS there is outflow, because the radio emission often starts sometime during the rising phase. The mass accretion rate in the outer disk is large and the inner radius of the disk is large. As the mass flow rate subsides the disk collapses to be optically thick, or at least it is optically thick farther in toward the center. At a lower rate, there are also radio injections associated with the transitions to the HS and out of it to the IS (e.g. Hjellming *et al.*, 2000) from comparison of the radio and X-ray transitions (for XTE J1748–288 and XTE J1859+226). This scenario would imply that spectral fits obtain larger inner radii for the disk in the VHS than in the HS, but only a factor of 2 seems to have been observed. Much needs to be done to understand the evolution.

5. Explanations for the Low Frequency QPO

The flux that is oscillating appears to be the hard power-law component of the spectrum, the candidate for indicating the presence of a hot corona. The thermal photons from the disk would be the dominant source of the seed photons to be up-scattered in the corona. Whether the corona is spherical and central or extended over the inner disk, or whether there is simultaneously spherical infall is not clear. Radio observations are indicating (See Fender, this volume) outflow and synchrotron radiation. A number of models have been suggested and would explain some of the observations. I only mention some of the ideas.

Disk-corona oscillations are probably not like the possible coronal radial infall oscillations suggested for the normal branch oscillations at about 6 Hz in the neutron star Z sources. Neither observations or simulations show the large amplitude and variability.

It has been suggested that the HBO in neutron stars are due to the Lense-Thirring frame dragging in the strong gravitational field close to rather heavy neutron stars (twice the nodal precession frquency of slightly tilted and elliptical orbits.) Systematic similarity of the neutron stars and BHCs has been claimed, as mentioned above. In some cases the approximately linear relation between two frequencies in the BHCs and the HBO and lower kilohertz frequencies for the neutron stars is impressive. Psaltis and Norman (2000) have made a case that the frequencies observed are the resonant frequencies in the disk, whatever the mechanism for modulating the X-ray flux.

One idea that appears to generate the right order of magnitude of frequencies and similar dependences to those observed is that of shocks at the inner disk and oscillation of the shock position (Molteni *et al.*, 1996; Manickam and Chakrabarti, 1999), but there have been objections that such shocks cannot be generated. The correlations between duration of the dips in GRS 1915+105 with the frequency of the QPO are difficult to distinguish from the correlations expected in a viscous refilling model (Trudolyubov *et al.*, 2000).

Finally, Tagger and Pellat (1999) make a case that Rossby waves of the Accretion Injection Instability are a natural way to carry gas and energy from a disk to the corona and thence into a wind or jet and to generate oscillations in the flux.

The data available so far has enough detail to provide many guideposts and constraints to the theories. These low frequency disk-corona oscillations are very common in BHCs and are a challenging physics problem which involves important aspects of accretion onto black holes.

References

Belloni, T., Migliari, S. and Fender, R.P.: 2000, *Astron. Astrophys.* **358**, L29.

Chakrabarti, S.K and Manickham, S.G.: 2000, *Mon. Not. R. Astron. Soc.* **531**, L41.

Chen, X., Taam, R.E. and Swank, J.H.: 1997, in: C. Dermer, M. Strickman and J. Kurfess (eds.), *Proceedings of the Fourth Compton Symposium*, AIP, New York, p. 995.

Cui, W., Shrader, C.R., Haswell, C.A. and Hynes, R.I.: 2000, *Astrophys. J.* **535**, L123.

Cui, W., Zhang, S.N., Focke, W. and Swank, J.H.: 1997, *Astrophys. J.* **484**, 383.

Cui, W., Zhang, S.N., Chen, W. and Morgan, E.H.: 1999, *Astrophys. J.* **512**, L43.

Dieters, S.W., *et al.*: 2000, *Astrophys. J.* **538**, 307.

Hjellming, R.M., Rupen, M.P. and Mioduszewski, A.J.: 2000, in: *Conference Rossi 2000*, March 2000.

Markwardt, C.B., Marshall, F.E. and Swank, J.H.: 1999, IAUC **7274**,

Markwardt, C.B., Swank, J.H. and Taam, R.E.: 1999, *Astrophys. J.* **513**, L37.

Merloni, A., Fabian, A.C. and Ross, R.R.: 2000, *Mon. Not. R. Astron. Soc.* **313**, 193.

Miyamoto, S., Kimura, K., Kitamoto, S., Dotani, T. and Ebisawa, K.: 1991, *Astrophys. J.* **383**, 784.

Molteni, D., Sponholz, H. and Chakrabarti, S.K.: 1996, *Astrophys. J.* **457**, 805.

Morgan, E.H., Remillard, R.A. and Greiner, J.: 1997, *Astrophys. J.* **482**, 993.

Muno, M.P., Morgan, E.H. and Remillard, R.A.: 1999, *Astrophys. J.* **527**, 321.

Psaltis, D., Belloni, T. and van der Klis, M.: 1999, *Astrophys. J.* **520**, 262.

Psaltis, D. and Norman, C.: 2000, *Astrophys. J.*, submitted (astro-ph/0001391).

Remillard, R.A., McClintock, J.E., Sobczak, G.J., Bailyn, C.D., Orosz, J.A., Morgan, E.H. and Levine, A.M.: 1999, *Astrophys. J.* **517**, L127.

Remillard, R.A., Morgan, E.H., McClintock, J.E., Bailyn, C.D. and Orosz, J.A.: 1999, *Astrophys. J.* **522**, 397.

Rodriguez, J., Varniere, P., Tagger, M. and Durouchoux, Ph.: 2000, *Astron. Astrophys.*, in press.

Smith, D.M., Heindl, W.A., Swank, J., Leventhal, M., Mirabel, I.F. and Rodriguez, L.F.: 1997, *Astrophys. J.* **489**, L51.

Sobczak, G.J., McClintock, J.E., Remillard, R.A., Cui, W., Levine, A.M., Morgan, E.H., Orosz, J.A. and Bailyn, C.D.: 2000, *Astrophys. J.* **531**, 537.

Strohmayer, T.E.: 2001, *Astrophys. J.* **552**, L49.

Tagger, M. and Pellat, R.: 1999, *Astron. Astrophys.* **349**, 1003.

Takizawa, M., *et al.*: 1997, *Astrophys. J.* **489**, 272.

Trudolyubov, S., Churazov, E. and Gilfanov, M.: 2000, *Astron. Astrophys.* **351**, L51.

Wijnands, R. and van der Klis, M.: 1999, *Astrophys. J.* **514**, 939.

RXTE OBSERVATIONS OF THE X-RAY TRANSIENT XTE J1859 +226

CRAIG MARKWARDT
University of Maryland, NASA/GSFC
E-mail: craigm@lheamail.gsfc.nasa.gov

Abstract. We present results of RXTE observations of XTE J1859+226, a transient and probable microquasar discovered in October, 1999. The light curve approximately followed a fast-rise exponential-decay, with a peak flux of about 1.2 Crab (2–10 keV). The source is at a galactic latitude of $+8.5°$, which makes it difficult to estimate distance and luminosity precisely. QPOs were detected in the 150–200 Hz and 0.5–10 Hz ranges. The high frequency QPOs and the results of spectral fitting can be used to loosely constrain the central object mass.

1. Introduction

Microquasars, as originally embodied in the prototype object GRS 1915+105, are probable stellar-mass black holes with relativistic jet outflows (Mirabel and Rodriguez, 1998). Since the discovery of GRS 1915+105, it seems that microquasars may be ubiquitous. Here we present RXTE X-ray observations of the new transient XTE J1859+226, which may indeed be another microquasar.

The galactic source XTE J1859+226 was discovered as an X-ray transient by the RXTE All-Sky Monitor in October 1999 (Wood *et al.*, 1999). RXTE observations were initiated and continued for several subsequent months. The X-ray light curve bears some resemblance to the well-known fast-rise exponential-decay profile (Figure 1), with a rise time to peak of a few days, followed by a somewhat chaotic X-ray plateau lasting a few weeks, and then a gradual decline over several hundred days. The decline phase contained a secondary plateau $\sim$ 60 days after maximum.

The peak flux was achieved on October 16 at a level of 5×10^{-8} erg s^{-1} cm^{-2} in the 2–80 keV band, which corresponds to a luminosity of 6×10^{38} d_{10}^2 erg s^{-1} where d_{10} is the distance in units of 10 kpc. Spectrally, the source began its rise in a very hard state, softened as it peaked in intensity, and continued to soften in its decline. This evolution is remarkably similar to that of GRS 1915+105 in its low-hard state, and the transient outburst of 4U 1630–47, another soft X-ray transient and black hole candidate (Dieters *et al.*, 2000).

Extensive optical and radio monitoring also took place (summarized by Hjellming at `http://www.aoc.nrao.edu/~rhjellmi/x1859.html`). The observations suggest at least two relativistic ejection episodes occurred over the course of the plateau phase: two flares occurred on approximately October 16.5 and 27, with

Astrophysics and Space Science is the original source of publication of this article. It is recommended that this article is cited as: *Astrophysics and Space Science* **276** (suppl.): 209–212, 2001.

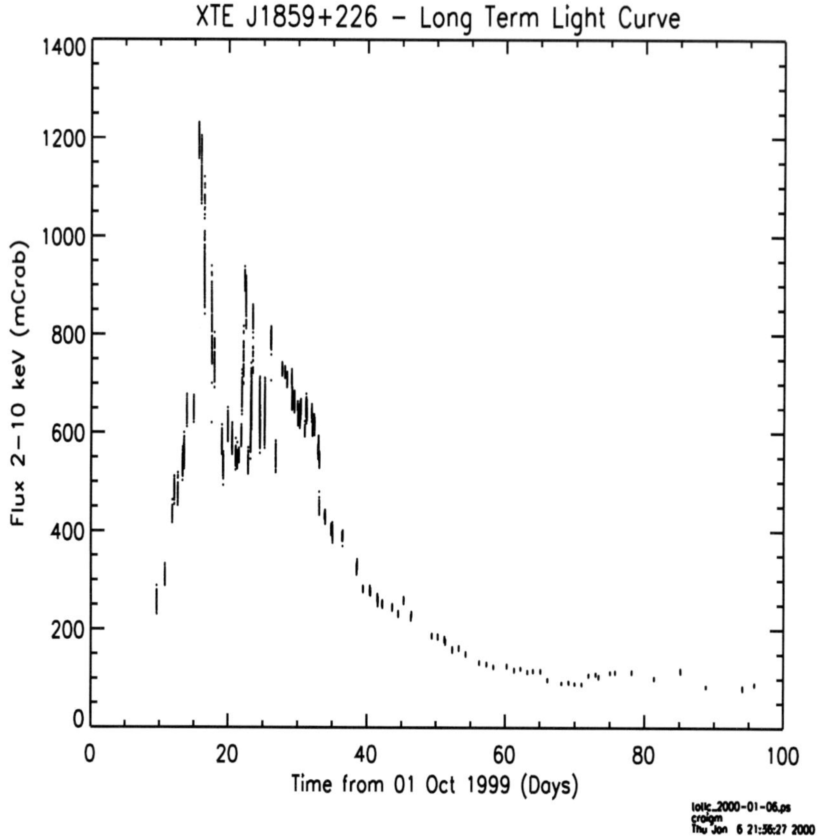

Figure 1. RXTE PCA light light curve of the October, 1999, outburst of XTE J1859+226 (2–10 keV).

subsequent power law declines and progressive spectral steepening. This evolution is indeed reminiscent of the expansion of a relativistic fireball, as first advanced Mirabel *et al.* (1998) for GRS 1915+105. In the case of XTE J1859+226, no actual ejecta were spatially resolved.

Continued optical and RXTE PCA observations of the source as it passed into quiescence have revealed low level activity in the summer of 2000 (Casares *et al.*, 2000; Tomsick and Heindl, 2000). Optical modulations have been detected at several periods, with the consensus settling on a multiple of 9.1 hr (e.g., Sanchez-Fernandez *et al.*, 2000).

The source is located at galactic latitude $+8.5°$. Optical extinction measurements of $E(B-V) = 0.8 \pm 0.4$ (Wagner *et al.*, 1999) and 0.58 ± 0.07 (Hynes *et al.*, 1999) suggest a minimum distance of a 1–3 kpc, which is consistent with the fitted value of neutral hydrogen absorption in the PCA/HEXTE spectrum, $N_H = 1.1 \times 10^{22}$ cm^{-2}. However, as the source is out of the galactic plane it is difficult to place an upper limit to the distance.

2. Fourier and Energy Spectra

Both high and low frequency quasiperiodic oscillations have been detected in the X-ray light curves of the source (Cui *et al.*, 2000; Markwardt *et al.*, 2000). In an average of the highest intensity power spectra, we find a broad high frequency QPO with a centroid frequency of 175 Hz and a full-width half-maximum of 120 Hz. Such high frequency oscillations have been seen in a number of black hole candidates (in the range 67–450 Hz) and at very least suggest they derive from phenomena close to the black hole horizon. Even when a distinct peak is not visible, the harder observations contain a broadband noise component which extends detectably to at least 250 Hz. Cui *et al.* (2000) report marginally distinct QPO frequencies of 187 ± 14 Hz and 150 ± 28 Hz. These features suggest that the QPO centroid frequency is variable, but reaches a maximum frequency of 200–300 Hz, as might be expected if the oscillation reached an inner boundary such as the innermost radius of stable orbits.

Lower frequency QPOs were detected intermittently throughout the outburst in the approximate range of 0.5–10 Hz. Often triplet peaks were observed, having the ratio of centroid frequencies of 1–2–4. As the center peak was usually the strongest, this arrangement strongly implies a subharmonic relationship. Such subharmonics have been seen recently in another black hole candidate XTE J1550–564 (Sobczak *et al.*, 2000). Both low and high frequency QPOs have been seen simultaneously.

Energy spectra were generated from the combined PCA and HEXTE observations, and satisfactorily fitted by a model consisting of a multicolor disk and a power law with Compton reflection. During the plateau the inner disk temperature was typically 0.85 keV but the power law component dominated (photon index ~ 2.5). Around November 5 the disk component became dominant, and the source began its slow decline in temperature and intensity.

The inner radius of the disk may be estimated from the X-ray flux and temperature, and also considering the color correction factor X that takes into account X-ray scattering and inclination effects (Zhang, Cui and Chen, 1997). Using a value of $X \sim 3$, we arrive at $R_{\mathrm{in}} \simeq 80 - 100$ km d_{10}, based on a reasonable range in minimum X-ray fluxes during the plateau phase. The mass and radius can also be constrained by considering that the maximum frequency observed, 200–250 Hz is bounded by the frequency at the radius of marginally stable circular orbits. Using these limits, we derive the approximate mass limit $M < 11 M_{\odot}$, and the distance lies in the range 5–13 kpc. For a maximally rotating black hole, $M < 30 M_{\odot}$. The main uncertainties lie in the distance to the source and the color correction factor.

There may be a weak disk-jet connection in this source. Of the two radio/IR ejection events, the first on October 16.5 involved the temporary disappearance of the low frequency QPO, and the second on October 27 involved the disappearance of both low and high frequency QPOs (the high frequency QPO did not resume subsequently). While there were no particularly strong drops in the disk flux (which might be tied to the loss of the inner accretion disk, as in GRS 1915+105), the disappearances of the QPOs suggests that there may indeed have been some mild disruptions in the inner accretion disk dynamics caused by the ejection events.

Acknowledgements

This work was produced in collaboration with Warren Focke and Jean Swank. I also thank Rob Hynes and Tariq Shahbaz for useful discussions during the Workshop.

References

Casares, J., *et al.*: 2000, IAUC No. 7451.

Cui, W., Shrader, C.R., Haswell, C.A. and Hynes, R.I.: 2000, *Astrophys. J.* **535**, L123.

Dieters, S.W., *et al.*: 2000, *Astrophys. J.* **538**, 307.

Hynes, R.I., Haswell, C.A., Norton, A.J. and Chaty, S.: 1999, IAUC No. 7294.

Markwardt, C.B., Focke, W.B., Swank, J.H. and Taam, R.E.: 1999, *Bull. Am. Astron. Soc.* **31(5)**, 1555, No. 126.03.

Mirabel, I.F., *et al.*: 1998, *Astron. Astrophys.* **330**, L9.

Mirabel, I.F. and Rodríguez, L.F.: 1998, *Nature* **392**, 673.

Sánchez-Fernández, C., Zurita, C., Casares, J., Shahbaz, T. and Castro-Tirado, A.: 2000, IAUC No. 7506.

Sobczak, G.J., Remillard, R.A., Muno, M.P. and McClintock, J.E.: 2000, *Astrophys. J.*, submitted (astro-ph/0004215).

Tomsick, J.A. and Heindl, W.A.: 2000, IAUC No. 7456.

Wagner, R.M., Smith, P.S., Schmidt, G.D. and Shrader, C.R.: 1999, IAUC No. 7279.

Wood, A., Smith, D.A., Marshall, F.E. and Swank, J.: 1999, IAUC No. 7274.

Zhang, S.N., Cui, W. and Chen, W.: 1997, *Astrophys. J.* **482**, L155.

XTE J1118+480: CLUES ON THE NATURE OF THE ACCRETION FLOW FROM THE OPTICAL VARIABILITY

A. MERLONI[1], T. DI MATTEO[2*] and A.C. FABIAN[1]
[1] *Institute of Astronomy, Madingley Road, Cambridge, CB3 0HA, U.K.*
[2] *Harvard-Smithsonian Center for Astrophysics, 60 Golden St., Cambridge, MA 02138, U.S.A.*

Abstract. We show how the simultaneous presence of a strong quasi periodic oscillation (QPO) of period $\sim$ 10 seconds in the optical and X-ray lightcurves of the X-ray transient XTE J1118+480 can be used to obtain information about the nature of the accretion flow around the source. The unusually high optical-to-X-ray flux ratio and the QPO observed simultaneously in both energy bands suggest that a significant fraction of the optical flux might originate close to the central source, where most of the X-rays are produced and be indicative of a magnetically dominated corona. We also show how the temporal evolution of the QPO can provide us with information on both the inner radius and the viscous properties of the optically thick accretion disc.

1. Introduction: The Magnetic Corona Model

The transient X-ray source XTE J1118+480 was discovered with the RXTE All-Sky Monitor on March 29, 2000 (Remillard *et al.*, 2000). At the peak of its outburst, it showed an X-ray spectrum typical of black hole candidates (BHC) in their *hard* state, with a prominent power-law component (photon index of about 1.8). The 13th magnitude optical counterpart, discovered by Uemura, Kato and Yamaoka (2000), exhibited a spectrum fairly typical of X-ray Novae in outburst (Dubus *et al.*, 2000 and references therein). A strong QPO with frequency $\nu_{\rm QPO} \sim 0.1$ Hz has been found in the X-ray power density spectrum (PDS) of the source (Wood *et al.*, 2000); more surprisingly, the optical light curves also showed a prominent QPO at the same frequency (Haswell *et al.*, 2000). The QPO frequency drifted to higher values systematically in both bands for about two months.

As discussed in detail in Merloni, Di Matteo and Fabian (2000) (MDF), the high optical-to-X-ray flux ratio and the optical/UV variability argues for the presence of significant cyclo-synchrotron (CS) radiation from active regions in a magnetically dominated corona in this source. Assuming that a fraction of the observed optical flux corresponding to its oscillating part ($\sim$ 30%) is due to CS emission from a magnetized corona, we modeled the spectrum taking into account reprocessing of coronal radiation in the accretion disc (according to the height of the active regions) and all the relevant radiative processes (see MDF for details). The disc

* Chandra fellow

Astrophysics and Space Science is the original source of publication of this article. It is recommended that this article is cited as: *Astrophysics and Space Science* **276** (suppl.): 213–216, 2001.

was assumed to extend down to the innermost stable orbit of a non-rotating black hole ($R_{in} = 6GM/c^2 = 3R_S$). We found that, in order to match the high (compared to X-rays) optical flux, both a low accretion rate ($\dot{M}/\dot{M}_{Edd} = \dot{m} \simeq 0.01$), and a high fraction of accretion power dissipated in the corona ($f > 0.9$) are needed; also the active regions need to be high above the disc (at about $10R_S$)*.

Hynes *et al.* (2000) reporting on *EUVE* observations of the source showed that the low measured EUV flux is inconsistent with an optically thick disc extending down to radii $R < 1000R_S$ (but see Dubus *et al.*, 2000 for a different estimate of the EUV flux). Based on this, they favor an inner advection-dominated flow (ADAF) surrounded by a geometrically thin disc truncated at hundreds of R_S. In the next section we show how to use the variability information to put constraints on the inner disc location.

2. The QPO Viscous Evolution

Beside the peaked QPO around 0.1 Hz, the observed PDS of XTE J1118+480 exhibits a broader feature around 1 Hz; plotted one against the other, these frequencies fall remarkably well on the correlation among QPOs discovered by Psaltis, Belloni and van der Klis (1999) in the PDS of BHC and neutron stars. Such a correlation is likely to reflect a general property of the accretion flows around these objects. Here we assume that the QPO observed in XTE J1118+480 is produced at a specific distance ($R_t = r_t R_S$) where a discontinuity of the flow occurs, as in Psaltis and Norman (2000). The QPO frequency must be essentially determined by the relativistic dynamical frequencies of the system (Psaltis and Norman, 2000; Stella, Vietri and Morsink, 1999).

XTE J1118+480 observations offer the opportunity to track the QPO evolution over a long period of time (Wood *et al.*, 2000). To model such evolution, we calculate $R_t(t)$ integrating the equation of the viscous evolution of a standard Shakura-Sunyaev disc:

$$\frac{\partial R_t}{\partial t} = -\frac{3}{2}\alpha \left(\frac{H}{R_t}\right)^2 \left(\frac{GM}{R_t^3}\right)^{1/2} J^{-1}, \tag{1}$$

where $J = 1 - \sqrt{3/r_t}$, α is the viscous parameter and the disc scaleheight $H/R_t \simeq 9\dot{m}J(1-f)r_t^{-1}$ for a radiation pressure dominated disc and $H/R_t \simeq 0.02(\alpha m)^{-1/10}$ $(\dot{m}J)^{-1/5}(1-f)^{-1/10}r_t^{1/20}$ for a gas pressure dominated one (Merloni, Fabian and Ross, 2000). We test two different hypothesis on the origin of the QPO (and the accretion flow geometry):

(a) According to the relativistic precession interpretation of the observed correlation (obeyed by XTE J1118+480) ν_{QPO} is twice the nodal precession frequency:

* Alternatively, a good agreement with the data could be achieved if the active regions move relativistically away from the disc (MDF, see also Markoff, Falcke and Fender, 2000, for a jet model of the SED)

(a) $\nu_{\rm QPO}$=2$\nu_{\rm LT}$; rad. p.; χ^2/dof=0.21; $\alpha[\dot{m}(1-f)]^2 \simeq 4.5 \times 10^{-8} a^{7/6} m^{-1/6}$ | (b) $\nu_{\rm QPO}$=$\nu_{\rm Kep}$; gas p.; χ^2/dof=0.80; $\alpha^{4/5}[\dot{m}^2(1-f)]^{1/5} \simeq 2.5 \times 10^{-4} m^{4/15}$

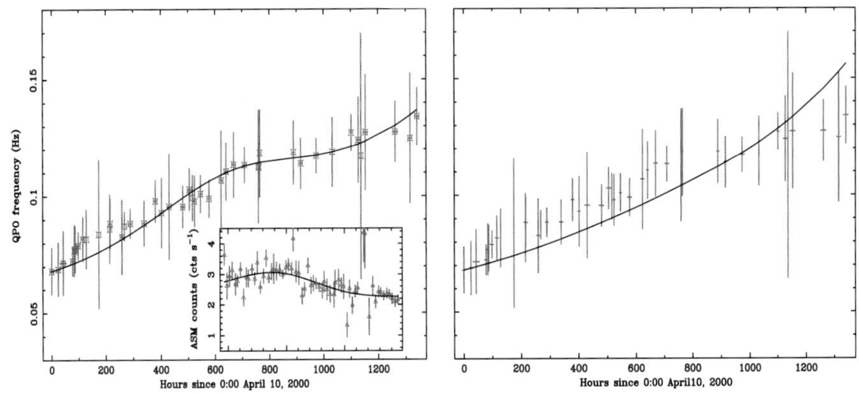

Figure 1. Fit to the observed QPO evolution (RXTE and USA data, Wood *et al.*, 2000) for (a) a QPO produced by relativistic LT precession in a radiation pressure dominated disc and (b) a QPO produced by modulation with Keplerian frequency in a gas pressure dominated disc. The error bars are given by the QPO FWHM, which is likely to reflect the uncertainties in the location where the oscillation are excited. In the inset the ASM lightcurve of XTE J1118+480 (triangles): the continuous line is a SPLINE fit to its smooth variation (flares are excluded), which is assumed to reflect long term variations of the total accretion rate.

$\nu_{\rm QPO} = 2\nu_{\rm nod} \simeq 1.62 \times 10^4 (a/m) r_{\rm t}^{-3}$ Hz, implying $r_{\rm t} \sim 50(a/m)^{1/3}$ (a is the BH angular momentum, m its mass in Solar mass units). In this case the radiation pressure solution, appropriate for the inner part of an optically thick disc, should be used in Equation (1).

(b) Alternatively, the observed low frequency QPO can be due to a Keplerian modulation of the flow: $\nu_{\rm QPO}$=$\nu_{\rm Kep} \simeq 1.14 \times 10^4 m^{-1} r_{\rm t}^{-3/2}$ Hz, so that $r_{\rm t} \sim 2000\, m^{-2/3}$, and the gas pressure solution must be used in this case.

Furthermore, we assume that the smooth variations of the X-ray ASM luminosity reflects a modulation of the total accretion rate so that $\dot{m}(t) = g(t)\dot{m}_0$, where $g(t)$ is the continuous function shown in the inset in Figure 1. We integrate Equation (1) and fitted the temporal evolution of the QPO frequency with the resulting $\nu_{\rm QPO}(R_{\rm t}(t))$ for the two cases and show the results in Figure 1. Both fits are acceptable, but the relativistic precession model (with $r_{\rm t} = O(10)$) explains the QPO evolution better than the Keplerian one (with $r_{\rm t} = O(1000)$). Finally, from such fits we can obtain an estimate of the viscous parameters of the optically thick accretion disc (α, $\dot{m}$, f, through the products shown in Figure 1) independent on any spectral analysis. The estimate from MDF ($\dot{m} f \simeq 10^{-3}$) based on the magnetic corona model is also consistent with case (a) (with $\alpha \sim 0.01$), while for case (b) it would imply an unreasonably low viscosity.

3. Conclusions

The main rationale for our model is the simultaneous presence of a strong QPO in the optical and X-ray lightcurves of XTE J1118+480, suggesting that the fluxes in the two bands both originate from the same region in the inner part of the accretion flow. Self-absorbed CS emission is the natural candidate to explain the optical variability. Such emission is expected in any magnetic corona model, and the inferred magnetic field value ($B \approx 2 \times 10^6$ G) is the one predicted to arise when the source is in the *hard* state (MDF). We have also shown how the evolution of the QPO can be satisfactory accounted for if the QPO is produced by a modulation whose frequency is equal to twice the relativistic nodal precession in a radiation pressure dominated disc and the location of such modulation (that is perhaps where the inner accretion disc undergoes a transition and is *at most* a few tens of Schwarzschild radii) viscously drift inwards.

References

Dubus, G., *et al.*: 2000, *Astrophys. J.* submitted (astro-ph/0009148).
Haswell, *et al.*: 2000, IAUC 7427.
Hynes, R.I., *et al.*: 2000, *Astrophys. J.* **539**, L37.
Markoff, S., Falcke, H. and Fender, R.P.: 2000, *Astrophys. J.*, submitted.
Merloni, A., Di Matteo, T. and Fabian, A.: 2000, *Mon. Not. R. Astron. Soc.* **318**, L15 (MDF).
Merloni, A., Fabian, A.C. and Ross, R.R.: 2000, *Mon. Not. R. Astron. Soc.* **313**, 193.
Psaltis, D., Belloni, T. and van der Klis, M.: 1999, *Astrophys. J.* **520**, 262.
Psaltis, D. and Norman, C.: 2000, *Astrophys. J.*, in press (astro-ph/0001391).
Stella, L., Vietri, M. and Morsink, S.: 1999, *Astrophys. J.* **624**, L63.
Remillard, R.R., *et al.*: 2000, IAUC 7390.
Uemura, M., Kato, T. and Yamaoka, H.: 2000, IAUC 7390.
Wood, K.S., *et al.*: 2000, *Astrophys. J. Lett.*, in press (astro-ph/0006234).

COMPTONIZATION AND PHASE LAG CORRELATIONS IN GRS 1915+105

L. NOBILI[1], T. BELLONI[2], R. TUROLLA[1] and L. ZAMPIERI[1]
[1] *Dipartimento di Fisica 'G.Galilei', Università di Padova, Italy*
[2] *Osservatorio Astronomico di Brera, Merate, Italy*

Abstract. We suggest a simple thermal Comptonization model to account for the observed phase lags-hardness ratio correlation of GRS1915+105.

Some distinctive features make GRS 1915+105 atypical with respect to Galactic BHCs (Belloni *et al.*, 2000 and reference therein). The dramatic drop in luminosity seen at certain stages and the simultaneous hardening of the spectrum plays in favor of the onset of an instability that in a very short time disrupts the inner part of the disk. This hypothesis is partially supported by theoretical calculations in which a thermal-viscous instability blows off the inner, radiation-pressure supported region of the disk (Szuszkiewicz and Miller, 1998). This event marks the transition from the so called high/soft state B, characterized in large part by a multi–temperature black body emission from a standard disk (high count rate in the soft X–ray band), to a low/hard state C, in which the primary emission is presumably due to a semi-evacuated disk lacking of its most luminous inner region. A relevant feature displayed by GRS 1915+105 during state C is the tight correlation between the 0.5–8 Hz QPO and the spectral and variability properties (Reig *et al.*, 2000). However, while a number of properties are easily interpreted in terms of the disk refilling model, the presence of time lags between hard and soft photons is not well understood yet. Even more intriguing is the observed gradual decrease of the time–lags from positive to negative values during the late state C. Because of this unusual behaviour, it is conceivable that the working mechanism could be slightly different from that usually assumed in other Galactic X–ray binary system. In fact, models based on a disk surrounded by a hot comptonizing corona cannot explain the timing properties of GRS 1915+105 and always produce positive lags.

To overcome these difficulties, we suggest here a different scenario for the accretion flow in the central region of GRS 1915+105, still within the framework of thermal Comptonization. The starting point is based on the following considerations: *i)* the strong correlations among the relevant spectral and timing properties of the source and their nearly regular evolution imply that a unique mechanism (i.e. the comptonizing cloud) is at work for the entire duration of state C. Then the continuous variation of the physical parameters of the cloud modulate the emerging flux. *ii)* The regular transition from positive to negative lags requires that the relative temperature of the cloud and the incident photons changes in time, starting

Astrophysics and Space Science is the original source of publication of this article. It is recommended that this article is cited as: *Astrophysics and Space Science* **276** (suppl.): 217–220, 2001.

from a value which is initially higher than the incident mean photon temperature T_γ , evolving to lower values at later time.

Taking into account these two basic facts, we have assumed that during state C a cloud formed e.g. by the debris of the puffed-up inner region or by the evaporation of the disk itself, fills the central region of the configuration. During the replenishment of the semi–evacuated part of the disk the cloud, assumed not uniform in temperature, becomes more and more compact, following the inward motion of the disk edge.

Since the cloud shrouds the inner part of the disk, in absence of an efficient heating mechanism electrons should be quickly cooled down by soft photons. However, if cooling of ions is inefficient, like in advection–dominated flows, protons decouple from electrons and reach a temperature that is a fraction $\eta \lesssim 1$ of the virial temperature, $KT_i \approx \eta GMm_p/r$. The electron thermal balance will be then fixed by the competition between ion heating and Compton cooling. A simple computation shows that, even if not complete virial equilibrium is achieved, the electron temperature T_H of the inner part of the cloud can be as high as 10 keV or more, depending weakly on the location of the inner edge of the disk, r_{in}. Moreover, the temperature $T(r)$ is one order of magnitude smaller at greater distance (see for details Nobili *et al.*, 2000). Since our aim is simply to test the capability of the model to reproduce the correct trend of the evolution and to estimate the amplitude of the spectral variations of GRS 1915+105, we have approximated the temperature stratification by assuming that the cloud is formed by an inner hot sphere at uniform temperature $T_H \approx 15$ keV, and an external cold shell at temperature $T_W \approx 1.5$ keV. The scattering depths of these two regions are τ_W and $\tau_H > \tau_W$ respectively. Initially the cloud is extended and rarefied so that soft photons are preferentially scattered in the inner and very hot part, where $\tau_H \gtrsim 1$, while the cold external shell does not play any role since $\tau_W \ll 1$ there. Under such conditions radiation gains energy producing a positive lag.

As density increases, the Compton y-parameter of the corona increases, until it exceeds unity below a certain radius r_{sc}. Photons emitted by the inner part of the disk undergo repeated scatterings with the hot electrons and start to fill the Wien peak at T_H. Then the primary radiation is formed by the flux emitted by the disk plus the hot, dense part of the corona. The resulting spectrum is therefore the superposition of a multitemperature blackbody (BB) generated by the disk at $r > r_{sc}$, and a Bose-Einstein distribution at T_H (BE). Both components contribute to the total spectrum in a variable way, depending on the extension of the region between r_{sc} and the disk inner edge r_{in}. In particular, for $r_{in} > r_{sc}$ (i.e. the Compton parameter is below 1 everywhere) the initial photon occupation number n is that of a pure multitemperature blackbody of a disk truncated at r_{in}. However, when $r_{in} < r_{sc}$ the BB+BE radiation has to traverse the warm part of the corona before escaping to infinity. In doing so, photons are downscattered producing now a negative lag.

The theoretical calculation of the emerging spectrum proceeds in two steps. First, we compute the time-dependent comptonized spectrum for a given seed

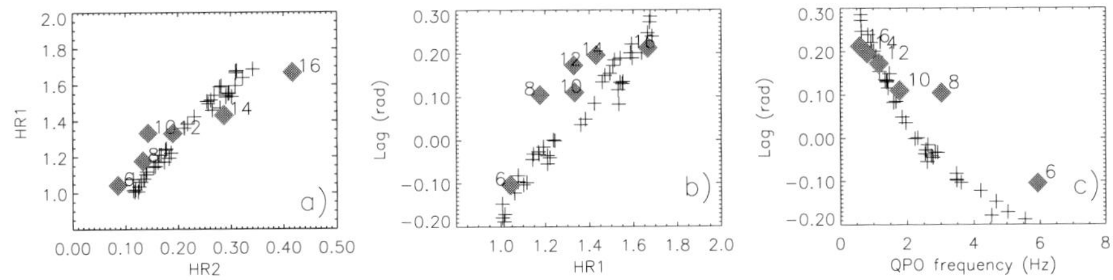

Figure 1. Comparison between observed data (plus sign) and calculated values (filled diamonds). a) color-color diagram; b) phase lags of the continuum versus HR1; c) phase lags versus QPO frequency. Each model is labeled by the value of r_{in} in units of GM/c^2.

photon distribution by solving the Kompaneets equation in a semi-infinite homogeneous medium and then we use an escape probability algorithm to obtain the time evolution of the spectrum emerging from a medium of finite depth.

The results presented here have been obtained for $T_H = 15$ keV, $T_W = 1.5$ keV, $r_{sc} = 20$ and a black hole mass $M = 6M_\odot$. The model with $r_{in} = 16$ (in units of GM/c^2) corresponds to the usual picture in which a hot corona ($T = T_H$) of moderate depth ($\tau = \tau_H \sim 3$) upscatters the soft photons produced by the disk; under these conditions the energy exchange in the warm region can be reasonably assumed to be negligible. The opposite situation is represented by the model with $r_{in} = 6$. Now the disk inner region lies entirely within the hot part of the corona which, because of the increase in the density, has $\tau_H \gtrsim 100$. Escaping photons are then reprocessed in the outer, warm part of the corona ($T = T_W$, $\tau = \tau_W \sim 10$).

The comparison of our results with observational data is illustrated in Figure 1. Panel 1a shows the color-color evolution, with colors defined as the ratios HR1 = (5–13 keV)/(2–5 keV) and HR2 = (13–60 keV)/(2–5 keV). The observed hardness ratios are quite well reproduced by the model for a value of the column density $N_H = 2 \times 10^{22}$ cm^{-2}, consistent with observations (Belloni *et al.*, 2000). When the inner edge of the disk is far out, the spectrum is a power-law produced by thermal Comptonization of soft disk photons. As r_{in} decreases, downscattering becomes progressively more important with respect to upscattering in the hot corona, the spectrum softens and, at the same time, the phase lag of the continuum decreases. Eventually, systematic downscattering dominates and soft 2–5 keV photons, which underwent more scatterings, escape later than hard 5–13 keV photons, *causing an inversion in the sign of the lag*. Panels 1b and 1c show the lag of the continuum as a function of the color HR1 = (5–13 keV)/(2–5 keV) and QPO frequency (the scattering of the computed points is partially due to the poor sampling of the parameter space). Time lags have been computed assuming that the typical size r_c of the comptonizing cloud is directly correlated with the inner edge of the disk, and varies with time in the range $20 \lesssim r_c c^2/GM \lesssim 400$. Finally, we derived a relation between ν_{QPO} and r_{in} associating the model with $r_{in} = 6$ (16) to the observation with the largest (smallest) QPO frequency, 6 (0.6) Hz. This yields $r_{in} \approx 13\nu_{QPO}^{-0.43}$. It is interesting to note that this relation is similar to that found in a quite independent way by di Matteo & Psaltis (1999) combining an empirical relation with the upper bound provided by the Keplerian orbital frequency.

Ultimately, despite the drastic simplification introduced in our calculations, a simple Compton model can account for the main observational properties of GRS 1915+105 in state C. In particular, we have adopted here an *ad hoc* picture for the accretion flow, but the basic results are largely independent of the specific choice for the geometry. Within a Comptonization framework the main ingredients to account for the observed correlations are the presence of a non isothermal cloud of varying optical depth and a source of energetic primary photons at small radii. However, we must consider also the possibility that dynamical Comptonization, due e.g. by a not uniformly expanding cloud or jet, could play a relevant role in producing the observed phenomenology.

References

Belloni, T., *et al.*: 2000, *Astron. Astrophys.* **355**, 271.
di Matteo, T. and Psaltis, D.: 1999, *Astrophys. J. Lett.* **526**, L101.
Nobili, L., Turolla, R., Zampieri, L. and Belloni, T.: 2000, *Astrophys. J.* **538**, L137.
Reig, P., *et al.*: 2000, *Astrophys. J.* **541**, 883.
Szuszkiewicz, E. and Miller, J.C.: 1998, *Mon. Not. R. Astron. Soc.* **298**, 888.

X-RAY FLARES/BURSTS FROM GRS 1915+105 AND THE TWO COMPONENT ACCRETION FLOW

J.S. YADAV and A.R. RAO
Tata Institute of Fundamental Research, Homi Bhabha Road, Mumbai, 400 005, India
E-mail: jsyadav@tifr.res.in

Abstract. We present results of our analysis of a set of RXTE/PCA observations of X-ray flares/bursts with burst cycle ranging from 30 to 1300 s from the Galactic X-ray transient source GRS 1915+105 during last four years. These flares/bursts can be classified into four types: 1) regular bursts with short burst phase of $\sim$ 20 s (R1), 2) regular bursts with short quiescent phase of $\sim$ 20 s (R2), 3) long regular bursts (R3) with burst cycle $\geq$ 1000 s, and 4) irregular bursts (IR). For all the observed bursts, the duration of the quiescent phase is inversely proportional to the square of the QPO frequency (2–10 Hz).

1. Introduction

The Galactic X-ray transient source GRS 1915+105 has shown spectacular X-ray variability during last four years of its observation by RXTE and other satellites (Greiner *et al.*, 1996; Yadav *et al.*, 1999; Belloni *et al.*, 2000). This source was discovered in 1992 with the WATCH all sky X-ray monitor onboard the GRANAT satellite (Castro-Tirado *et al.*, 1994). The X-ray intensity was found to vary on a variety of time scales and the light curve showed a complicated pattern of dips and rapid transitions between high and low intensity (Belloni *et al.*, 1997; Taam *et al.*, 1997). Recently, Yadav *et al.* (1999) have made a detailed study of various types of X-ray bursts seen in GRS 1915+105 from IXAE/PPCs observations during 1997 June–August and attempted to explain these bursts in the light of the recent theories of advective accretion disks. In X-ray astronomy, the terminology X-ray burst is used for type I and type II classical bursts seen in LMXBs. The timing and spectral properties of the flares/bursts described here are completely different than that of the classical bursts (Yadav and Rao, 2000).

In Table I, we list details of RXTE/PCA observations discussed here along with some of the properties of the observed X-ray flares/bursts. Each burst cycle consists of a low flux quiescent phase followed by a high flux burst phase and the fast transition in less than 10 s. The dips or the quiescent phase in these observations are spectrally hard while the brighter portions (the burst phase) are soft (Yadav, 2001). For details of these observations as well as of our analysis please see Yadav and Rao (2000).

Astrophysics and Space Science is the original source of publication of this article. It is recommended that this article is cited as: *Astrophysics and Space Science* **276** (suppl.): 221–224, 2001.

TABLE I
Summary of selected observations of GRS 1915+105

Date (UT)	Exposure (s)	Type of Bursts	ASM[a] Flux	Rec. Time (s)	Av. Q. Flux
1996 April 06 05:40	5600	Regular (R2)	99.4	280[b]	6200
1996 Oct 07 05:44	7000	Regular (R3)	98.8	1150[b]	3200
1997 May 26 12:25	3215	Regular (R1b)	47.6	105[b]	7500
1997 June 18 14:17	3472	Irregular (IR)	61.5	var.	var.
1997 June 22 19:27	2550	Regular (R1a)	59.7	55[b]	8700

Q. Flux = average quiescent time flux (c/s), var. = variable, [a] Mean ASM flux (c/s) for a day, [b] Mean burst recurrence time.

2. Results and Discussion

On the basis of timing properties, these flares/bursts can broadly be put into two classes: regular bursts centered around a fixed period with low dispersion ($\delta P/P \sim 1 - 50\%$) and irregular bursts with no fixed periodicity ($\delta P/P > 50\%$). These flares/bursts can be classified into four types: 1) regular bursts with short burst phase of $\sim$ 20 s (R1) with $\delta P/P \sim 1 - 10\%$, 2) regular bursts with short quiescent phase of $\sim$ 20 s (R2) with $\delta P/P$ up to 50%, 3) long regular bursts (R3) with burst cycle $\geq$ 1000 s and $\delta P/P \sim 1 - 15\%$, and 4) irregular bursts (IR) (see Figure 1). We have measured the quiescent time and the burst duration for all types of bursts and results are shown in Figure 1. The data of IR bursts seen on 1997 June 18 fall diagonally; clearly showing a strong correlation between the burst duration and the quiescent time. On the other hand, data for different types of regular bursts fall on either horizontal or vertical branches implying no such correlation for the regular bursts (Yadav *et al.*, 1999). The R1a and R1b regular bursts fall on the horizontal branch with the burst time $\sim$ 20 s. The R2 regular bursts fall on the vertical branch with the quiescent time $\sim$ 20 s while the R3 regular bursts fall on vertical branch with the quiescent time $\sim$ 320 s (this is shown in the inset of Figure 1).

The most striking features of these flares/bursts are slow exponential rise, sharp linear decay and hardening of spectrum as burst progresses (Paul *et al.*, 1998). The decay time scales are shorter than the rise time scales. In sharp contrast, the decay time is longer than the rise time in classical bursts and spectrum is initially hard and becomes softer as the burst decays (Lewin *et al.*, 1995). Yadav *et al.* (1999) have suggested that the source is in a high-soft state during the burst phase and in a low-hard state during the quiescent phase on the basis of available spectral observations and derived disk parameters of GRS 1915+105. The fast time scale for the transition of the state is explained by invoking the appearance and disappearance of the advective disk in its viscous time scale. Such fast changes of states are possible in the Two Component Accretion Flows (TCAF) where the advective disk

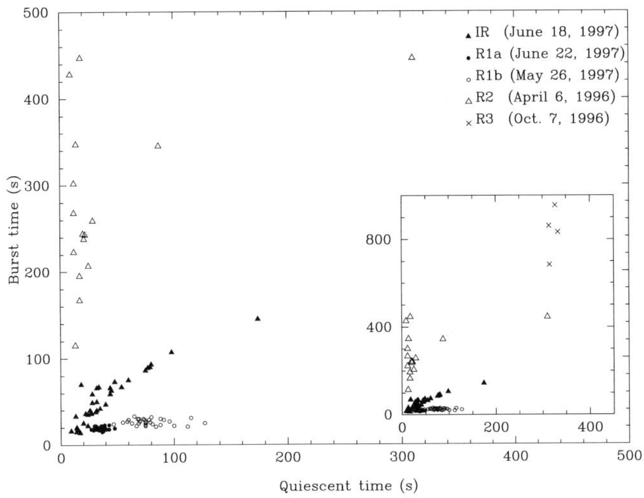

Figure 1. Plot of the quiescent time and the burst duration for all the bursts.

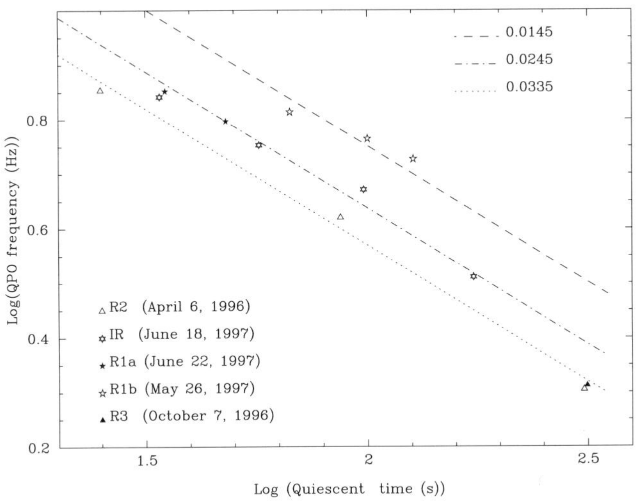

Figure 2. Variation of QPO frequency ν_I (minimum) with the quiescent time for different types of X-ray flares/bursts (data points). Plotted lines are the quiescent time α ν_I^{-2} for different values of $\Theta_{\dot{M}}$.

covers the standard thin disk (Chakrabarti, 1996; Chakrabarti and Titarchuk, 1995). Chakrabarti and Manickam (2000) have provided a relation; $t_{off} = f(\Theta_{\dot{M}})\nu_I^{-2}$ based on TCAF. Where t_{off} is the duration of off state (duration in which the sonic sphere becomes ready for catastrophic Compton cooling), ν_I is the intermediate QPO frequency between 2–10 Hz, and $\Theta_{\dot{M}}$ is a dimensionless parameter defined as $\Theta_{\dot{M}} = (\Theta_{out}/\Theta_{in}) \times \dot{m}_d$ where Θ_{in} and Θ_{out} are the solid angles of the inflow and outflow respectively and $\dot{m}_d$ is the disk accretion rate in units of Eddington accretion rate.

In Figure 2 we plot the above relation in the log-log scale taking t_{off} as the quiescent time for $\Theta_{\dot{M}} = 0.0145$, 0.0245 and 0.0335 alongwith our QPO results (points). These results are in good agreement. The data points of R2 and R3 bursts when ASM flux was 99.4 & 98.8 c/s respectively fall along the dotted line with $\Theta_{\dot{M}} = 0.0335$ (see Table I). The data points of R1a and IR bursts when ASM flux was 59.7 and 61.5 c/s respectively fall along the dashed-dotted line ($\Theta_{\dot{M}} = 0.0245$). The data points of R1b bursts during which ASM flux has lowest value of 47.6 c/s fall along the dashed line ($\Theta_{\dot{M}} = 0.0145$). It may be noted here that the $\Theta_{\dot{M}}$ and the ASM flux though determined independently agree well for different types of flares/bursts as both of these are related to the disk accretion rate $\dot{m}_d$. Our results in Figure 2 suggest that $t_{\rm off}$ represents the quiescent time of the flares/bursts which may or may not be of the order of the viscous time scales of a thin accretion disk.

References

Belloni, T., *et al.*: 1997, *Astrophys. J.* **488**, L109.

Belloni, T., *et al.*: 2000, *Astron. Astrophys.* **355**, 271.

Castro-Tirado, A.J., *et al.*: 1994, *Astrophys. J. Suppl.* **92**, 469.

Chakrabarti, S.K.: 1996, *Phys. Rep.* **266**, 229.

Chakrabarti, S.K. and Manickam, S.G.: 2000, *Astrophys. J.* **531**, L41.

Chakrabarti, S.K. and Titarchuk, L.G.: 1995, *Astrophys. J.* **623**, 623.

Greiner, J., Morgan E.H. and Remillard: 1996, *Astrophys. J.* **473**, L107.

Lewin, W.H.G., Van Paradijs, Jan and Taam, R.E.: 1995, in: W.H.G. Lewin, Jan Van Paradijs and van den Heuvel (eds.), *X-ray Binaries*, Cambridge University Press, p. 175.

Taam, R.E., Chen, X. and Swank, J.H.: 1997, *Astrophys. J.* **485**, L83.

Paul, B., *et al.*: 1998, *Astrophys. J.* **492**, L63.

Yadav, J.S.: 2001 *Astrophys. J.* **548**, 876.

Yadav, J.S., *et al.*: 1999, *Astrophys. J.* **517**, 935.

Yadav, J.S. and Rao, A.R.: 2000, Proc. 33rd COSPAR, Warsaw, July 16–22, 2000.

BROADBAND X-RAY SPECTROSCOPY OF GRS 1915+105

A. RAU and J. GREINER
Astrophysical Institute Potsdam, 14482 Potsdam, Germany

Abstract. We analyzed RXTE data of GRS 1915+105 of X-ray low states between 1996 and 1998. The X-ray spectrum is dominated by the power law component. We found (i) that the power law is pivoting at 10–30 keV, (ii) that the power law slope correlates with radio flux, (iii) three different groups of 3–10 keV residuals, (iv) that one of these residual groups correlates with the power law slope.

The prototypical microquasar GRS 1915+105 was discovered by Granat (Castro-Tirado, Brandt and Lund, 1992) as a transient X-ray source. Observations with RXTE PCA revealed astonishing X-ray variability (Greiner *et al.*, 1996) which has been classified into a dozen different X-ray states (Belloni *et al.*, 2000).

We analyzed 68 observations from the RXTE public archive of GRS 1915+105 during X-ray low states, so called χ-states (Belloni *et al.*, 2000; no large amplitude variations nor obvious structured variability) from November 1996 to June 1998. For each observation we used PCU 0 data of the PCA and binned HEXTE cluster A data and fitted these with a model consisting of cold absorption (wabs), disk blackbody (diskbb) and reflection (refsch, including a power law) in *XSPEC* with a fixed folding energy of 400 keV and solar abundances. Energies between 4 and 8.5 keV were ignored during the fit.

All reduced X-ray spectra can be divided into three groups depending on the residuals in the energy range 3-10 keV. The first group shows a small hump at 6.4 keV and a dip at 4.5 keV, and the second group has two broad humps at 4.5 and 6 keV. Both groups are mainly seen during the long 1997 low state. The third group has a deep dip at 4.5 keV as the only feature and occurs only at times of radio flares. A deepening of the 'dip'-residuals with steeper power law and increasing radio flux is seen, whereas no such trend is visible for the other groups. This correlation between dip depth and radio flares suggests that the depth is connected with the strengths of a possibly nearly-spherical outflow or accretion disk wind (inclination is 70° from radio jet characteristics, e.g. nearly edge-on) where a part may be collimated into a jet.

During χ-states (independent of the three residual groups) an increasing PCA count rate is related to a decreasing HEXTE count rate and an increasing spectral photon index. Interestingly, the power law pivotes at 10–30 keV, while the disk component is always negligible. Therefore a factor of 3–4 in PCA count rate can be explained by a varying power law flux by the same factor (Figure 1a). Thus, the power law producing corona of hot electrons can be the origin of the observed

Astrophysics and Space Science is the original source of publication of this article. It is recommended that this article is cited as: *Astrophysics and Space Science* **276** (suppl.): 225–226, 2001.

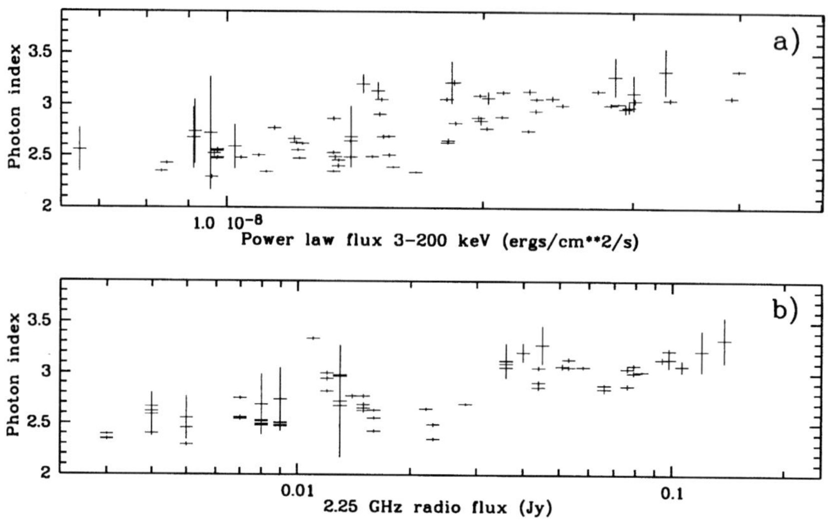

Figure 1. X-ray power law photon index as function of a) the 3–200 keV power law flux and b) the 2.25 GHz Green Bank Interferometer (GBI) radio flux.

count rate variation. An increasing photon index is caused by a decreasing electron temperature in the corona and the observed variation implies a temperature variation between 25 and 50 keV (Sazonov and Sunyaev, 2000).

Also correlated with the power law slope is the 2.25 GHz radio flux as measured with the GBI such that a higher photon index (lower electron temperature) is related to higher radio flux (Figure 1b). This could be due to a shift of the synchrotron radiation to longer wavelengths with decreasing electron temperature. This would be in agreement to a correlation of the K band flux with the X-ray hardness ratio (Greiner *et al.*, 2001).

References

Belloni, T., Klein-Wolt, M., Mendez, M., *et al.*: 2000, *Astron. Astrophys.* **355**, 271.
Castro-Tirado, A.J., Brandt, S. and Lund, N.: 1992, *IAU Circ.*, 5590.
Greiner, J., Morgan, E. and Remillard, R.A.: 1996, *Astrophys. J.* **473**, L107.
Greiner, J., Vrba, F.J., Henden, A.A., *et al.*: 2001, these Proceedings.
Sazonov, S.Y. and Sunyaev, R.A.: 2000, *Astron. Lett.* **26**, 494.

CORRELATIONS BETWEEN THE RADIO EMISSION AND THE X-RAY TIMING PROPERTIES IN GRS 1915+105

M. MUNO, R. REMILLARD and E. MORGAN
MIT Center for Space Research; E-mail: muno@mit.edu

E. WALTMAN
Naval Research Laboratory

R. HJELLMING
National Radio Astronomy Observatory

G. POOLEY
Mullard Radio Astronomy Laboratory, Cavendish Laboratory

GRS 1915+105 is noted for exhibiting X-ray, radio, and infrared emission which is variable on time scales of seconds to minutes (Mirabel *et al.*, 1998; Eikenberry *et al.*, 1998; Greiner *et al.*, 1996; Belloni *et al.*, 1997). However, the source also exhibits steady emission which is reminiscent of black hole candidates such as Cyg X−1 and GX 339−4. Both the variable and steady X-ray conditions can be seen as transitions between two basic modes of X-ray emission (Muno *et al.*, 1999). The soft state exhibits little X-ray emission above 40 keV and no 0.5–10 Hz quasi-periodic oscillation (QPO) in its power density spectrum (PDS). The hard state exhibits a power law energy spectrum up to at least 100 keV, and a 0.5–10 Hz QPO. Bright (100 mJy at 15 GHz), optically thick radio emission often accompanies the hard state of GRS 1915+105 (Foster *et al.*, 1996), similar to the hard states of Cyg X−1 (Brocksopp *et al.*, 1999) and GX 339−4 (Corbel *et al.*, 2000). This so-called 'radio plateau' emission (e.g. MJD 50730-50750 in Figure 1) originates from a compact jet (Dhawan *et al.*, 2000). However, other hard X-ray states are sometimes accompanied by radio emission which is a factor of 10 fainter (see MJD 50440-50560 and MJD 51053-51096 in Figure 1).

During periods of optically thick radio emission, the power contained in the radio-emitting jets may be a significant fraction of the accretion energy which can be inferred from the X-ray emission. This has led Fender *et al.* (1999) to speculate that the Compton corona responsible for the X-ray power law represents the base of the compact radio jet. This motivates us to examine whether the X-ray spectral and timing properties during the hard state of GRS 1915+105 are related to the intensity of the radio emission.

We have analyzed all 111 *RXTE* PCA and HEXTE observations of GRS 1915 + 105 during the hard-steady state taken between 1996 April 15 (MJD 50188) and 1999 March 22 (MJD 51259). Radio flux densities at 15 GHz from

Astrophysics and Space Science is the original source of publication of this article. It is recommended that this article is cited as: *Astrophysics and Space Science* **276** (suppl.): 227–230, 2001.

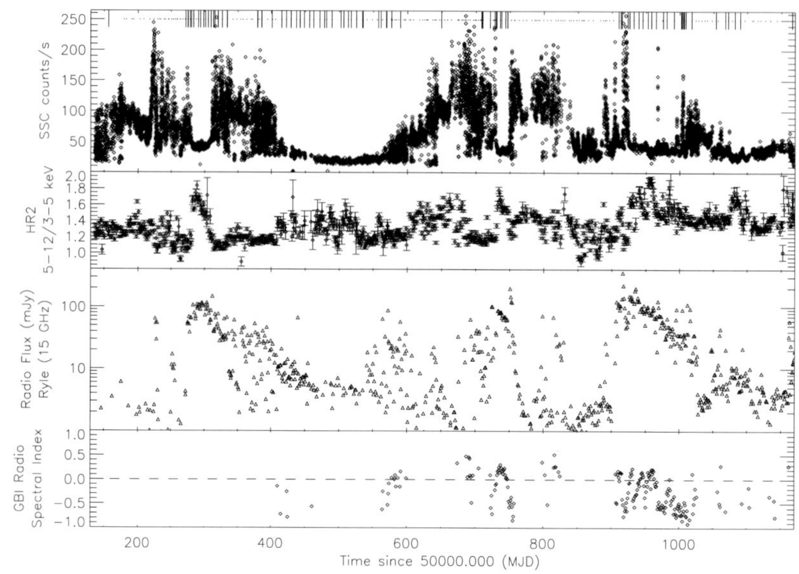

Figure 1. Monitoring data for GRS 1915+105. *Top panel*: 1.5–12 keV light curve from the *RXTE* All-Sky Monitor (ASM). Vertical bars across the top represent hard-steady pointed *RXTE* observations, and dots the remainder of the pointed observations. *Second panel*: *RXTE* ASM hardness ratio HR2 (5–12 keV / 3–5 keV). *Third panel*: Radio flux from the Ryle telescope at 15 GHz. *Bottom panel*: GBI radio spectral index ($\alpha = \Delta \log S_\nu / \Delta \log \nu$) using $\nu = 8$ and 2 GHz; points are plotted only if the flux density at both frequencies was larger than 20 mJy.

the Ryle telescope occur within a half-day of 98 of these observations, and public measurements of the radio spectral index between 2 and 8 GHz from the NSF-NRAO-NASA Green Bank Interferometer are available for 25 observations. For each *RXTE* observation, we have computed the PDS, Fourier phase lag spectrum, and the coherence function from the PCA data (Nowak *et al.*, 1999). Representative examples of these X-ray timing properties of GRS 1915+105 are displayed in Figure 2.

We find that radio emission always accompanies the hard-steady state of GRS 1915+105, but that the radio flux density at 15 GHz and the X-ray flux between 2–200 keV are not correlated. However, in Figure 3 we demonstrate that as the radio emission becomes brighter and more optically thick, 1) the frequency of a ubiquitous 0.5–10 Hz X-ray quasi-periodic oscillation decreases, 2) Fourier phase lags measured between between hard (11.5–60 keV) and soft (2.5–4 keV) photons in the frequency range of 0.01–10 Hz change sign from negative to positive, 3) the coherence between hard and soft photons at low frequencies decreases, and 4) the relative amount of low frequency power in hard photons compared to soft photons decreases.

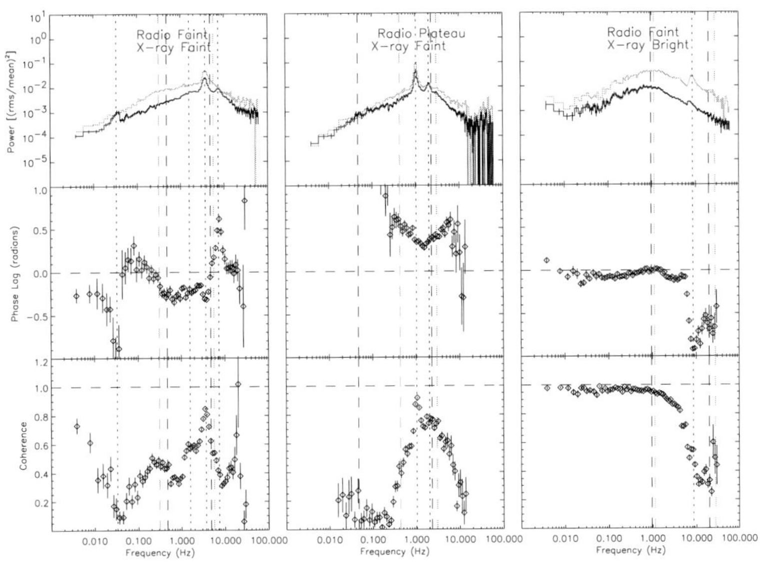

Figure 2. Timing properties of the three variants of the hard X-ray state, from observations on MJD 50488, 50737, and 50708 from left to right. *Top panel*: PDS for the 2–4.3 keV energy band (black lines), and the 11.5–60 keV band (grey lines). *Middle panel*: the phase lags between the 2–4.3 keV and the 11.5–60 keV bands, where positive values indicate hard photons lag soft. *Bottom panel*: the coherence function between the 2–4.3 keV and the 11.5–60 keV bands (a measure of the linearity of the transfer function relating soft and hard signals; see e.g. Nowak *et al.*, 1999). In all panels, the frequencies of QPOs are shown with black short-dashed lines, and break frequencies with black long-dashed lines. The break frequency in the 11.5–60 keV band is shown with grey long-dashed lines.

These correlations provide compelling evidence of a link between the X-ray and radio emitting regions, but it is difficult to interpret the sign shifts in the phase lags and the loss of coherence in the X-ray bands when the radio emission is faint. The radio plateau state of GRS 1915+105 resembles the hard state in Cyg X−1 in that the X-ray phase lags are positive and the radio emission is bright and optically thick, but the low observed coherence in Figure 2 stands in contrast to the high coherence observed in Cyg X−1 (Nowak *et al.*, 1999). Nevertheless, the search for similar trends in other sources may provide insights into the connection between the radio and X-ray emission from binary black hole candidates.

References

Belloni, T., Méndez, M., King, A.R., van der Klis, M. and van Paradijs, J.: 1997, *Astrophys. J.* **479**, L145.

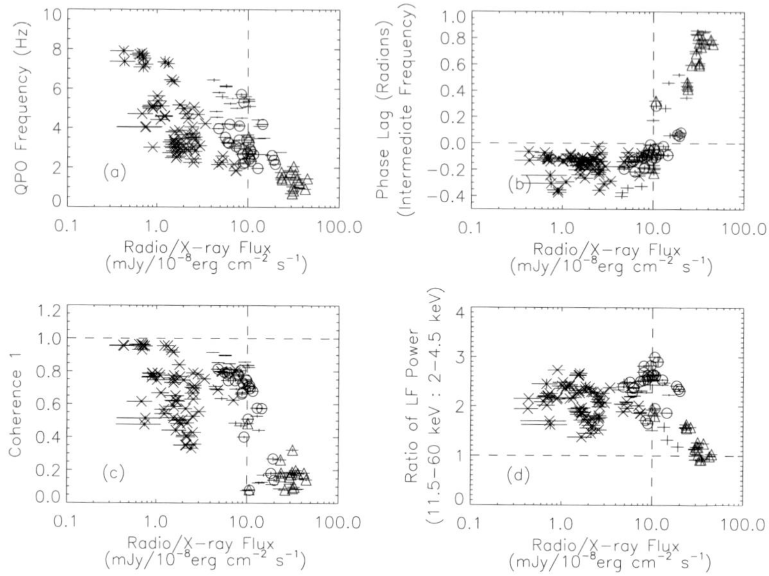

Figure 3. X-ray timing properties as a function of the ratio of radio to X-ray flux. Triangles represent bright (> 15 mJy at 15 GHz) optically thick ($\alpha < -0.2$) conditions, circles bright optically thin conditions, and $\times$'s radio-faint conditions. a) The QPO frequency. b) Phase lags (as in Figure 2) averaged between the low frequency break and the QPO. c) The coherence function (as in Figure 2) averaged over frequencies below the low frequency break. d) The ratio of the integrated broad-band power below the QPO frequency in high (11.5–60 keV) to low (2–4.5 keV) energy bands.

Brocksopp, C., Fender, R.P., Larionov, V., Lyuty, V.M., Tarasov, A.E., Pooley, G.G., Paciesas, W.S. and Roche, P.: 1999, *Mon. Not. R. Astron. Soc.* **309**, 1063.

Corbel, S., Fender, R.P., Tzioumis, A.K., Nowak, M., McIntyre, V., Duroudhoux, P. and Sood, R.: 2000, *Astron. Astrophys.* **359**, 251.

Dhawan, V., Mirabel, I.F. and Rodríguez, L.F.: 2000, *Astrophys. J.* **543**, 373.

Eikenberry, S.S., Matthews, K., Morgan, E.H., Remillard, R.A. and Nelson, R.W.: 1998, *Astrophys. J.* **494**, L61.

Fender, R.P., *et al.*: 1999, *Astrophys. J.* **519**, L165.

Foster, R.S., Waltman, E.B., Tavani, M., Harmon, B.A., Zhang, S.N., Paciesas, W.S. and Ghigo, F.D.: 1996, *Astrophys. J.* **467**, L81.

Greiner, J., Morgan, E.H. and Remillard, R.A.: 1996, *Astrophys. J.* **473**, L107.

Mirabel, I.F., Dhawan, V., Chaty, S., Rodríguez, L.F., Marti, J., Robinson, C.R., Swank, J. and Geballe, T.: 1998, *Astron. Astrophys.* **300**, L9.

Muno, M.P., Morgan, E.H. and Remillard, R.A.: 1999, *Astrophys. J.* **527**, 321.

Nowak, M.A., Vaughan, B.A., Wilms, J., Dove, J.B. and Begelman, M.C.: 1999, *Astrophys. J.* **510**, 874.

VISCOUS EVOLUTION OF AN ACCRETION DISK IN A BINARY SYSTEM

GALINA V. LIPUNOVA and NIKOLAI I. SHAKURA
Sternberg Astronomical Institute, Moscow State University, Universitetskii pr. 13, Moscow, 119899 Russia; E-mail: galja@sai.msu.ru

Abstract. A new analytical solution describing a time-dependent disk in a binary system at outburst, which has a fixed tidally-truncated outer radius, is applied to 1975 X-ray nova burst of A0620-00. The X-ray flux of a multicolor (black-body) α-disk decays quasi-exponentially with time. Modeling the X-ray and the optical light curves of A0620-00, we can estimate α and obtain 'distance–mass of compact object' relation.

We consider the basic viscous diffusion equation of accretion (e.g., Lipunova and Shakura, 2000) as applied to accretion disks. This equation shows that the time behaviour of a disk depends on the relation between kinematic coefficient of turbulent viscosity ν, column density of a disk Σ, and radius r.

If ν is a function of radius alone, the basic differential equation of diffusion type is linear. This case was considered by Weizsäcker (1948), Lüst (1952), Lynden-Bell and Pringle (1974).

The general case, when ν is a power–law function of Σ and r, was considered by Filipov (1984; 1988), Lyubarskii and Shakura (1987).

We adopt the $\nu \propto \Sigma^l\, r^m$ relation and pioneer in considering a confined accretion disk. We assume specific outer boundary conditions for a disk in a binary system (Lipunova and Shakura, 2000) and obtain a new solution. The standard α-model of the viscosity is assumed.

The bolometric luminosity is found to vary with time as a power low with an index of $-5/2$, if Thomson scattering dominates the opacity, and $-10/3$, if free-free and free-bound transitions do.

Assuming the spectrum to be a multi-color black body one, the X-ray luminosity is obtained to vary as $\exp{(-t^{5/6})}$. This is quite close to the exponential behaviour.

To illustrate how the parameters of the disk can be obtained, we model the 1975 burst of A0620-00 (Figure 1). We used the X-ray data in the 3–6 keV band of *Ariel-5* (Kaluzienski *et al.*, 1977) and the data in B and V optical bands (Kuulkers 1998, and references therein).

We see that the model is consistent with the idea of a black hole in this binary and with the distance estimates available in literature (Barret *et al.*, 1996; Chen *et al.*, 1997). We choose the models with E(B$-$V)=0.39 (Wu *et al.*, 1976). From Kuulkers (1998) we have B$-$V (27 d)= 0.24 ($\pm$10%), then BH mass lies in the range 9–14 $M_{\odot}$.

Astrophysics and Space Science is the original source of publication of this article. It is recommended that this article is cited as: *Astrophysics and Space Science* **276** (suppl.): 231–232, 2001.

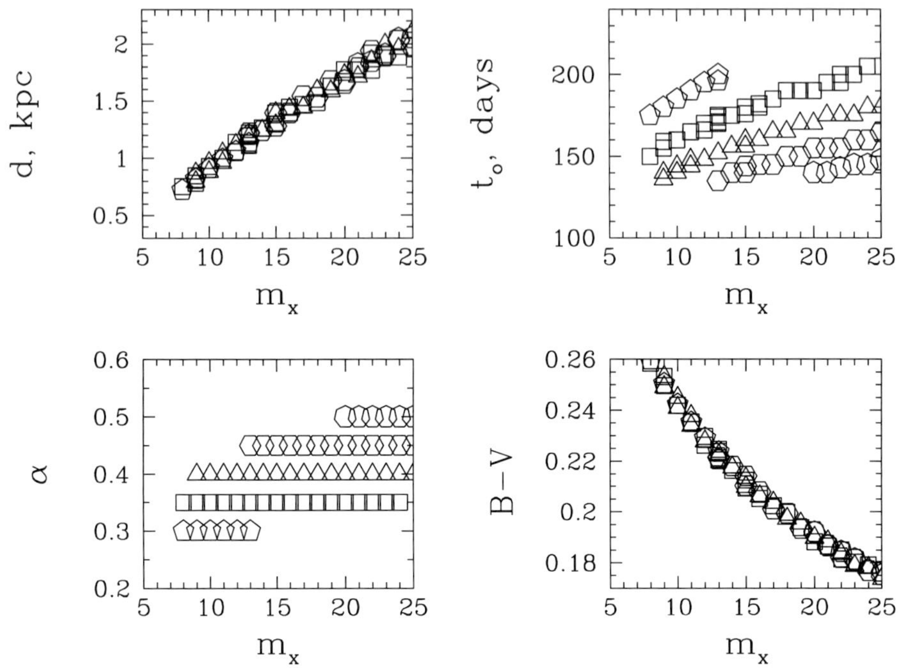

Figure 1. A multitude of models is restricted to the following values of parameters: $m_X = M_{BH}/M_\odot = 5 \div 25$, B−V (27th day after peak) = $0.17 \div 0.26$. We checked the values of α: 0.2, 0.25, 0.3, 0.35, 0.4, 0.45, 0.5, 0.55, 0.6. d is the distance to the source. The right upper panel displays the corresponding values of t_0 (see Lipunova and Shakura, 2000) if we set t to be the time after the peak of the outburst, MJD 42638.4 = 13/08/1975.

Acknowledgements

The work is partially supported by the RFBR grant 00–02–17164, the program 'Universitety Rossii' (grant 5559), and NTP 'Astronomy' grant 1.4.4.1 (Russia).

GVL is thankful to Instituto de Astrofísica de Andalucía for supporting her visit to Granada. She is also supported by ISSEP.

References

Filipov, L.G.: 1984, *Adv. Space Res.* **3**, 305.
Filipov, L.G.: 1988, *Adv. Space Res.* **8**, 141.
Kaluzienski, L.J., Holt, S.S., Boldt, E.A. and Serlemitsos, P.J.: 1977, *Astrophys. J.* **212**, 203.
Kuulkers, E.: 1998, *New Astron. Rev.* **42**, 1.
Lipunova, G.V. and Shakura, N.I.: 2000, *Astron. Astrophys.* **356**, 363.
Lüst, R.: 1952, *Naturforschg* **7(a)**, 87.
Lynden-Bell, D. and Pringle, J.E.: 1974, *Mon. Not. R. Astron. Soc.* **168**, 603.
Weizsäcker, C.F.: 1948, *Naturforschg* **3(a)**, 524.

THE ACCRETION-EJECTION INSTABILITY IN X-RAY BINARIES

P. VARNIÈRE and M. TAGGER
Service d'Astrophysique (CNRS URA 2052), CEA/Saclay 91191 Gif sur Yvette
E-mail: varniere@cea.fr

Abstract. The Accretion-Ejection Instability (AEI), which can occur in magnetized disks near equipartition, is a good candidate to explain the low-frequency QPO in black-hole binaries. Here we present analytical work concerning the behavior of QPO frequency and the emission of Alfvén waves from the disk to the corona.

1. Introduction

The AEI is a spiral instability, similar to galatic spirals but driven by magnetic stress rather than self-gravity. It occurs in the inner region of an accretion disk threaded by a vertical magnetic field of the order of equipartition with the gas pressure. The spiral extracts energy and angular momentum from the disk, causing accretion, and stores them in a Rossby vortex at its corotation radius (see also the contributions of Tagger and Caunt, these proceedings). This vortex then leaks energy and angular momentum as Alfvén waves to the corona, where it can power a wind or a jet. For more details see Tagger and Pellat (1999).

2. QPO Frequency in Pseudo-Newtonian Potential

Relativistic effects change the rotation curve of the disk near the last stable orbit. They allow the existence of an Inner Lindblad Resonance (ILR) for the $m = 1$ mode. This changes the properties of the 1-armed spiral (best candidate to explain the QPO), and therefore its frequency.

We have studied, using a pseudo-newtonian potential, the properties of the instability when the disk inner radius r_{int} approaches the Last Stable Orbit at r_{LSO}. When r_{int} is large the QPO frequency varies as $\omega \propto r^{-3/2}$; but as r_{int} approaches r_{LSO} the correlation changes and becomes *positive* (the QPO frequency decreasing with a decreasing radius) when $r_{int} < 1.4 r_{LSO}$, *i.e.* when the $m = 1$ mode has an Inner Lindblad Resonance in the disk (Varnière, Rodriguez and Tagger, 2001). We consider this as a possible explanation for the positive correlation found (contrary to other sources) in GRO J1655 as shown in these proceedings and Rodriguez, Varnière and Tagger (2001).

Astrophysics and Space Science is the original source of publication of this article. It is recommended that this article is cited as: *Astrophysics and Space Science* **276** (suppl.): 233–234, 2001.

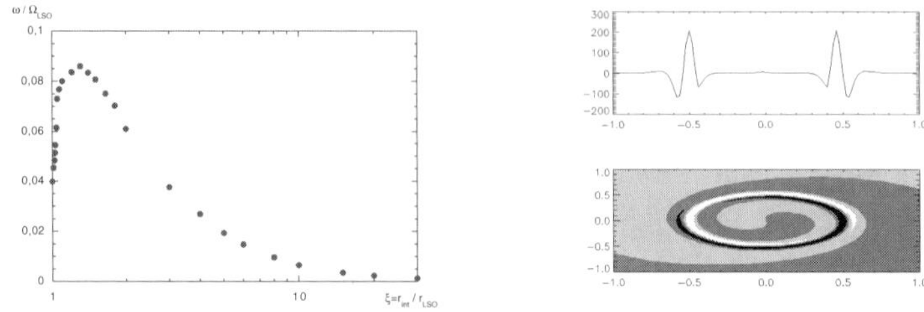

Figure 1. (left) the QPO frequency as a function of the disk inner radius r_{int}. The frequency-radius correlation is reversed when $r_{int} \lesssim 1.4 r_{LSO}$. (right) radial profile and spatial structure of the flux of Alfvén waves emitted to the corona.

3. Emission of Alfvén Waves

The Rossby vortex twists the footpoint of the field lines threading the disk. If the disk has a low density corona this twisting will be propagated upward as Alfvén waves. The energy and angular momentum extracted from the disk will thus be transfered to the corona where they can power a wind or jet. We study this with a variational form

$$F = [\text{energy of the waves}] + i \ [\text{outgoing spiral} + \text{vortex} \\ + k_z \text{ Alfvén Waves}]$$

where imaginary terms correspond to amplification or damping of the instability. From the numerical solution we compute the flux of the emitted Alfvén waves. The result, plotted in Figure 1, shows that it peaks at the corotation radius, where the Rossby vortex is localized.

4. Conclusions

The frequency of the AEI, and its variations with the disk inner radius, make it a very good candidate to explain the ‘ubiquitous’ QPO. The accretion energy and momentum extracted from the disk are propagated to the corona as Alfvén waves.

References

Rodriguez, J., Varnière, P. and Tagger, M.: 2001, *Astron. Astrophys.*, submitted.
Tagger, M. and Pellat, R.: 1999, *Astron. Astrophys.* **2**, 155.
Varnière, P., Rodriguez, J. and Tagger, M.: 2001 *Astron. Astrophys.*, submitted.

THE ACCRETION-EJECTION INSTABILITY: OBSERVATIONAL TESTS

J. RODRIGUEZ, P. VARNIÈRE, M. TAGGER and PH. DUROUCHOUX
Service d'Astrophysique (CNRS URA 2052), CEA/Saclay 91191 Gif sur Yvette
E-mail: jrodriguez@cea.fr

1. First Steps: GRO J1655-40

QPOs are commonly considered to originate in the disks of X-ray binaries. One thus expects, and most often observes, that the QPO frequency decreases as the disk inner radius increases. However Sobczak *et al.* (1999) pointed out an opposite behavior in the case of GRO J1655-40. This might prove usefull in testing the theory of the Accretion-Ejection Instability (see contributions by Caunt, Tagger and Varnière, these proceedings). At the same time, Merloni, Fabian, and Ross (1999) showed that, when fitting the sources spectra with the standard model (multicolor blackbody+ power law), severe distorsions could be due to the vertical structure of the disk. In that view we re-did the spectral analysis of the observations in which Sobczak *et al.* found QPO's.

Using the criterion of Merloni *et al.*, we find that all the data points from GRO J1655-40 where the fits give an anomalously low inner radius must be excluded. The remaining points confirm the inverse correlation.

2. Another Good Candidate: GRS 1915+105

Because of its high and rapid variability, and following the results found by Markwardt *et al.*, 1999, and Muno *et al.*, 1999, we then focussed on GRS 1915+105. In particular we studied very precisely two 30 min cycles present in the observation of September 9, 1997, in which the QPO frequency had been seen to vary (Markwardt *et al.*, 1999; Swank *et al.*, 1997), correlated to the flux, the color temperature and the color radius itself.

We extracted, during two $\sim$ 30 min dips, every 20-s a 16-s spectrum and a lightcurve, extracting both the QPO frequency and the spectral parameters (Rodriguez *et al.*, 2000).

We have also analyzed recent observations of GRS 1915+105, pointed as a target of opportunity, on April 17, 22, 23 by RXTE. Although only the first date showed high variability, in the three observations a strong QPO, together with its first harmonic are highly correlated to the flux (work in progress).

Astrophysics and Space Science is the original source of publication of this article. It is recommended that this article is cited as: *Astrophysics and Space Science* **276** (suppl.): 235–236, 2001.

3. Results and Interpretation

- Applying more stringent criteria leads us to exclude some of the data points observed by Sobczak *et al.* in GRO J1655-40, but the points we retain do show the inverse correlation between the QPO frequency and the disk inner radius.
- Though highly variable GRS 1915+105 does not present this inverse correlation, during the 30 min cycles.
- These observations are compatible with the theory of the Accretion-Ejection Instability (see Varnière and Tagger, 2001), if the disk of GRO J1655-40 is always near the Last Stable Orbit when QPO are observed, and the disk of GRS 1915+105 never approaches it during its 30 min cycles.
- The criterion of Merloni *et al.* (2000) leads us to exclude the data points of GRO J1655-40 where the spectral fits return a very low color radius ($\sim$ very few kms). But we find that these points are not compatible with the predictions of Merloni, as they systematically show a higher temperature when the color radius decreases. On the other hand, data points at the onset of the low state, in GRS 1915+105, are compatible with the predictions of Merloni *et al.*.
- We note that in GRO J1655-40, as well as during the hard-steady state of GRS 1915+105, the very low color radius actually means a small size of the emitting region. This and the higher temperature would be compatible with the presence in the disk of a spiral shock or hot point, where accretion energy would be dissipated.

References

Markwardt, C.B., Swank, J.H. and Taam R.E.: 1999, *Astrophys. J.* **513**, L37.

Merloni, A., Fabian, A.C. and Ross, R.R.: 2000, *Mon. Not. R. Astron. Soc.* **313**, 193.

Rodriguez, J., Varnière, P., Tagger, M. and Durouchoux, Ph.: 2000, *Astron. Astrophys.*, submitted.

Sobczak, G.J., McClintock, J.E., Remillard, R.A., Cui, W., Levine, A.M., Morgan, E.H., Orosz, J.A. and Bailyn, C.D.: 2000, *Astrophys. J.* **531**, 537.

Tagger, M. and Pellat, R.: 1999, *Astron. Astrophys.* **349**, 1003.

Varnière, P., Rodriguez, J. and Tagger, M.: 2000, *Astron. Astrophys.*, submitted.

Varnière, P. and Tagger, M.: 2001, These Proceedings.

ANALYTICAL SOLUTIONS OF ACCRETION DISKS WITH OUTFLOWS

W. KLUŹNIAK
Copernicus Astronomical Center, ul. Bartycka 18, 00-716 Warszawa, Poland

Abstract. Analytical solutions of accretion flow in three dimensions are discussed, with the flow restricted to be axisymmetric. Steady outflows are a generic feature for many classes of solutions. Many two-dimensional numerical solutions of accretion exhibit non-trivial flow structure in agreement with our results. Another mechanism of driving outflow is the well-known propeller mechanism, could it be responsible for the behaviour of GRS 1915+105?

1. On Accretion Disk Solutions

Analytical considerations in the literature are usually essentially one-dimensional (e.g. the radial structure of accretion disks is determined by solving a set of height-averaged equations, while the vertical structure is found at a fixed radius).

Hydrodynamic flow in accretion disks is intrinsically two-dimensional or even three-dimensional. Perhaps for this reason, it is fashionable and useful to model accretion flows and jet outflows numerically. Most of these simulations are two-dimensional, they are carried out in a limited domain (i.e., the results may depend on the boundary conditions adopted), and frequently it is not clear whether a steady state has been reached, or whether the solution is transient.

It would, then, be worthwile to expend more effort on obtaining analytical results in two or three dimensions. In fact, nearly exact analytical results can be obtained, and the effort expended is immediately rewarded with novel results which only subsequently are also uncovered numerically. For lack of space, I will give but one example: analytical work has clearly shown outflow in the plane of the disk (Urpin, 1985; Kluźniak and Kita, 2000), such outflow is discerned also in many numerical simulations (e.g., Igumenshchev and Abramowicz, 1999; Lee, 2000).

2. GRS 1915+105, An Accreting Pulsar?

It is almost universally assumed that GRS 1915+105, the micro-quasar with bizarre light-curves, is a black hole. Thus, work on understanding the outflows associated with radio emission, usually focuses on possible ejection of material from the inner parts of a black-hole accretion disk. But the mass of the source is not known, and it is quite possible that the powerhouse is a rotating magnetized neutron star in a state of intermittent accretion.

Astrophysics and Space Science is the original source of publication of this article. It is recommended that this article is cited as: *Astrophysics and Space Science* **276** (suppl.): 237–238, 2001.

The unusually steady (in frequency) quasi-periodic oscillation is strongly suggestive of a pulsar. Several contributions to this volume discuss the characteristic transitions between two or three accretion states of GRS 1915+105. It seems to me, that these are about what would be expected of an accreting pulsar at a period which is the inverse of the QPO frequency. It is accepted in neutron star theory, that a rotating magnetized neutron star can accrete matter only if the mass penetrates the corotation radius (where, by definition, the keplerian period is equal to the stellar period), this is possible only above a certain mass accretion rate. Otherwise matter is expelled.

The corotation radius for such a pulsar, as well as the radius of the light cylinder, comes out to be very close to the inner radius of the disk (inferred from observations and reported in several contributions in this volume, e.g., Belloni, 2000) of GRS 1915, in that state which is accompanied by strong radio emission, while the disk radius in the radio-quiet state is much smaller [although other spectral models do not support this interpretation, Vilhu, 2000].

Obviously, this model has the great advantage that it connects one observed number (the QPO frequency) with another (inner disk radius), and correlates in a natural way one state of the system with expulsion of matter and radio activity (which is quenched in the other state by accretion onto the neutron star). Were the black hole models equally successful! The QPO frequency should be monitored over the years – it is likely to decrease on the pulsar model, as much of the source luminosity is powered by the kinetic energy of the rotating star.

Acknowledgement

This work was supported in part by KBN grant 2P03D 004 18.

References

Belloni, T.: 2000, These Proceedings.
Igumenshchev, I.V. and Abramowicz, M.A.: 1999, *Mon. Not. R. Astron. Soc.* **303**, 30.
Kluźniak, W. and Kita, D.: 2000, submitted (astro-ph/0006266).
Lee, W.: 2000, These Proceedings.
Urpin, V.A.: 1984, *Astron. Zh.* **61**, 84 [SvA, 28, 50].
Vilhu, O.: 2000, These Proceedings.

ADVECTION DOMINATED FLOWS AROUND BLACK HOLES IN TWO DIMENSIONS

WILLIAM H. LEE
Instituto de Astronomía, UNAM, Apdo. Postal 70–264, Cd. Universitaria, México D.F. 04510, México

Abstract. We show the results of two-dimensional, azimuthally symmetric simulations of advection dominated fluid flows around black holes. We use the SPH method, and the α-prescription for the viscosity, including all terms in the stress tensor. We have performed calculations with $0.001 < \alpha < 0.2$ and find strong circulation patterns in the flow, on spatial scales that depend on the magnitude of the viscosity.

1. Introduction, Method and Initial Conditions

The astrophysical relevance of accretion flows in which radiative cooling is inefficient has been recently emphasized (Narayan and Yi, 1994). These flows show complicated hydrodynamical behavior that is very different from that obtained with the thin disk approach (Shakura and Sunyaev, 1973) for efficient cooling.

We use the Smooth Particle Hydrodynamics (SPH) method (Monaghan, 1992), in azimuthal symmetry. The equation of state is $P = \rho u(\gamma - 1)$ with $\gamma = 4/3$. Viscosity is computed using an α prescription ($\mu = \alpha \rho c_s^2 / \Omega_k$), including all the terms from the viscous stress tensor. The black hole is modeled as a Newtonian point mass with an absorbing boundary at $r_{in} = 1.3 r_G = 1.3(2GM_{BH}/c^2)$. We use two kinds of initial conditions: 1) equilibrium hydrostatic tori with constant specific angular momentum; 2) matter injection at a steady rate at a radius $r = 300 r_G$. We assume no radiative energy losses.

2. Results

For initially hydrostatic tori, the presence of a hot, high-entropy bubble near the equatorial plane is easily seen (Figure 1). This moves outward in the disk, on top of a background accretion flow. There is a large-scale circulation pattern, on a scale $z \simeq r$. For a lower viscosity $\alpha = 0.001$ there are small-scale vortices in the disk, and the overall behavior is more turbulent. A background accretion flow is also present.

For the case of matter injection, the flow structure is governed by the magnitude of the viscosity, with small-scale eddies at low values of α and large-scale circu-

Astrophysics and Space Science is the original source of publication of this article. It is recommended that this article is cited as: *Astrophysics and Space Science* **276** (suppl.): 239–240, 2001.

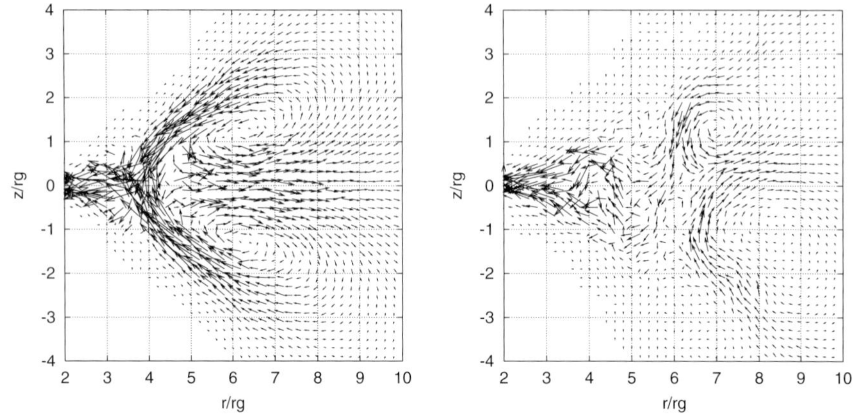

Figure 1. Momentum vector field $\rho\vec{v}$ for $\alpha = 0.01$.

lations when the viscosity is high. The fluid shows significant oscillations in the vertical direction as it moves radially inward.

With these simulations we have tested our code with previously published results obtained with a different numerical scheme (Igumenshchev *et al.*, 1996; Igumenshchev and Abramowicz, 1999), and we have found excellent agreement both in a qualitative and quantitative manner. In future work we will extend our results to study the parameter space (α, γ) and increase the resolution.

Acknowledgements

I acknowledge support from CONACyT (27987E), DGAPA (IN-119998) and helpful conversations with M. Abramowicz and W. Kluźniak.

References

Igumenshchev, I.V., Chen, X. and Abramowicz, M.A.: 1996, *Mon. Not. R. Astron. Soc.* **278**, 236.
Igumenshchev, I.V. and Abramowicz, M.A.: 1999, *Mon. Not. R. Astron. Soc.* **303**, 309.
Monaghan, J.J.: 1992, *Annu. Rev. Astron. Astrophys.* **30**, 543.
Narayan, R. and Yi, I.: 1994, *Astrophys. J.* **428**, L13.
Shakura, I.N. and Sunyaev, R.A.: 1973, *Astron. Astrophys.* **24**, 337.

FORMATION AND DYNAMICS OF NEUTRON HALOES IN DISK ACCRETING BLACK HOLES

A.A. BELYANIN and E.V. DERISHEV
Institute of Applied Physics, 46 Ulyanov St., 603950 Nizhny Novgorod, Russia

There are many theoretical and observational indications that the hot plasma of ion temperature T_i exceeding several MeV can exist in the inner accretion disks around black holes. Formation of two-temperature region with $T_i > T_e$ in accretion disks has been predicted as long ago as in 1976 (Shapiro *et al.*, 1976). When the energy transfer from ions to electrons is slow as compared with accretion timescale, the accretion flow becomes nearly adiabatic, and the ion temperature can reach virial values. This idea has led to the prediction of advection-dominated accretion flows which seems to find support in observations of several X-ray transients (Narayan and Yi, 1996).

The presence of hot ions in accretion disks leads to the rich variety of inelastic nuclear interactions. The most prominent of them is neutron production which primarily occurs through dissociation of helium nuclei. This process has potentially important observational implications arising e.g. from possibility of radiative neutron capture with subsequent formation of gamma-ray lines (Aharonian and Sunyaev, 1984).

Neutron production in accretion disks was considered several times (Aharonian and Sunyaev, 1984; Guessoum and Kazanas, 1989) under the assumption that neutrons are quickly thermalized in the disk and are advected to the black hole with the same rate as protons. In this case the neutron fraction in the accretion flow is inevitably small and the radiative-capture gamma-ray lines are weak and smeared out by hot rapidly rotating disk. However, the above assumption is not true if the proton density in the disk is sufficiently low. In this case, as shown in Derishev and Belyanin (2000), the neutron and ion components in the disk effectively decouple, allowing for the accumulation of neutrons. Such a pile-up can lead to much larger neutron number than in previously considered cases.

It is important that, once formed, such a halo becomes self-sustained. This means that, even if the ion temperature in the disk falls below the threshold for He dissociation by protons, the neutron production is supported by collisions of energetic neutrons from the halo with helium nuclei. The self-sustained mode implies that at least one of three neutrons resulting from helium dissociation breaks another helium nucleus before decaying or being advected into the black hole. These two possibilities provide, respectively, lower and upper bounds on the range of accretion rates in which the self-sustained neutron halo may exist.

Astrophysics and Space Science is the original source of publication of this article. It is recommended that this article is cited as: *Astrophysics and Space Science* **276** (suppl.): 241–242, 2001.

Our analysis shows (Derishev and Belyanin, 2000) that neutron halo exists in a broad range of disk parameters and can lead to a number of important observational and dynamical effects. In particular, neutron haloes can be the natural source of relativistic electrons and positrons produced in proton-neutron and neutron-neutron collisions. In the presence of abundant soft X-rays from disk, relativistic electron-positron pairs are quickly cooled, and the snapshot of electron distribution would demonstrate that the majority of them has energies between 50–500 keV. The resulting spectrum above 100 keV has a typical comptonization pattern. Pair cascade, if developed, leads to a hard power-law tail with index around 2. Both features are frequently observed in the hard state of accreting black-hole candidates. In addition, very hard radiation should be present with photon energies up to 70 MeV. However, it is expected to contain much less energy than hard X-rays and is difficult to observe. Deuterium gamma-ray line at 2.2 MeV resulting from neutron capture is also expected at a level detectable by future INTEGRAL mission.

Furthermore, the presence of neutron halo strongly affects the dynamics of accretion and leads to the rich variety of transient dynamical regimes.

Acknowledgements

This work has been supported by Russian Foundation for Basic Research (the project 99-02-18244) and by the Russian Academy of Science through a grant for young scientists.

References

Aharonyan, F.A. and Sunyaev, R.A.: 1984, *Mon. Not. R. Astron. Soc.* **210**, 257.
Derishev, E.V. and Belyanin, A.A.: 2000, *Astron. Astrophys.*, submitted.
Guessoum, N. and Kazanas, D.: 1989, *Astrophys. J.* **345**, 356.
Narayan, R., McClintock, J. and Yi, I.: 1996, *Astrophys. J.* **457**, 821.
Shapiro, S.L., Lightman, A.P. and Eardley, D.M.: 1976, *Astrophys. J.* **204**, 187.

Chapter V
JET FORMATION AND DYNAMICS

RELATIVISTIC JET FORMATION IN MICROQUASARS

DAVID L. MEIER
Jet Propulsion Laboratory, California Institute of Technology

Abstract. Microquasars produce collimated, relativistic flow at speeds up to a Lorentz factor of ~ 3 (0.94c). It is generally believed that such flow velocities indicate that jet acceleration and collimation occurs in the relativistic environment of a supermassive black hole. Recently, several groups around the world have begun to test theories of jet formation using magnetohydrodynamic (MHD) simulations of magnetized gas flow around black holes. This paper reviews the field of simulations of MHD jet formation, with an emphasis on producing the observed jet speeds and on the role that black hole angular momentum might play. It is shown that the most powerful jets occur when the black hole is rotating rapidly *and* when the accretion disk is geometrically thick and hot. This model may explain the association of jet ejection with the hard state in microquasars.

1. Introduction

The distinguishing feature that separates microquasars from other binary X-ray sources, and links them with the giant extragalactic radio galaxies and quasars, is the presence of twin relativistic jets emanating from the source center. Two important observations indicate that the jets are ejected from deep within the potential well of the accreting compact object and, therefore, are closely related to the accretion process itself. First of all, the speed of the jet is of order the escape velocity from the vicinity of the compact object – 0.26 c for SS433 (Margon, 1984), which is believed to harbor a neutron star, and 0.92 c (Lorentz factor $\Gamma = (1 - v^2/c^2)^{-1/2} \approx 2.5$) for GRS 1915+105 and GRO J1655-40 (Mirabel and Rodriguez, 1999), which are believed to contain black holes of several solar masses. Secondly, jet ejection does not take place unless the accretion disk has entered a hard X-ray state (Fender *et al.*, 1999), and then the actual ejection event often is accompanied by a temporary disappearance of the hard X-ray emission, a possible indication that the hot disk material has been ejected in the jet.

The leading model for producing and collimating relativistic jets is the magnetohydrodynamic (MHD) acceleration model (Lovelace, 1976; Blandford, 1976). This model is able to account for jet collimation, acceleration to relativistic velocities, and the magnetic fields associated with detected synchrotron emission, all while operating within the accretion paradigm for microquasars. In addition, recent results have shown that it can account for the observed association of jet production with the black hole accretion disk states (Meier, 2001b).

Astrophysics and Space Science is the original source of publication of this article. It is recommended that this article is cited as: *Astrophysics and Space Science* **276** (suppl.): 245–252, 2001.

This paper reviews some of the important properties of numerical simulations of MHD jet acceleration and discusses analytic work that combines theories of accretion disk structure, disk dynamos, and jet production to explain microquasar behavior.

2. Review of Numerical Simulations of Jet Production by Magnetized Accretion Disks

Recent results on numerical simulations of jet production by magnetized accretion disks also have been reviewed by Meier and Koide (2000), Meier (2001a), and Meier, Koide and Uchida (2001). The simulations fall into two general categories: a) time-dependent, useful for studying violent and transient effects, and b) steady-state, useful for studying more benign, longer-lived phenomena.

2.1. Non-relativistic time-dependent simulations: The basic acceleration and collimation mechanism

The basic mechanism for acceleration and collimation of jets by rotating magnetic fields was described in detail by Blandford and Payne (1982) for accretion disks and Blandford and Znajek (1977) for rotating black holes. The first authors to demonstrate this mechanism with numerical simulations were Uchida and Shibata (1985). Some of the best modern illustrations of their 'sweeping pinch' mechanism are the three-dimensional renderings of the 2.5-dimensional simulations of Uchida *et al.* (2000). These simulations begin with a *strong* vertical magnetic field threading a differentially-rotating, thick accretion torus. The differential rotation winds up the field into a toroidal component (B_ϕ). The vertical magnetic pressure gradient ($\partial B_\phi^2/\partial Z$) quickly becomes strong enough to accelerate the disk surface layers upward, and the radial hoop stress ($-B_\phi^2/R$) pinches and collimates the outflow into a jet, with and outflow speed of order the escape velocity from the inner edge of the disk or torus. Such an outburst is a very dynamical event, occurring in only a few tens of milliseconds for a 10 $M_\odot$ black hole disk; a violent microquasar outburst lasting hours or days might involve many such events strung together in rapid-fire fashion.

2.2. Non-relativistic steady-state simulations

In low-luminosity, quiescent sources, the accretion and jet ejection process may reach a steady or quasi-steady state. Recently many papers have studied steady jet ejection numerically (e.g., Ustyugoga *et al.*, 1995; Ouyed and Pudritz, 1997; Krasnopolsky *et al.*, 1999). In these calculations the thick disk is replaced by an infinitely-thin disk boundary condition on the equatorial plane, with a poloidal (R, Z) magnetic field anchored in the differentially-rotating disk and matter injected from the lower boundary along the field lines at a speed slower than the escape speed. While the axial field and velocity (B_Z and V_Z) and azimuthal velocity (V_ϕ) are fixed on the lower boundary, the radial components (B_R and V_R) and azimuthal field are allowed to change as the solution is obtained. The final numerical results are indeed very similar to those of Blandford and Payne (1982). The field lines must have a polar angle of $\theta > 30°$ in order for magnetically-driven outflow to commence, but well above the disk the field lines bend toward the axis, resulting in a slowly-collimated wind or jet, again with a terminal velocity of order the escape speed at the foot of the jet. Such a scenario appears to be a good model for the steady compact radio cores seen continuously in sources like Cyg X-1 and GX 339-4, or in GRS 1915+105 between outbursts (Dhawan *et al.*, 2000).

2.3. Pseudo-Relativistic Simulations of Black Hole Accretion Systems

To study the dynamical strong field problem in very tenuous coronae, the author and his colleagues have employed non-relativistic simulations, also with an infinitely thin disk boundary condition, and interpreted them in a pseudo-relativistic manner (Lind *et al.*, 1994; Meier *et al.*, 1997). That is, as long as the inertia of the magnetic field and internal energy are small ($c_{sound} < c$; $V_A < c$), the non-relativistic MHD equations have the same form as the relativistic ones if the velocity V is replaced with the proper velocity $U \equiv \Gamma V$. When U exceeds c (unity here), it can be identified with the Lorentz factor Γ. Thus, these pseudo-relativistic simulations account for the inertia of the kinetic energy, which will be important even for a cold gas.

The initial conditions for these simulations are very similar to those of the steady-state ones, except that the disk is truncated interior to the last stable orbit at $6\,r_g$, where $r_g \equiv GM/c^2$, and B_R, V_R, and B_ϕ are all *fixed* on the lower boundary, as would be the case in an actual accretion disk. These conditions result in a new magnetic field structure during the simulation: the anchored field lines at the inner edge of the disk are bent inward, creating a substantial B_R in the corona, which is then wound up by the differential rotation into B_ϕ, expelling plasma upward and pinching it into a collimated jet. The simulation at late times is shown in Figure 1 and details of the magnetic structure near the disk inner edge are shown in Figure 2. This highly-collimated inner jet appears to compete with the outer, loosely-collimated wind found in the steady state simulations: in the ~ 80 simulations we have run, we see either the inner jet or wind, but rarely both at the same time.

The pseudo-relativistic simulations have been useful for studying a possible mechanism for producing the highly relativistic jets ($\Gamma \gtrsim 10$) observed in radio galaxies and quasars – the 'magnetic switch' – which achieves jet velocities higher than the escape speed by accelerating a very tenuous disk corona rather than the dense disk material. However, as the speeds of the fastest microquasar jets agree well with the escape velocity from the ergosphere of a rotating black hole, it ap-

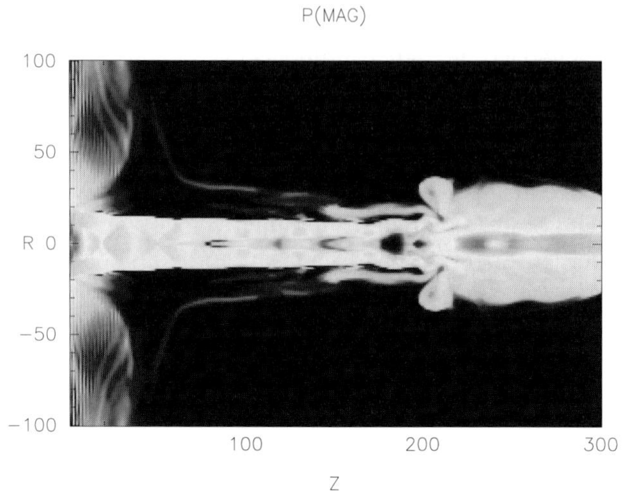

Figure 1. Magnetic pressure intensity plot at time $t = 700\tau_g$ for a simulated jet produced by a Keplerian accretion disk (on the equator at left) around a black hole (at [0,0]). Red indicates the highest magnetic pressure and dark blue the lowest. Such a plot is similar to the appearance of a total intensity radio image (if the field/particle ratio is uniform). Distances are in units of r_g.

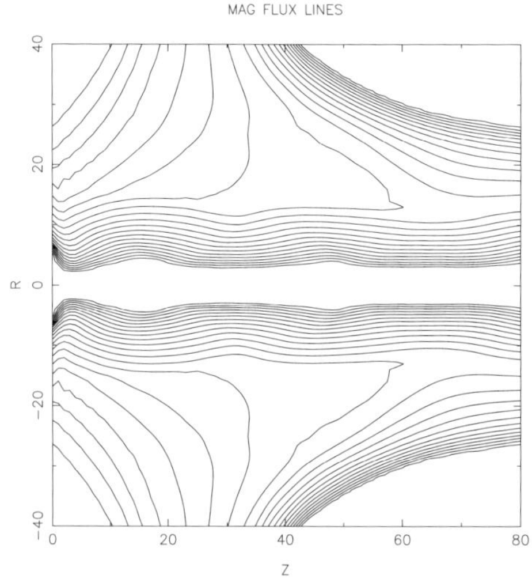

Figure 2. Field structure of the inner disk edge for the simulation in Figure 1. The black hole is at the origin and the expanded scale is again in r_g.

pears unnecessary at this time to invoke such a mechanism for microquasars. On the other hand, the discovery of objects with high Lorentz factors ('microblazars') could change this situation.

2.4. Fully General Relativistic Jet Production near Rotating Black Holes

The full jet acceleration problem involves MHD processes in curved spacetime near the black hole horizon. Our fully relativistic simulations begin with a rather thick accretion disk, with inner edge at $R = 6\,r_g$, threaded by a *weak* axial magnetic field (internal Alfven velocity of $V_A = 0.01c$). In addition to the disk, a freely-falling halo fills the remainder of the computational grid, which extends from $1.5\,r_g$ to $40\,r_g$ in spherical radius and from 0 to $\pi/2$ in polar angle. Five different scenarios have been simulated in a series of papers involving different combinations of rotating ($j \equiv J/(GM^2/c) = 0.95$) and non-rotating holes and rotating, non-rotating, and counter-rotating disks. The simulations were followed for a fairly short period of time – on the order of $\sim 100\,\tau_g \equiv 100\,GM/c^3$ – but long enough to allow the rotation of the disk or geometry and MHD processes to run their course.

The Schwarzschild cases had no significant MHD-driven outflow develop unless the disk was rotating. The outflow velocity was of order the escape velocity from the last stable orbit of $6\,r_g$ ($0.4c$) (Koide *et al.*, 1998). In the Kerr cases, however (Koide *et al.*, 2000), a powerful jet resulted *even when the disk was not rotating or was counter-rotating against the black hole*. The main driver of the jet was the rotating geometry itself, rather than the rotation of the accretion disk. The disk fell rapidly into the ergospheric region ($R < 2r_g$), where the accreting matter must rotate with the black hole geometry. This rotation drove an MHD jet in a manner similar to the non-relativistic rotating disk simulations. The highest jet speeds achieved so far in these simulations are of order the escape velocity from the edge of the ergosphere ($\Gamma < 3$).

3. Applications to Microquasars

3.1. Jet Power

One of the most important properties of jets is the total amount of energy per unit time carried away from the central engine by the jet. It is shown in Meier (2001b) that the jet power from a rotating, magnetized accretion flow can be computed from the following expression

$$L_{jet} = B_\phi^2\, H^2\, R^2\, \Omega^2 \,/\, 4\,c \tag{1}$$

where B_ϕ is the strength of the azimuthal magnetic field responsible for much of the viscous stress, H the disk scale height, R the radius of the jet-emitting region, and Ω the angular velocity of the magnetic field as measured by an observer at infinity. Ω actually is composed of two parts: rotation of the metric space surrounding the (possibly-spinning) black hole and rotation of the accretion disk within that metric

$$\Omega = \Omega_m + \Omega_d \tag{2}$$

The factor $B_\phi H/R$ represents the poloidal magnetic field strength protruding from the disk surface – only a fraction of the toroidal component – as suggested by dynamo theory arguments (Livio *et al.*, 1999; Vishniac, private communication) and simulations (Khanna, 1999). With a given accretion model, the jet power can be computed from (1).

3.2. Specific Examples

3.2.1. *Poor jet-producing quality of thin, cool accretion disks*

We first consider standard, Shakura and Sunyaev (1973) disks around Kerr and Schwarzschild black holes. (For more details see Meier, 2001b.) With $\Omega = \Omega_d = \Omega_{K(epler)}$ at $R \approx 7\,r_g$ in the Schwarzschild case, and $\Omega = \Omega_m + \Omega_K$ at $R \approx 2\,r_g$ in the Kerr case

$$L_{jet}^{Sch-SS} = 10^{35.4}\,\mathrm{erg\,s^{-1}}\,m_1^{0.9}\,\dot{m}_{-1}^{1.2} \tag{3}$$

$$L_{jet}^{Kerr-SS} = 10^{36.3}\,\mathrm{erg\,s^{-1}}\,m_1^{0.9}\,\dot{m}_{-1}^{1.2}\,(1 + 1.1j + 0.29j^2) \tag{4}$$

where m_1 is the black hole mass in $10M_\odot$ units, $\dot{m}_{-1}$ is the accretion rate in units of the Eddington rate of $2.2 \times 10^{-7}\,M_\odot\,yr^{-1} m_1$. The jet powers in Equations (3) and (4) are weaker by orders of magnitude than those observed in sources like GRS 1915+105 and others ($10^{39-40}\,\mathrm{erg\,s^{-1}}$). The chief reason for the predicted jet powers being so small is the weak poloidal magnetic field, due to the cool, geometrically thin ($H << R$) nature of the disks.

As several authors have pointed out, black hole spin in the thin disk case makes little difference in the jet power, regardless of whether one is comparing non-rotating systems with rotating ones ($10^{35.4}$ vs. $10^{36.4}\,\mathrm{erg\,s^{-1}}$) or the contribution of disk power *vs.* hole power within the same system ($10^{36.5}$ vs. $10^{36.2}\,\mathrm{erg\,s^{-1}}$). The most noticeable effect of hole rotation in these cases is that the jet is produced at a smaller radius, resulting in somewhat larger field strengths and rotation rates.

3.2.2. *Importance of black hole spin and hot, thick accretion flows*

The situation is quite different when the disk is hot and geometrically thick: $B_p \sim B_\phi$ should produce much stronger jets. However, thick disks generally rotate more slowly than the Keplerian rate, so the jet power still should be rather small, *unless the metric rotates as well.*

At the present time the theory of hot, geometrically thick accretion flow is in a state of flux; new models (CDAFs, Narayan *et al.*, 2000; and ADIOSs, Begelman and Blandford, 1999) are being proposed to replace the standard advection-dominated ones (Narayan and Yi, 1995). Because the ADAFs are the most developed, they are used here to illustrate the main points, but the computations should be re-done when the theory of hot accretion flows has stabilized. Inserting ADAF solutions for $H(\sim R)$, B_ϕ, and $\Omega_d \approx \Omega_K$ into Equation (1) we obtain

$$L_{jet}^{Sch-AD} = 10^{36.5}\,\mathrm{erg\,s^{-1}}\,m_1\,\dot{m}_{-1}\,f^2 \tag{5}$$

$$L_{jet}^{Kerr-AD} = 10^{38.0}\,\mathrm{erg\,s^{-1}}\,m_1\,\dot{m}_{-1}\,(0.5f^2 + 1.5fj + 1.0j^2)\,g^2 \tag{6}$$

where $f = \Omega_d/0.4\Omega_K$ (possibly much less than unity) and $g = B_\phi/B_\phi^{AD}$ (possible greater than unity because of shear in the metric not yet taken into account). In the hot Kerr case, most of the jet power is derived from the rotational energy of the black hole, not from the released binding energy of the presently-accreting matter. (The implicit dependence on $\dot{m}$ enters through the disk magnetic field strength.) The situation for hot, thick flows, therefore, is quite different from that for cool geometrically thin accretion flows. The poloidal field is much stronger, and, when the black hole also rotates, very strong jets are produced.

4. Discussion and Conclusions

The above model explains a number of features of microquasars:

- Jet powers up to 10^{39-40} erg s^{-1} for a 10 $M_\odot$ black hole
- A suppression factor of $63\, m_1^{0.1}\, \dot{m}_{-1}^{-0.2}\, g^2$ in the jet power when the accretion transitions from a hard to a soft state (which compares well with the > 35 factor observed)
- Jet speeds of order $\Gamma_{escape} = \alpha_{lapse}^{-1} = \{[1+(1+2r_g/r)(jr_g/r)^2]/[1-2r_g/r+(jr_g/r)^2]\}^{1/2} \approx 2.4$ if the jet is ejected from the ergosphere ($r = 2r_g$) of a Kerr hole at the local escape speed.

Chief among the things the model cannot explain is why GRO J1655-40 began as a radio loud microquasar but evolved to a radio quiet one, despite repeated accretion events that appear very similar to the one that triggered the initial jet (Tavani *et al.*, 1996).

Acknowledgements

Part of this research was carried out at the Jet Propulsion Laboratory, California Institute of Technology, under contract to NASA.

References

Begelman, M.C. and Blandford, R.D.: 1999, *Mon. Not. R. Astron. Soc.* **303**, L1.
Blandford, R.D.: 1976, *Mon. Not. R. Astron. Soc.* **176**, 465.
Blandford, R.D. and Payne, D.G.: 1982, *Mon. Not. R. Astron. Soc.* **199**, 883.
Blandford, R.D. and Znajek, R.: 1977, *Mon. Not. R. Astron. Soc.* **179**, 433.
Dhawan, V., Mirabel, I.F. and Rodriguez, L.F.: 2000, *Astrophys. J.* **543**, 373.
Fender, R., *et al.*: 1999, *Astrophys. J.* **519**, L165.
Khanna, R.: 1999, in: A.F. Moz and M. Nunez (eds.), *ASP Conf. Series* **178**, p. 57.
Koide, S., Shibata, K. and Kudoh, T.: 1998, *Astrophys. J.* **495**, L63.
Koide, S., Meier, D., Shibata, K. and Kudoh, T.: 2000, *Astrophys. J.* **536**, 668.
Krasnopolsky, R., Li, Z.-Y. and Blandford, R.: 1999, *Astrophys. J.* **526**, 631.

Lind, K., Meier, D. and Payne, D.: 1994, *Bull. Am. Astron. Soc.* **184**, 40.03.
Livio, M., Ogilvie, G.I. and Pringle, J.E.: 1999, *Astrophys. J.* **512**, 100.
Lovelace, R.V.E.: 1976, *Nature* **262**, 649.
Margon, B.: 1984, *ARAstron. Astrophys.* **22**, 507.
Meier, D.: 2001, *Astrophys. J. Lett.*, in press.
Meier, D.L.: 2001, in: R. Laing and K. Blundell (eds.), *Particles and Fields in Radio Galaxies*, ASP Press, in press.
Meier, D.L. and Koide, S.: 2000, in: H. Hirabayashi, P.G. Edwards and D.W. Murphy (eds.), *Proceedings of the First VSOP Symposium*, Sagamihara, Japan.
Meier, D.L., Koide, S. and Uchida, Y.: 2001, *Science*, in press.
Meier, D., *et al.*: 1997, *Nature* **388**, 350.
Mirabel, I.F. and Rodriguez, L.F.: 1999, *Annu. Rev. Astron. Astrophys.* **37**, 409.
Narayan, R. and Yi, I.: 1995, *Astrophys. J.* **444**, 231.
Narayan, R., *et al.*: 2000, *Astrophys. J.* **539**, 793.
Ouyed, R. and Pudritz, R.: 1997, *Astrophys. J.* **482**, 712.
Shakura, N.I. and Sunyaev, R.A.: 1973, *Astron. Astrophys.* **24**, 337.
Tavani, M., *et al.*: 1996, *Astrophys. J.* **473**, L103.
Uchida, Y. and Shibata, K.: 1985, *Publ. Astron. Soc. Jpn.* **37**, 515.
Uchida, Y., *et al.*: 2000, in: P.C.H. Martens, S. Tsuruta and M.A. Weber (eds.), *ASP Conf. Series* **195**, p. 213.
Ustyugova, G., *et al.*: 1995, *Astrophys. J.* **439**, L39.

THE CONTINUOUS JETS OF CYGNUS X-1

R. SPENCER and C. DE LA FORCE
Jodrell Bank Observatory, University of Manchester

A. STIRLING, M. GARRETT, R. FENDER and R. OGLEY
University of Central Lancashire, UK; JIVE, The Netherlands; University of Amsterdam; SACLAY, Paris

Abstract. A resolved jet-like radio-emitting structure on mas scales has been found in Cygnus X-1. The jet emission is relatively smooth, and does not show the knots and blobs seen in Cygnus X-3, SS433 and GRS1915+105. This suggests that the jet production mechanism in Cyg X-1 is continuous, with the steady production of the energetic electrons and magnetic field required for the production of the radio emission. This is unlike the episodic jets seen in the other sources. The implications of these results for jet formation mechanisms will be discussed.

1. Introduction

Cygnus X-1 (V1357 Cygni) was one of the earliest stars to be detected in the radio band (Braes and Miley, 1971; Hjellming and Wade, 1971). Its radio flux density has occasionally reached levels of 50 mJy in the 1970's, however it is mostly at the 12–15 mJy level with a flat ($\alpha = 0$) spectrum to mm wave frequencies (Fender *et al.*, 2000a). The source also has long periods in the low/hard state (Brocksopp *et al.*, 1999), with transitions to the soft state being associated with changes in the radio emission. It is one of the best candidates for a black hole compact companion. However its radio structure has remained a mystery until recent VLBI observations.

About a dozen of the radio emitting X-ray binary stars have been found to possess radio jets. Some sources show superluminal motion (e.g. GRS 1915+105; Fender *et al.*, 1999), others have lower but still relativistic velocities (e.g. SS433, Paragi *et al.*, this conference). They typically have very 'blobby' structures where the ejection of knots of emission can in some cases be related to changes in X-ray state (see review by Mirabel and Rodriguez, 1999). The flat spectrum and steady radio emission in the low/hard state of Cyg X-1 suggests that its radio properties may be somewhat different from these sporadic and transient sources.

Cyg X-1 is used as part of the astrometric reference frame, and the hint of resolution effects in trans-Atlantic VLBI measurements by Lestrade (priv. comm.) led us to propose VLBA observations on this source.

Astrophysics and Space Science is the original source of publication of this article. It is recommended that this article is cited as: *Astrophysics and Space Science* **276** (suppl.): 255–258, 2001.

2. VLBA Observations

Initial observations were undertaken in March 1997 at 15 GHz. These showed the presence of a jet-like structure on the $\sim$ 1 mas scale (Stirling *et al.*, 1998). More comprehensive follow-up observations were made in August 1998, over three epochs (10, 12 and 14 August) and at two frequencies (8.4 and 15 GHz). The results found previously at 15 GHz were confirmed, though the possible existence of emission in a direction perpendicular to the jet , again on the 1 mas scale, cannot be ruled out.

The 8.4 GHz observations however show a clear one-sided jet $\sim$ 15 mas (30 au if at 2 kpc) at position angle ~ -20 degrees present at each epoch. The emission along the jet varies smoothly, and is unlike the blobby jets seen in GRS1915+105 and SS433 (Stirling *et al.*, 2000). However it should be noted that a compact jets exist in GRS 1915+105 when in the 'plateau' state (Dhawan *et al.*, 2000). The properties of steady jets produced by both persistent and transient sources in the low/hard state are discussed in Fender (2000b).

A kink in the jet (see image presented by de la Force *et al.*, this conference) stays consistent between the epochs, indicating a slow (< 3000 km s^{-1}) jet, a backward moving shock feature or precession with non-ballistic flow. The smooth emission suggests a continuous jet, as described by the models of Blandford and Königl (1979) and where the fading of brightness down the jet is caused by adiabatic expansion rather than energy losses of the radiating electrons.

3. Jet Energetics

The brightness ratio of the jet to the counter-jet region in the map exceeds 64 to 1. If we interpret this as being due to Doppler boosting of a relativistic jet then $\beta \times cos(\theta) = 0.6$ for a spectral index of 0. The inclination angle for the binary has been proposed to be at 37 degrees, so if the jet is at the same angle then $\beta = 0.78$, $\gamma = 1.59$, and $\beta_{app} = 1.25$. We would therefore expect to see an apparent motion of 110 mas/day of the components in the jet.

Assuming synchrotron radiation, a cylindrical jet, unity filling factor, and ignoring the contribution from protons, we find a minimum energy of 10^{39} ergs. If the time to form the jet is equal to the time for material to travel down the length of jet then the power in the 15 mas jet is $\sim 3 \times 10^{34}$ ergs sec^{-1}. This is rather less than the minimum power required for discrete blob ejection in GRS1915+105 (10^{39} ergs sec^{-1}). However if the flat spectrum extends to infra-red frequencies the power increases to 10^{37} ergs sec^{-1} for the whole jet, which is comparable to the bolometric X-ray luminosity in this state.

The power required for continuous jets in persistent sources may be much less than that for sporadic jets in transient sources. We might therefore expect less dramatic physical changes in persistent sources as a result, consistent with long periods in the low/hard state.

The inner core of the jet (as shown in the 15 GHz images) is $\sim 1\times < 0.5$ mas^2 in size, or 2 au in length. The equipartition magnetic field B is then ~ 300 mG, resulting in an electron lifetime, $t_{yr} = \frac{1}{(B^3 \nu_{GHz})^{0.5}}$, of 1.6 years at 15 GHz and 22 days at IR. These are very much greater than the travel times down the jet if the bulk motion is relativistic and show that the losses down the jet are primarily adiabatic. The typical energy of the electrons radiating at 15 GHz in this field is 100 Mev.

The expected magnetic field at the inner disk ($\sim 6r_g$) is 2×10^{11} G if $B \propto r^{-2}$ or 3×10^5 G if $B \propto r^{-1}$ as in the Hjellming and Johnston (1988) models. The jet power for the core region can be estimated from $L_j > \pi r^2 \Gamma^2 \beta c U_{min}$ where U_{min} is the minimum energy density. Ignoring the protons again gives 9×10^{34} ergs sec^{-1} for the inner jet, comparable with the figure found above for the jet as a whole.

Several mechanisms for the production of steady jets have been proposed. Ones which may be relevant here are the Blandford and Payne (1982) (BP) method which uses the twisting of magnetic fields generated in a disk to accelerate a jet, and the Blandford and Znajek (1977) (BZ) mechanism where the energy is extracted from a rotating black hole.

Expressions for the jet luminosity (in ergs/sec) in the two cases are:

$$BR : L = 1.6 \times 10^{38} (\frac{r_0}{r_g})^{3/2} B_8^2$$

$$BZ : L = 1.0 \times 10^{35} (\frac{J}{J_{max}})^2 (\frac{M}{M_\odot})^2 B_8^2$$

where r_0 is the inner radius of the disk, J the angular momentum of the black hole ($= J_{max}$ for a maximally rotating black hole) and B_8 the magnetic field at the inner disk radius in units of 10^8 G. For the given jet power and a maximally rotating 10 $M_\odot$ black hole we get $B = 6 \times 10^5$ G for BP and 3×10^7 G for BZ, both within range of expected fields derived earlier for the base of the jet. Other mechanisms may be important (e.g. Meier, this conference) especially if an unstable state is reached with sporadic jets produced when the system changes to high/soft state.

4. Conclusions

We have found an apparently one-sided jet ~ 30 au long in Cygnus X-1. The appearance is consistent with a Doppler boosted twin jet at 0.78c at around 40 degrees to the line of sight. A large proper motion $\gg$ beam size during the few hours of the observing runs is expected for such a velocity, so the jet image must be smeared. Multiple short snapshot observations are required to see if features do in fact move through the jet at the expected high proper motion. However these images will have lower signal to noise and have poorer mapping fidelity.

Standard jet production mechanisms for steady jets can produce the power required for a relativistic jet.

We note that the kink in the jet stays at a constant position; if this indeed indicates lack of motion along the jet then the jet is slow and must be intrinsically one-sided, and all the calculations above would have to be revised. Further observations are required to verify this feature.

It would be interesting to find the status of the jet when the source flares in the radio and in the transition to the high/soft X-ray state. Comparison with other episodic jet sources suggests that the jet should appear blobby and have a steep spectrum during this phase.

References

Blandford, R.D. and Königl, A.: 1979, *Astrophys. J.* **232**, 34.
Blandford, R.D. and Payne, D.G.: 1982, *Mon. Not. R. Astron. Soc.* **199**, 883.
Blandford, R.D. and Znajek, R.L.: 1977, *Mon. Not. R. Astron. Soc.* **179**, 433.
Braes, L.L.E. and Miley, G.K.: 1971, *Nature* **232**, 246.
Brocksopp, C., *et al.*: 1999, *Mon. Not. R. Astron. Soc.* **309**, 1063.
Dhawan, V., Mirabel, I.F. and Rodriguez, L.F.: 2000, *Astrophys. J.*, in press, astro-ph/0006086.
Fender, R.P., *et al.*: 1999, *Mon. Not. R. Astron. Soc.* **304**, 865.
Fender, R.P., *et al.*: 2000a, *Mon. Not. R. Astron. Soc.* **312**, 853.
Fender, R.P.: 2000b, *Mon. Not. R. Astron. Soc.*, in press, astro-ph/0008447.
Hjellming, R.M. and Johnston, K.J.: 1988, *Astrophys. J.* **328**, 600.
Hjellming, R.M. and Wade, C.M.: 1971, *Astrophys. J. Lett.* **168**, L21.
Mirabel, I.F. and Rodriguez, L.F.: 1999, *Annu. Rev. Astron. Astrophys.* **37**, 409.
Stirling, A.M., Spencer, R.E. and Garrett, M.A.: 1998, *New Astron. Rev.* **42**, 657.
Stirling, A.M., *et al.*: 2000, *Mon. Not. R. Astron. Soc.*, submitted.

MAGNETO-ROTATIONAL MECHANISMS OF JET FORMATION

G.S. BISNOVATYI-KOGAN
Space Research Institute, Profsoyuznaya str. 84/32, Moscow 117810, Russia
E-mail: gkogan@mx.iki.rssi.ru

Abstract. The magneto-rotational mechanism of jet formation in stationary and non-stationary events is analyzed. Collapse of a rotating magnetized body during star formation or supernovae explosion may lead to a jet-like mass ejection for certain angular velocity and magnetic field distributions at the beginning of the collapse.

1. Introduction

Matter outflow is observed in most astrophysical objects in the form of stellar wind, or collimated outflow (ejection) from young stellar objects, AGN and quasars, microquasars (galactic X-ray sources). Mechanisms of mass loss are connected with a radiative and/or electromagnetic acceleration. The last one, which is probably the best for producing collimated outflows, may work quasi-stationary, or may be connected with explosive events. Rotation is always present in compact objects, and formation of collimated jets results from the action of magneto-rotational phenomena.

2. Problem of Collimation

During accretion and outflow of matter the relative dynamical action of magnetic field increases for a given element of matter. As was shown by Schwartzman (1971), the conservation of a magnetic flux during stationary accretion implies a dependence $B_r \sim r^{-2}$. At constant mass flux $\dot{M} = 4\pi\rho v_r r^2$ and free-fall velocity $v_r \sim r^{-1/2}$, the density increases as $\rho \sim r^{-3/2}$, and kinetic energy density $E_k \sim r^{-5/2}$. The growth of the magnetic energy density is faster $E_m \sim B_r^2 \sim r^{-4}$. After equipartition $E_k \sim E_M$ is reached, it cannot be violated during subsequent accretion.

The outflow from rotating magnetized object contain azimuthal component of the magnetic field, which energy density decreases not faster than the specific kinetic energy of matter. For the outflow with constant velocity v_r in stationary spherical outflow the density $\rho \sim r^{-2}$, $B_\phi \sim r^{-1}$. In this condition $E_k \sim E_m$, and the trajectories of mass outflow become more tightly spiraled, and finally the

Astrophysics and Space Science is the original source of publication of this article. It is recommended that this article is cited as: *Astrophysics and Space Science* **276** (suppl.): 259–262, 2001.

flow is becoming highly collimated with direction of the flow along the rotational axis (Heyvaerts and Norman, 1989; Bogovalov, 1998). We may say here about the universal magnetic collimation in the outflows in vacuum from rotating magnetized objects.

The outbursts may be collimated at the very beginning. When jets are separated from the object of its origin the problem appears of preservation of the jet against its spherization during a motion in a rarefied medium. One of the plausible mechanism of jet preservation is a magnetic pinch collimation produced by an axial electrical current, suggested by Bisnovatyi-Kogan *et al.* (1969).

3. Jet Formation by Matter Outflow from Magnetized Accretion Disk

Accretion of matter with a large scale magnetic field into a black hole leads to formation of an accretion disk with a strong poloidal magnetic field. Models of non-rotating accretion disks, supported by magnetic field had been constructed by Bisnovatyi-Kogan and Ruzmaikin (1974, 1976). Rotation of matter in such a disk is accompanied by generation of a strong electrical field, leading to particle acceleration and ejection of matter along magnetic field lines. Qualitative and phenomenological models had been considered by Bisnovatyi-Kogan and Blinnikov (1976), Blandford (1976), Lovelace (1976). Analytical self-similar solutions for jets and winds produced by centrifugally and magnetically driven mechanisms had been obtained by Blandford and Payne (1982), Lovelace *et al.* (1991),

Extensive numerical simulations have been done for construction of more realistic models of quasi-stationary jet formation from magnetized accretion disks. In the paper of Romanova *et al.* (1998) have been done numerical simulations of dynamics of magnetic loops in the coronae of accretion disks. It was obtained that in presence of differential rotation the loops are opened in the inner parts of the disk, and opening of loops is followed by magnetically driven outflow from the disk. The outflow may be transient, consisting of small-scale outbursts from different loop opening, or may be steady, corresponding to outflow along the open field lines at the inner part of the disk, and showing some collimation. Computations of magnetocentrifugally driven winds had been performed by Ustyugova *et al.* (1999). The stationary regime of the outflow was obtained by solution of time-dependent MHD equations, with a split-monopole poloidal field configuration frozen into the disk. Close to the disk the outflow is driven by by the centrifugal force, while at all larger distances the flow is driven by the magnetic force, which is proportional to $-\nabla(rB_\phi)^2$, where B_ϕ is the toroidal field. The collimation distance over which the flow becomes collimated is much larger than the size of the sumulation region, so the obtained outflows are approximately spherical.

MHD simulation have been performed by Ustyugova *et al.* (2000) of Pointing outflows, where the mass flux is negligible and energy and angular momentum are carried predominantly by electromagnetic field in the form of MHD waves.

As a result of time-dependent simulations a quasi-stationary collimated Pointing jet arises from the inner part of the disk and a steady uncollimated hydromagnetic outflow from the outer disk for a case Keplerian disk initially threaded by a dipole like poloidal magnetic field.

4. Magnetorotational Explosions

A mechanism of producing magnetorotational explosion was suggested by Bisnovatyi-Kogan (1970), where amplification of the toroidal component of the magnetic field due to differential rotation leads to the transformation of a part of the rotational energy to the energy of the ejection. First numerical simulations devoted to the magnetorotational mechanism were made by Le Blanc and Wilson (1970). and investigated analytically by Meier et.al. 1976.

One of the most important parameters for this problem is relation of magnetic energy to the gravitational energy of the star: $\alpha = E_{\rm mag}/|E_{\rm grav}|$. Simulations have been done by Ardeljan *et al.* (2000). for the initial values of the $\alpha = 10^{-2}$, 10^{-4}, 10^{-6}. Details of the definition of the initial magnetic field are described in the paper by Ardeljan *et al.* (2000).

Initial magnetic field chosen for our simulations has quadrupole-like kind of symmetry (i.e. its 'z'-component is equal to zero at the equatorial plane). As a result an ejection predominantly in the equatorial plane was obtained. The toroidal magnetic field grows with time and produces MHD shock which moves to the boundary of the star. Part of the matter of the envelope of the star (about 7% of the mass of the star) has radial kinetic energy larger than its potential energy and can be ejected. This ejected matter carries about 3.3% of the total energy of the star (see also poster PV-06). When the initial magnetic field have a dipole-like symmetry jet formation is possible as was shown in calculations of Le Blanc and Wilson (1970).

There is a possibility of asymmetric MHD explosion, when we have asymmetric magnetic field amplification. One example of development of asymmetric picture was considered by Bisnovatyi-Kogan and Moiseenko (1992). When the collapsing and exploding star has initially toroidal and poloidal fields of different symmetry: dipole poloidal and symmetric poloidal, or quadrupole poloidal and antisymmetric toroidal fields, formation of an additional toroidal field from the existing poloidal due to differential rotation leads to spontaneous breaking of the symmetry. Due to the asymmetry of the toroidal field during the explosion the outbursts could also become asymmetric, giving observational asymmetric or even one-side jets often observed in extragalactic jets, and may be in microquasars (Mirabel *et al.*, 1992; Fender, 1999).

Acknowledgements

I am grateful to Dr A.J. Castro-Tirado for hospitality, RFBR for partial support by grants 98-01-00352, 99-02-18180 and ESA-INTAS grant No.120.

References

Ardeljan, N.V., Bisnovatyi-Kogan, G.S., Moiseenko, S.G.: 2000, *Astron. Astrophys.* **355**, 1181.
Bisnovatyi-Kogan, G.S.: 1970, *Astron. Zh.* **47**, 813.
Bisnovatyi-Kogan, G.S. and Blinnikov, S.I.: 1976, *Pisma Astron. Zh.* **2**, 489.
Bisnovatyi-Kogan, G.S., Komberg, B.V. and Friedman, A.M.: 1969, *Astron. Zh.* **46**, 465.
Bisnovatyi-Kogan, G.S. and Moiseenko, S.G.: 1992, *Astron. Zh.* **69**, 563.
Bisnovatyi-Kogan, G.S. and Ruzmaikin, A.A.: 1974, *Astrophys. Space Sci.* **28**, 45.
Bisnovatyi-Kogan, G.S. and Ruzmaikin, A.A.: 1976, *Astrophys. Space Sci.* **42**, 401.
Blandford, R.D.: 1976, *Mon. Not. R. Astron. Soc.* **176**, 465.
Blandford, R.D. and Payne, D.G.: 1982, *Mon. Not. R. Astron. Soc.* **199**, 883.
Bogovalov, S.V.: 1998, *Pisma Astron. Zh.* **24**, 381.
Fender, R.P.: 1999, in: *Astrophysics and Cosmology: A collection of critical thoughts*, Springer Lectures Notes in Physics, in press (astro-ph/9907050).
Heyvaerts, J. and Norman, C.A.: 1989, *Astrophys. J.* **347**, 1055.
Le Blanc, L.M. and Wilson, J.R.: 1970, *Astrophys. J.* **161**, 541.
Lovelace, R.V.E.: 1976, *Nature* **262**, 649.
Lovelace, R.V.E., Berk, H.L. and Contopulos, J.: 1991, *Astrophys. J.* **379**, 696.
Mirabel, I.F., Rodriguez, L.F., Cordier, B., Paul, J. and Lebrun, F.: 1992, *Nature* **358**, 215.
Meier, D.L., Epstein, R.I., Arnett, W.D. and Schramm, D.N.: 1976, *Astrophys. J.* **204**, 869.
Romanova, M.M., Ustyugova, G.V., Koldoba, A.V., Chechetkin, V.M. and Lovelace, R.V.E.: 1998, *Astrophys. J.* **500**, 703.
Schwartzman, V.F.: 1971, *Astron. Zh.* **48**, 479.
Ustyugova, G.V., Koldoba, A.V., Romanova, M.M., Chechetkin, V.M. and Lovelace, R.V.E.: 1999, *Astrophys. J.* **516**, 221.
Ustyugova, G.V., Lovelace, R.V.E., Romanova, M.M., Li, H. and Colgate, S.A.: 2000, *Astrophys. J.* **516**, 221.

MHD JETS AROUND GALACTIC OBJECTS

FABIEN CASSE and JONATHAN FERREIRA
Laboratoire d'Astrophysique de l'Observatoire de Grenoble,
BP 53 F-38041 Grenoble cedex 9, France
E-mail: Fabien.Casse@obs.ujf-grenoble.fr

Abstract. We present a self-similar, steady-state model describing both a magnetized accretion disc and a magnetohydrodynamic jet. We focus on the role of a hot corona in such a structure. This corona enables the disc to launch various types of jets. By considering the energy conservation, we also present a diagnostic of the luminosity of the magnetized disc, which could explain some observational signatures of galactic objects.

1. The Disc-Jet Connection

In the universe, jets of matter are observed around almost every kind of systems where plasma accretes onto a central object. The central object can be a young protostar, a compact galactic object or a supermassive black hole. The jet velocity is always found to be a few times the escaping velocity of the central object (e.g., Livio, 1997). The most successful model, accounting for all known properties of observed jets, relies on a large scale magnetic field anchored on an accretion disc (Blandford and Payne, 1982). The field extracts the disc angular momentum, thereby allowing accretion, and accelerates a fraction of disc material to high speeds. Moreover, the magnetic field confines the accelerating jet matter, leading to a cylindrical collimation of the jet. Hereafter, such systems are called Magnetized Accretion-Ejection Structures (MAES).

In a MAES, the bipolar magnetic field pinches *the disc*. Thus, the only force able to counteract both the magnetic and gravitational forces, *whithin the disc*, is the vertical plasma pressure gradient. As a consequence, the mass loading in the jet depends critically on this vertical equilibrium. This equilibrium is the connection between the accretion motion and the jet. A more detailed explanation of this phenomenon is displayed in Casse and Ferreira (2000a).

2. Entropy Generation

In all previous studies dealing with magnetized accretion disc steadily launching jets, the energy equation was simply disregarded. It was replaced by the assumption of isothermal or adiabatic magnetic surfaces. In the turbulent disc, however, some

Astrophysics and Space Science is the original source of publication of this article. It is recommended that this article is cited as: *Astrophysics and Space Science* **276** (suppl.): 263–266, 2001.

physical mechanisms (convection, turbulent thermal conductivity) could provide a non-local energy transport, more efficiently than by photons. We can mimic these effects by the following equation

$$\rho T \frac{dS}{dt} = Q$$

where Q is the difference between local heating and cooling rates of the plasma. This function Q is prescribed according to the energy conservation (Casse and Ferreira, 2000b). In a steady-state framework, this equation can be written as

$$Q = \frac{\gamma}{\gamma - 1} \frac{k_B}{m_p} \rho \vec{u_p} \cdot \vec{\nabla} T - \vec{u_p} \cdot \vec{\nabla} P .$$

where T is the temperature, P the plasma pressure, u_p the poloidal velocity of matter and γ the polytropic index of plasma. Because the gradient of plasma characteristics are almost vertical in a thin disc, one can see that entropy generation will have a strong impact on the vertical mass flux when the poloidal velocity is vertical (i.e. in the corona taking place at the disc surface). In the disc, the entropy generation will be advected with the flow onto the central object.

This is an important point since advection-dominated accretion flow (ADAF) only occurs in thick or slim discs ($h/r > 0.3$). Here, most of the entropy is generated in the disc atmosphere and taken in the jet, allowing the disc to be thin. Using this approximate form of the energy equation, we solve the full set of magnetohydrodynamics (MHD) equations from the disc midplane (resistive MHD regime) to the jet asymptotics (ideal MHD regime).

3. Global Energy Conservation

The global energy conservation enables us to know that the power released by accretion P_{acc} is shared between the total jet power P_{jet} (MHD Poynting flux, kinetic energy and enthalpy) and the disc luminosity L_D. We measure the amount of entropy generated with a parameter f whose value is determined by the disc turbulence ($0 < f \leq 1$). It is noteworthy that, according to the amount of entropy, the resulting jet is magnetically-driven ('cold' jet $f \sim 0$) or magneto-thermally-driven ('hot' jet $f \sim 1$). The energy conservation becomes (Casse and Ferreira, 2000b)

$$2L_D = (1 - f)P_{diss} = P_{acc} - 2P_{jet}$$

where P_{diss} is the viscous and ohmic heating integrated all over the disc. One can easily see that if most of the heat is deposited as entropy in the corona, a thin magnetized disc could launch jets and have a weak luminosity. Assuming the disc

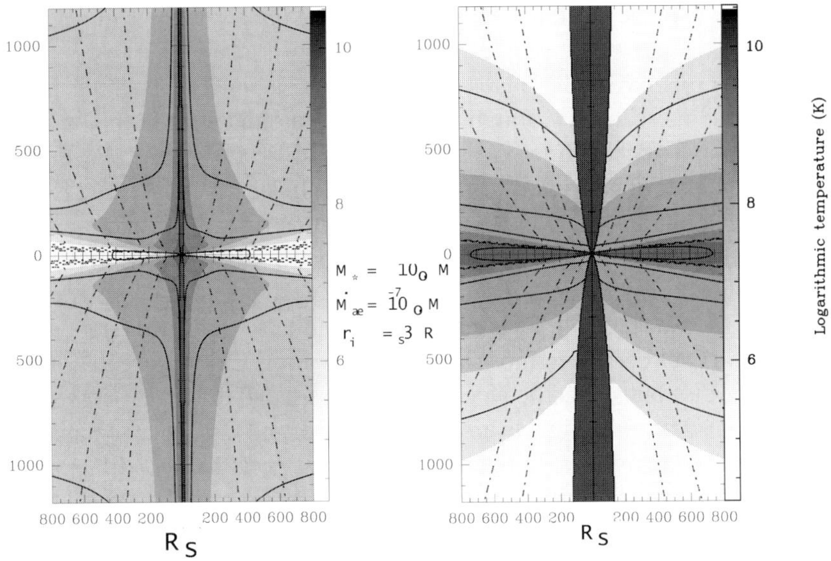

a. X-ray low/hard state ($f \rightarrow 1$)
Low disc emission & hot corona
Dense hot emitting jet

b. X-ray high/soft state ($f \rightarrow 0$)
Disc radiation & no coronal emission
Tenuous, cold and low emitting jet

Figure 1. Poloidal cross-section of MAES. In grey are shown the logarithmic temperature levels, the solid lines are the logarithmic plasma density levels ($\log n_H$ = 12, 12.5, 13, 15 in cm^{-3} for left pannel and $\log n_H$ = 9, 10, 11, 15 for right pannel). The denser level is closer to the disc. The dashed lines represent plasma streamlines coming from the disc to the jet. The poloidal magnetic field lines are not displayed here but are strictly parallel, in the jet, to the poloidal streamlines, according to MHD equations. The left figure displays a structure that might be able to describe the X-ray low/hard state of some microquasars (see Fender *et al.*, 1999). The right pannel could correspond to a X-ray high/soft state.

to be optically thick, we can calculate its luminosity L_D and effective temperature T_{eff}, namely

$$\frac{L_D}{L_{EDD}} \simeq 0.36 \left(\frac{1-f}{1+\Lambda}\right)\left(\frac{\dot{M}_{ae}}{10^{-7}M_\odot/yr}\right)\left(\frac{M_*}{10M_\odot}\right)\left(\frac{3R_s}{r_i}\right)$$

$$kT_{\text{eff}} \simeq 1\text{keV}\left(\frac{1-f}{1+\Lambda}\right)^{\frac{1}{4}}\left(\frac{\dot{M}_a(r)}{10^{-7}M_\odot/yr}\right)^{\frac{1}{4}}\left(\frac{M_*}{10M_\odot}\right)^{-\frac{1}{2}}\left(\frac{r}{3R_s}\right)^{-\frac{3}{4}}$$

where we have assumed the external radius r_e of the MAES to be much larger than the internal radius r_i ($R_s = 2GM_*/c^2$ and L_{EDD} is the Eddington luminosity). The parameter Λ is the ratio of the magnetic to the viscous torque at the disc midplane and depends on the local MHD turbulence (Casse and Ferreira, 2000a).

An important consequence arises from these estimates. Indeed a magnetized disc driving jets could be less luminous than a standard accretion disc, even with $f \simeq 0$ (no corona). This will be the case if the turbulence is such that the magnetic torque is larger than the viscous one, the disc will be darker but keeping almost the same effective temperature.

In a stationary MAES, the characteristic timescale for jet launching is only a few dynamical timescales, i.e. roughly one second around a stellar black hole. So both low and high state of microquasars (e.g. Mirabel *et al.*, 1998, Fender *et al.*, 1999) might be described by two different MAES. Indeed the high state exhibits a strong thermal emission in X-rays (soft X-rays emitted from the disc, $f \rightarrow 0$) and a very small radio emission that could be accounted by a very tenuous, cold jet (or no jet at all). At the opposite, the low state is characterized by a non-thermal emission in hard X-rays (associated with a corona) and a 'radio' jet. This low state might be obtained with a MAES where most of the heat is not released as radiation by the optically thick disc but shared between coronal losses and thermal energy sent into the jet ($f \rightarrow 1$). The resulting jets are hot and dense, and may be associated with a radio and infra-red emissions. In Figure 1, we display these two MAES. Anyway, a disc instability allowing the MAES to shift between these two states remains to be found. Also, to clearly validate this scenario, a modelisation of spectrum emitted from MAES must be performed and compared to observations.

Acknowledgement

The authors would like to thank Rob Fender for fruitful discussions.

References

Blandford, R.D. and Payne, D.G.: 1982, *Mon. Not. R. Astron. Soc.* **199**, 883.
Casse, F. and Ferreira, J.: 2000a, *Astron. Astrophys.* **353**, 1115.
Casse, F. and Ferreira, J.L.: 2000b, *Astron. Astrophys.* **361**, 1178.
Fender, R., *et al.*: 1999, *Astrophys. J.* **519**, L165.
Livio, M.: 1997, in: Wickramasinghe, T., Ferrario, L. and Bicknel, G.V. (eds.), *The formation of astrophysical jets, ASP Conf. Ser.* **121**, p. 845.
Mirabel, I.F., *et al.*: 1998, *Astron. Astrophys.* **330**, L9.

ON THE FORMATION OF ACCRETION-POWERED GALACTIC AND EXTRA-GALACTIC OUTFLOWS

TAPAS K. DAS
IUCAA Post Bag 4 Ganeshkhind Pune 411 007, India; E-mail: tapas@iucaa.ernet.in

Abstract. Though widely observed to be emanating from a variety of astrophysical sources, the underlying physical mechanism behind the formation of galactic and extragalactic outflows is still enshrouded in a veil of mystery. In addition, it has not been possible to calculate accurately the amount of matter expelled in these events. In this article we present a *non-self-similar analytical model*, which, for the first time, we believe is able to explain the outflow formation phenomenon as well as compute the mass outflow rate by simultaneously solving the equations governing the exact transonic accretion and outflow. Our model predicts the dependence of this rate on various flow parameters as well as indicates the *exact* location from where the outflows are launched.

1. Introduction

It is now a well-established fact that quasars and microquasars suffer massloss through outflows and jets (Mirabel and Rodriguez, 1999; Ferrari, 1998; Begelman, Blandford and Rees, 1984). These galactic and extra-galactic jet sources are commonly believed to harbour accreting compact objects at their hearts as the prime movers for almost all non-stellar energetic activities around them including the production of bipolar outflows and relativistic jets. Unlike normal stellar bodies, compact objects do not have their own physical atmosphere from where matter could be ripped off as winds, hence outflows from the vicinity of these prime movers *have to be generated only from the accreting material.* Instead of separately investigating the jets and accretion processes as two disjoint issues around the dynamical centre of the galactic and extra-galactic jet sources, it is absolutely necessary to study these two phenomena within the same framework and any consistent theoretical model for jet production should explore the outflow formation only from the knowledge of accretion parameters. Note that while self-similar models are a valuable first step, they can never be the full answer, and indeed any model which works equally well at all radii is fairly unsatisfactory to prove its viability. Thus the preferred model for jet formation must be one which is able to select the *specific* region of jet formation. Keeping these basic facts in our mind, we propose a *non-self-similar analytical model* capable of self-consistently exploring the hydrodynamic origin of accretion powered jets/outflows emanating from galactic and extragalactic sources. Using this model it has been possible to simultaneously solve the accretion and wind equations to compute (from first prin-

Astrophysics and Space Science is the original source of publication of this article. It is recommended that this article is cited as: *Astrophysics and Space Science* **276** (suppl.): 267–272, 2001.

ciple) what fraction of accreting material is blown as wind and the exact location (distance from the central accretor) of the jet launching zone has been successfully pointed out.

Denoting $\dot{M}_{in}$ and $\dot{M}_{out}$ to be the instantaneous time rate of inflow and outflow around an accreting compact object, we define a quantity 'Mass Outflow Rate ($R_{\dot{M}}$)' as the ratio of $\dot{M}_{out}$ to $\dot{M}_{in}$, which is basically a measure of the fraction of barionic accreting material being expelled as outflows/jets. Our major aim was to compute the exact value of $R_{\dot{M}}$ in terms of *minimum* number of accretion parameters and to investigate the dependence of $R_{\dot{M}}$ on various flow parameters. We do our calculation for accretion with considerable intrinsic angular momentum (Disc-Outflow system, see section 2 for detail) and for spherical/quasi-spherical accretion with negligibly small angular momentum (see section 3 for detail).

Because of the absence of any matter donating real physical atmosphere around accreting compact objects, we first need to incorporate some virtual surfaces around these objects from where outflows may be generated. Formation of these surfaces are explained in section 2.1 (for the disc-outflow system) and in section 3.1 (for spherical/quasispherical accretion-outflow system).

In this article, we are basically interested in highlighting the conceptual ideas behind our model rather than presenting the mathematical details, thus we keep this article free from any formulae or equations. Interested readers may kindly refer respective works (Das, 1998, 1999, 1999a, 2000, 2000a; Das and Chakrabarti, 1999 (DC99 hereafter)) for detail mathematical formalism.

2. The Disc-Outflow System

2.1. Model Description and Prescription for the Formation of Outflow Generating Surface

We consider thin, axisymmetric polytropic inflows around a Scwarzschild Black hole in vertical equilibrium. We ignore the self-gravity of the flow and calculations are done using Paczyński-Wiita (Paczyński and Wiita, 1980) potential which mimics surroundings of the Schwarzschild black hole. Considering the inflow to be polytropic, we explore both the polytropic and the isothermal outflow.

Due to the fact that close to the BH the radial component of the infall velocity of accreting material would be enormously high, viscous time scale would be much longer than the infall time scale and a rotating inflow entering into a black hole will have almost constant specific angular momentum close to the black hole for any moderate viscous stress (Das, 1998 and DC99). Though at the outer edge of the accretion disk the angular momentum distribution may be Keplerian or even super-Keplerian, matter would be highly sub-Keplerian close to the black hole to satisfy the inner boundary condition at the event horizon of the BH (see Chakrabarti, this volume). This almost constant angular momentum produces a very strong centrifugal force which increases much faster compared to the gravitational force and becomes comparable at some specific radial distance, location of which is easy to compute. Here, (actually, a little farther out, due to thermal pressure) matter starts piling up and produces the centrifugal pressure supported boundary layer (CENBOL). Further close to the black hole, the gravity always wins and matter enters the horizon supersonically after passing through a sonic point. Formation of CENBOL may be attributed to the shock formation in accreting fluid or to the maximisation of polytropic pressure of the inflow. In CENBOL region the flow becomes hotter and denser and for all practical purposes behaves as the stellar atmosphere so far as the formation of outflows are concerned. A part of the hot and dense shock-compressed inflowing material is then 'squirt' as outflow from the CENBOL. In case where the shock does not form, regions around pressure maximum achieved just outside the inner sonic point of the *inflow* would also drive the flow outwards. The outflow is shown to be thermally and centrifugally accelerated but is assumed to be confined by external pressure of the ambient medium. Subsonic outflows originating from CENBOL would pass through sonic points and reach far distances as in wind solution.

2.2. THE OVERALL SOLUTION SCHEME

Let us suppose that matter first enters through the outer sonic point and passes through a shock. At the shock, part of the incoming matter, having higher entropy density is likely to return back as winds through a sonic point, other than the one it just entered. Thus a combination of topologies, one from the region of accretion and the other from the wind region is required to obtain a full solution. In the absence of the shocks, the flow is likely to bounce back at the pressure maximum of the inflow and since the outflow would be heated by photons, and thus have a smaller polytropic constant, the flow would leave the system through an outer sonic point different from that of the incoming solution. By simultaneously solving the proper set of equations in appropriate geometry (see Das, 1998, DC99 and Chapter 2.1 and 2.2 of Das, 2000a for detail solution scheme and flow geometry), we get the *combined* flow topologies (Figure 2 of DC99) where the value of $R_{\dot{M}}$ along with *all* other outflow parameters can be exactly computed only from the knowledge of the *minimum number of* accretion parameters namely specific energy ($\mathcal{E}$), specific angular momentum (λ), accretion rate (scaled in the unit of Eddington rate) ($\dot{M}_{in}$) and adiabatic indices (γ) of the flow. Also the dependence of $R_{\dot{M}}$ on *all possible* flow parameters has been investigated self-consistently (see Das, 1998 and DC99 for detail).

3. Outflows from Spherical/Quasi-Spherical Accretion

3.1. The outflow generating surface

For this class of accretion, absence of angular momentum rules out the possibility of formation of CENBOL. A novel mechanism is present in the literature (Kazanas and Ellison, 1986; Protheroe and Kazanas, 1983) where the kinetic energy of spherically accreting material has been randomized by proposing a steady, standing, collisionless, relativistic hadronic pressure supported spherically symmetric shock around a Schwarzschild black hole which produces a nonthermal spectrum of high energy (relativistic) protons. A fraction of the energy flux of infalling matter is assumed to be converted into radiation at the shock standoff distance through hadronic ($p - p$) collision and mesonic ($\pi^{\pm}, \pi^{0}$) decay. Pions generated by this process, decay into relativistic electrons, neutrinos/antineutrinos and produces high energy γ rays. These electrons produce the observed non-thermal radiation by Synchrotron and inverse Compton scattering. Shock accelerated relativistic protons are not readily captured by the black hole rather considerable high energy density of these relativistic protons would be maintained to make this shock self-supported (Protheroe and Kazanas, 1983). In this work, we take this pair-plasma pressure mediated shock surface as the alternative of the CENBOL from where the outflow could be launched.

Here we consider that a Schwarzschild type black hole quasi-spherically accretes fluid obeying polytropic equation of state. We also assume that for our model, the effective thickness of the shock is small enough compared to the shock standoff distance. We investigate polytropic as well as isothermal outflows from polytropic accretion.

3.2. The overall solution scheme

The *exact* value for the shock location and all relevant pre- and post-shock quantities can be computed in terms of $\mathcal{E}$, $\dot{M}_{in}$ and γ. Now using the the sloution scheme as described for the disc-outflow system, the set of equations are simultaneously solved in proper geometry to get the combined flow topology for polytropic and isothermal outflows (see Das, 1999, 1999a, 2000 for detail solution scheme and flow geometry) where $R_{\dot{M}}$ along with *all other* outflow parameters was calculated and its dependence on all flow parameters has been studied.

It is to be noted that along with a pseudo-Newtonian treatment discussed in section 3.1, accretion-powered spherical wind has also been studied very recently using a full general relatavistic framework. Equations governing the general relatavistic, transonic, spherically symmetric accretion has been explicitly solved in Schwarzschild metric (Das, 2001a) and the connection between the inflow and the outflow is then established (Das, 2001b).

4. Some Important Results and Directions for Future Work

4.1. Explanation of Quiescent States of X-ray Novae

An interesting situation arises in our model when the polytropic index of the outflow is large and the compression ratio is also very high. In this case, the flow virtually bounces back as the wind and the outflow rate can be equal to the inflow rate or even higher, thereby evacuating the disk completely (see Figure 6 of Das, 1998). These cases can cause runaway instabilities by rapidly evacuating the disk. It is possible that some of the black hole systems, including that in our own galactic centre, may have undergone such evacuation phase in the past and gone into quiescent phase. Thus our model could explain the quiescent states in X-ray novae systems like GS 2000+25 or GRS 1124-633 etc. (Tanaka, 1995; Ueda *et al.*, 1994) and also in some systems with massive black holes, especially the black hole at our galactic centre.

4.2. Spectral Properties of Our Galactic Centre

We suggest that (Das, 1998; Das, 2000a, DC99) a possible explanation for extreme low luminosity and low radiative efficiency (upper limit on the mass accretion rate of $SgrA^*$ is of the order of $\sim 8 \times 10^{-5}\ M_\odot\ Yr^{-1}$, Bondi accretion rate on it has been approximated as $\sim 3 \times 10^{-5}\ M_\odot\ Yr^{-1}$ (Quarteart, Narayan and Reid, 1999 and references therein)) could be due to the presence of profuse mass loss from the vicinity of $SgrA^*$. We found that for such low accretion rates the mass outflow rate is exorbitantly high, almost to the point of evacuating the disk, which prompted us to strongly speculate that the spectral properties of our galactic centre could be explained by inclusion of wind using our model. However, possibility that outflow in an accretion disk wind might explain the low accretion rate at the Galactic Center also has been suggested by Blandford and Begelman (1999).

4.3. Outflow Driven Contamination of Metalicity to the Outer Galaxies

A number of observational evidences suggest that the fluid accreting onto black hole has potential to generate appropriate temperature which supports significant nucleosynthesis to take place in accretion disks around black holes (Mukhopadhyaya and Chakrabarti, 1999 and references therein). It is interesting to investigate whether the fate of the shock induced nucleosynthesis generated heavier elements could be predicted by our disk-outflow model. One of the major speculations of our model (Das, 1998, DC99 and §5.1 of Chapter 2.2 of Das, 2000a) that outflows from the hot and dense CENBOL (where the composition change is much more significant) would carry away modified compositions and contaminate the atmosphere of the surrounding stars and galaxies in general. Strong indications of disk-evacuation by wind for some region of parameter space suggests (Das,

1998, DC99) that overall such contributions to metalicity must not be neglected. Significant work in this direction is in progress and is expected in near future.

Acknowledgements

I would like to thank the LOC of this workshop, and my special thanks goes to Dr A.J. Castro-Tirado for kindly providing the financial assistance for attending the workshop.

References

Begelman, M.C., Blandford, R.D. and Rees, M.J.: 1984, *Rev. Mod. Phys.* **56**, 255.
Blandford, R.D. and Begelman, M.C.: 1999, *Mon. Not. R. Astron. Soc.* **303**, L1.
Das, T.K.: 1998, in: S.K. Chakrabarti (ed.), *Observational Evidence for Black Holes in the Universe*, Kluwer Academic, Holland, p. 113.
Das, T.K.: 1999, *Ind. J. Phys.* **73(B)**, 1,1.
Das, T.K.: 1999a, *Mon. Not. R. Astron. Soc.* **308**, 20.
Das, T.K.: 2000, *Mon. Not. R. Astron. Soc.* **318**, 294.
Das, T.K.: 2000a, Ph. D. Thesis.
Das, T.K.: 2001a, submitted to *MNRAS*.
Das, T.K.: 2001b, submitted to *A&A*.
Das, T.K. and Chakrabarti, S.K.: 1999, *Class. Quantum Grav.* **16**, 3879 (DC99).
Ferrari, A.: 1998, *Annu. Rev. Astron. Astrophys.* **36**, 539.
Kazanas, D. and Ellison, D.C.: 1986, *Astrophys. J.* **304**, 178.
Mirabel, I.F. and Rodriguez, L.F.: 1999, *Annu. Rev. Astron. Astrophys.* **37**, 409.
Mukhopadhyay, B. and Chakrabarti, S.K.: 2000, *Astron. Astrophys.* **353**, 1029.
Paczyński, B. and Wiita, P.J.: 1980, *Astron. Astrophys.* **88**, 23.
Protheroe, R.J. and Kazanas, D.: 1983, *Astrophys. J.* **256**, 620.
Tanaka, Y.: 1995, *Nature* **375**, 659.
Ueda, Y., Ebisawam K. and Done, C.: 1994, *Publ. Astron. Soc. Jpn.* **46**, 107.

JET FORMATION AND DYNAMICS: COMPARISON OF QUASARS AND MICROQUASARS

WOLFGANG KUNDT
IfA of Bonn University, Auf dem Hügel 71, D-53121 Bonn

Abstract. Quasars and Microquasars share the following properties: (i) They have similar, elongated morphologies – reminiscent of being driven by supersonic beams – consisting of cores, knots, and heads, with jet-opening angles $\lesssim 10^{-2}$, and no beam branching; (ii) core/lobe power ratios of $10^{2\pm2}$; (iii) fluctuating, broad and hard core spectra; (iv) (occasional) sidedness; (v) (occasional) superluminal growth.

In all cases, the central engine is thought to be a rotating magnet whose reconnecting magnetic fields generate the relativistic pair plasma – of typical Lorentz factor $10^{3\pm2}$ – which rams the jet channels and blows the cocoons (subsonically) after having been stalled in a head. The supersonic jets form on passing a central deLaval nozzle, first proposed by Blandford and Rees in 1974, which forms naturally due to the huge density contrast of $10^{-8.3}T_4$ with respect to the ambient medium (of temperature T, $T_4 := T/10^4$ K). Beam stability and narrowness are likewise guaranteed by the density contrast (of jet fluid and CSM). Observed are both the (thermal) radiation of the rammed channel-wall material, and the synchrotron radiation of the deflected beam particles.

1. Properties of the Jet Sources

At this workshop, an opinion has been repeatedly expressed – e.g. by Spencer and by Durouchoux in their discussions – that the obvious analogy between the (extragalactic) quasars and the (stellar) microquasars was perhaps more fundamental than expected from their modeling which allows for quite different *central engines*: young stellar objects, neutron stars, stellar-mass black holes or supermassive black holes, and quite different *jet compositions*: relativistic pair plasma (Kundt and Gopal-Krishna, 1980; Begelman *et al.*, 1994), focussed stellar winds or fired bullets (plasmons). Kaiser has pointed out difficulties of the plasmon model for jets. Kluzniak and Merloni have highlighted details of the X-ray power spectrum of black-hole candidates (BHCs) which prefer an interpretation via a neutron star (with a well-defined spin frequency, and hard surface). Fender arrived at marginally tolerable beam energetics – using the measured spectra – by making the minimum assumption of pure pair plasma. And in his numerical modeling of jets, Meier had increasing success with an increasing Lorentz factor of the beam particles.

All these results agree with the work by Gopal-Krishna and myself (1980), also by Blome and myself (1988), in which we conclude that the beam fluid must be relativistically light (and magnetized at pressure balance) in order for the jets to

Astrophysics and Space Science is the original source of publication of this article. It is recommended that this article is cited as: *Astrophysics and Space Science* **276** (suppl.): 273–277, 2001.

stay narrow, multiply focussed, and never branching and in order to conserve 4-momentum and particle number in the heads of the flows. Independently, I have never seen a robust mechanism of *jet formation* proposed other than the buoyant escape of some quasi-weightless fluid out of the deep potential well of the central engine. Also, I have repeatedly argued that the *stellar-mass BHCs* are binaries containing a neutron star inside a massive accretion disk ($M \approx 5\ M_\odot$) (Kundt and Fischer, 1989; Kundt, 1998b, 1999). Black holes are in disharmony with:

- (i) the hard spectra (up to γ-ray energies),
- (ii) the (X-ray) luminosity maxima and minima,
- (iii) quasi periods,
- (iv) (noisy and asymmetric) lightcurves and line velocities,
- (v) pair-formation capability, and with various further details.

Microquasars received their name because their properties are quite similar to those of the quasistellar radio sources (quasars) and because their powers are over a million times weaker than those of the latter. This classification should include not only the jet-forming binary neutron stars and black-hole candidates (which were the main subjects of this workshop) but also the jet sources issuing from *binary forming white dwarfs*, and from *young stellar objects* (YSOs). The former are less obvious because they are hidden inside planetary nebulae. They were highlighted in Kundt (1996a) and have M1-92 (Trammell and Goodrich, 1996), He2-90, and CRL 618 as their presently clearest representatives. The latter include (young) expanding radio triples and one-sided nuclear radio jets (Kundt, 1996a, 1998a), and tend to have several successive heads so that they are much longer (several pc) than originally realized (Reipurth, 1997). Finally, more and more jets are detected even from high-mass X-ray and γ-ray binaries, cf. (Paredes *et al.*, 2000).

Talking about morphologies, a well-known dichotomy are the Fanaroff and Riley classes I and II of extragalactic radio sources with center- and edge-brightened cocoons, which have been recently redefined by Eilek as of *type A and type B* depending on whether or not their head still ploughs supersonically into its surrounding medium (Kundt, 2000). Class II is contained in type A. The new classification offers the prediction that all sources evolve from type A to type B as their ram pressure in the head tends to decrease with increasing source size (for constant injection power) whilst the ambient sound speed tends to increase (with increasing ambient temperature). Type A is statistically younger. This same dichotomy can be found in microquasars whereby a core-jet source can be easily misclassified as of type B before its distant heads have been identified.

2. Jet Fluid and Central Engine

Why do the microquasars deserve their name, i.e. behave like scaled-down quasars? How different are they from their mega counterparts? In this section I shall argue that indeed, their *formation mechanism* is basically the same: Magnetic reconnections – like on the Sun – generate relativistic pair plasma whose buoyant escape from the central potential well rams two antipodal escape channels, perpendicular to the feeding disk, seen as two knotty jets. The big difference between the two source classes – beyond their vastly different power – is the density of their ambient media, $10^{2\pm3}$ cm^{-3} versus $10^{-5\pm2}$ cm^{-3}, i.e. again a factor of $\gtrsim 10^6$, with the microquasars being embedded in *megadense* media.

Crucial in this comparison is the claim that all jet sources are driven by *pair plasma*, including the young neutron-star binary SS 433 (Kundt, 1996a, 1998b,c). Here are six reasons to support this long-standing claim:

- Pair formation via magnetic reconnections is a robust mechanism – demonstrated by flares on the sun – which is expected to take place abundantly around all magnetized rotators: young neutron stars, young white dwarfs, young stars, and the dense centers of galactic disks (s.b.).
- Pair-plasma jets can perform an ordered, loss-free $\vec{E} \times \vec{B}$ drift (at practically equal velocities of both charges) – apart from unavoidable inverse-Compton losses – so that the (dubious) assumption of any downstream in-situ post acceleration is avoided (Kundt, 1998c). Analogous examples from daily life are the electron beams illuminating a TV screen.
- Our Galactic disk is an abundant emitter of the 511 keV pair-annihilation line, some 10^{43} s^{-1} annihilations, a rate that exceeds radioactive interstellar positron productions by a factor of $\gtrsim 30$ but which is expected from all the Galactic microquasars (Kundt, 1997).
- When compared with hadronic jets, pair-plasma jets require distinctly lower powers, just low enough to be consistent with the Eddington limit for stellar sources.
- Pair-plasma jets can be destroyed by inverse-Compton losses on the photon flood from a nearby bright star (when unsuitably oriented), such as σ Orionis (Reipurth *et al.*, 1998), or the companion in Cygnus X-1 (R. Spencer, these Proceedings).
- Wardle *et al.* (1998) have concluded at a pair-plasma composition of quasar beams from the presence of some 1% circular polarization and simultaneous presence of strong linear polarization which imply equal numbers of low-energy ($\gamma < 10^2$) electrons and positrons.

How different are the *central engines* of quasars and microquasars? If spinning stellar magnets are the engines of the latter, as argued above, what are their mega cousins? Zsolt Paragi called them ‘black boxes’. In (1999b) I gave eight reasons of why I consider supermassive black holes inconsistent which I will repeat below; cf. (Hoyle *et al.*, 2000). My preference is for *burning disks*: mass densities $\rho(r)$ inside galactic disks grow as $r^{-2+2\epsilon}$ with decreasing distance r from the center, for a rotation curve $v(r) \sim r^{\epsilon}$, hence approach stellar values near the center. Even for an initially empty center, a typical galactic mass-spiral-in rate of $\lesssim M_{\odot}$ yr^{-1} guarantees that the innermost solar-system-sized parts of the disk fill up to nuclear-burning densities, on the timescale of Myrs. Such a central disk will be disrupted by

TABLE I

Comparison of QSOs with microQSOs: Θ := beam opening angle, L_c/L_l = core/lobe power ratio, $\beta_\perp$:= (apparent) $v_\perp/c$, $\Omega := \rho_{amb}/\rho_{jet}$. Note that $\gamma_{bulk} \geq \sqrt{1+\beta_\perp^2}$ holds rigorously, due to special-relativistic kinematics

source	Θ	$\frac{L_c}{L_l}$	spec-	-trum	γ	$\beta_\perp$	sided	C.E.	Ω	radio loud	radio quiet
AGN	10^{-2}	10^2	$\lesssim$TeV	cont	$10^{3\pm2}$	$\lesssim 30$	more	BD	10^8	quasar	QSO
n ⋆	10^{-2}	10^2	$\lesssim$MeV	line	$10^{3\pm2}$	$\lesssim 5$	less	n ⋆	10^{12}	μquasar	μQSO

nuclear detonations long (in density) before a black hole can form (1996b) and will be observed in the form of the BLR. Its vertical mass distribution reads $\rho(z)/\rho_c \approx 1/\cosh^2(z/z_h)$ with $\rho_c \lesssim 10^2$ g cm^{-3}, $z_h = v_{th}/(2\pi G\rho_c)^{1/2} \lesssim 10^{10}$ cm, and its Keplerian revolution velocities $\sqrt{\pi G\sigma r}$ can reach 0.1c (Pandey and Gupta, 1998; Kundt, 1999b). Sgr $A^\star$ may have a present nuclear-burning disk radius of $10^{13.6}$ cm.

Burning disks (BDs) might never have given supermassive BHs a chance as engine for an AGN had not Salpeter (1971) expressed an early warning considering their stability, assuming radiation-pressure support without nuclear burning. On the other hand, here are the eight BH difficulties expressed in Kundt (2000):

(i) a missing factor of 10^4 in the unresolved *core masses* of nearby galaxies (of $10^{6.5\pm3}$ $M_\odot$: (Kormendy and Richstone, 1995)), on account of the integrated QSO light;

(ii) on average, *mass-outflow* rates from AGN are not smaller than mass-inflow rates: $\dot{M}_{out}/\dot{M}_{in} \approx \beta_{-1} r_{16} N_{24.5}/L_{46}$;

(iii) a BH is too *γ-ray compact* for the relativistic jet plasma to escape, not only in the case of the radio-quiet sub-population;

(iv) BLR spectra are $\gtrsim 10^2$-times *metal enriched* over solar, like for the glowing ashes of nuclear fires (Turnshek, 1988);

(v) QSO *spectra* range all the way up to TeV energies and beyond, much too *hard* for a swallowing giant BH;

(vi) the putative BHs *evolve invertedly* to the QSO phenomenon (1999b; cf. Schmidt *et al.*, 1995; Richstone *et al.*, 1998);

(vii) the *nuclear-burning hurdle* has prevented BH formation, by a margin in surface-mass density of $\lesssim 10^2$;

(viii) the *universality* of the *bipolar-flow* phenomenon asks for a uniform engine: a spinning magnet which generates the indispensible non-thermal component (pair plasma).

Table I summarizes my understanding of the two discussed jet classes. Their differences may be solely due to the differences in source power and ambient density. To me, they look like the same physical phenomenon.

Acknowledgements

I should like to thank Hans Baumann, Ashok Singal, and Gernot Thuma for several joint seminars on the subject, Josep Paredes for constructive criticism, and Günter Lay for converting my write-up into something submittable.

References

Begelman, M.C., Rees, M.J. and Sikora, M.: 1994, *Astrophys. J.* **429**, L57.
Blandford, R.D. and Rees, M.J.: 1974, *Mon. Not. R. Astron. Soc.* **169**, 395.
Blome, H.-J. and Kundt, W.: 1988, *Astrophys. Space Sci.* **148**, 343.
Hoyle, F., Burbidge, G. and Narlikar, J.V.: 2000, in: *A different Approach to Cosmology*, Cambridge University Press.
Kormendy, J. and Richstone, D.: 1995, *Annu. Rev. Astron. Astrophys.* **33**, 581.
Kundt, W.: 1996a, in: *Jets from Stars and Galactic Nuclei, Lecture Notes in Physics* **471**, Springer.
Kundt, W.: 1996b, *Astrophys. Space Sci.* **235**, 319.
Kundt, W.: 1997, in: H. Lesch, R.-J. Dettmar, U. Mebold and R. Schlickeiser (eds.), *The Physics of Galactic Halos*, Akademie Verlag, Berlin, p. 255.
Kundt, W.: 1998a, in: S. Massaglia and G. Bodo (eds.), *Astrophysical Jets*, Gordon and Breach, p. 61.
Kundt, W.: 1998b, *Fundamentals of Cosmic Physics* **20(1)**, 1.
Kundt, W.: 1998c, in: A. Richter (ed.), *Understanding Physics*, Copernicus Gesellschaft, Katlenburg-Lindau, p. 187.
Kundt, W.: 1999 in: F. Giovannelli and L. Sabau-Graziati (eds.), *Multifrequency Behaviour of High-Energy Cosmic Sources*, *Mem. Soc. Astr. It.* **70**, N. 3/4, p. 1105.
Kundt, W.: 2000, in: J. Biretta and P. Leahy (eds.), *Life Cycles of Radio Galaxies*, STSci Baltimore proceedings, Elsevier (in press).
Kundt, W. and Fischer, D.: 1989, *J. Astrophys. Astron.* **10**, 119.
Kundt, W. and Gopal-Krishna: 1980, *Nature* **288**, 149.
Pandey, U.S. and Gupta, A.C.: 1998, *Bull. Astron. Soc. India* **26**, 731.
Paredes, J.M., Martí, J., Ribó, M. and Massi, M.: 2000, *Science* **288**, 2340.
Reipurth, B.: 1997, *The Messenger* **88**, 20.
Reipurth, B., Bally, J., Fesen, R.A. and Devine, D.: 1998, *Nature* **396**, 343.
Richstone, D., *et al.*: 1998, *Nature* **395A**, 14.
Salpeter, E.E.: 1971, *Nature Phys. Sci.* **233**, 5.
Schmidt, M., Schneider, D.P. and Gunn, J.E.: 1995, *Astron. J.* **110**, 68.
Trammell, S.R. and Goodrich, R.W.: 1996, *Astrophys. J.* **468**, L107.
Turnshek, D.A.: 1988, in: J.C. Blades, D.A. Turnshek and C.A. Norman (eds.), *QSO Absorption Lines: Probing the Universe*, Cambridge Univ. Press, p. 17.
Wardle, J.F.C., Homan, D.C., Ojha, R. and Roberts, D.H.: 1998, *Nature* **395**, 457.

JET DYNAMICS AND EMISSION

JOSÉ-LUIS GÓMEZ
Instituto de Astrofísica de Andalucía (CSIC), Apdo. 3004, Granada 18080, Spain

Abstract. Numerical relativistic hydrodynamical simulations have been recently used to study the dynamics of relativistic jets in active galactic nuclei and microquasars. Computation of the non-thermal emission from this hydrodynamical models have allowed a direct comparison with recent high resolution observations of these sources, revealing the important role this method can play in the understanding of the physical processes taking place in these sources. This paper reviews some of the latest results in numerical relativistic hydrodynamical and emission simulations of jets. Recent observations are also analyzed on the light of these simulations.

1. Introduction

Over the past two decades there has been a significant effort to obtain a better understanding of the fluid dynamics in astrophysical jets through numerical simulations. First simulations, restricted to classical fluids dynamics, explored with great success the propagation and morphology of jets (see the reviews of Norman, 1996 and Ferrari, 1998). Application to relativistic flows began in the nineties (Martí, Müller and Ibáñez, 1994; Duncan and Hughes, 1994; Martí *et al.*, 1995, 1997; Komissarov and Falle, 1998; Rosen *et al.*, 1999; Aloy *et al.*, 1999a), thanks to the development of new numerical techniques capable of solving the time-dependent equations of relativistic hydrodynamics for large flow Lorentz factors and strong shocks (see the review by Martí and Müller, 1999).

Some of the latest simulations have started to explore three dimensional relativistic jets (Aloy *et al.*, 1999a,b, 2000; Hardee, 2000), magnetized relativistic jets (Komissarov, 1999), as well as jet formation and collimation making use of the first general relativity magnetohydrodynamical codes (Koide, Shibata and Kudoh, 1999; see also the review by Meier, 2001). Computation of the non-thermal (synchrotron) radio emission from the hydrodynamical results (Gómez *et al.*, 1995, 1997; Mioduszewski, Hughes and Duncan, 1997; Komissarov and Falle, 1997; Aloy *et al.*, 1999b, 2000; Jones, Ryu and Engel, 1999) allows a direct comparison with the observations through which test our models.

Astrophysics and Space Science is the original source of publication of this article. It is recommended that this article is cited as: *Astrophysics and Space Science* **276** (suppl.): 281–288, 2001.

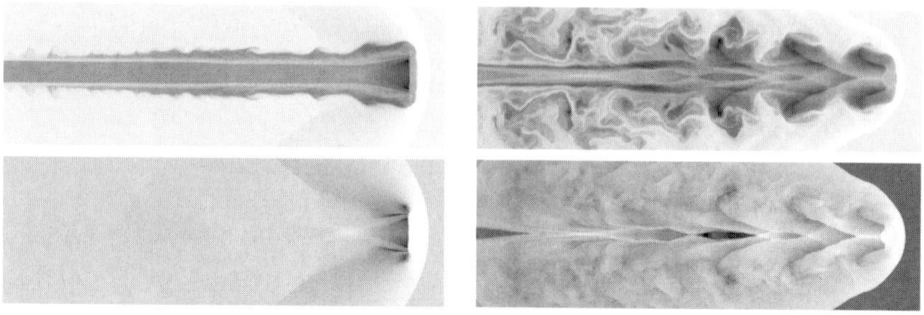

Figure 1. Numerical hydrodynamical simulations of a relativistically 'hot' (*left*) and 'cold' (*right*) jet (from Martí *et al.*, 1997). Panels show the logarithm of the proper rest-mass density (top) and pressure (bottom).

2. Morphology of Relativistic Jets

The main differences that are expected to arise between the morphology of classical and relativistic jets come from the increase in the effective inertial mass of relativistic jets (Martí *et al.*, 1995, 1997; Duncan and Hughes, 1994), as can be deduced from the momentum equation. Relativistic implications in the jet morphology may be caused by relativistic flow speeds or relativistic internal energies. Martí *et al.* (1997; see also the review by Martí, 2000) explored the main differences between hot jets (i.e., beams with internal energies comparable to the rest-mass energies), and highly supersonic ones (those in which the kinematic relativistic effects due to high jet Lorentz factors dominate).

Hot jets show little internal structure and relatively smooth cocoons (see Figure 1). The lack of complex internal structures allows the terminal shock to remain well defined during the jet propagation through the external medium. Therefore, hot jets are expected to present a terminal hot spots, similar to those observed in Faranoff-Riley AGN type II sources. On the other hand, highly supersonic jet models, in which the high Lorentz factors dominate the kinematic relativistic effects, present a richer internal structure, with multiple internal shocks, and extended overpressured cocoons (Figure 1).

Indications of the collision between the jet and external medium has been observed in several microquasars (Mirabel and Rodríguez, 1999). For the superluminal microquasars GRO J1655-40 and GRS 1915+105, only the latter has been found to show some evidence for the existence of lobes, although the most plausible interpretation for the observed features is in terms of regular HII regions (Rodríguez and Mirabel, 1999). Formation of terminal hot spots and lobes is determined by the jet momentum and the density contrast between the jet and external medium. Therefore, the apparently missing lobes in these two sources suggests that either both sources carry relatively small momentum, which would require cold and/or light jets (since they are the fastest jets observed in microquasars), or propagate through a light medium. Ejections in GRS 1915+105 and GRO J1655-

40 are quite episodic, which may be an indication for light jets in both sources, or even the lack of continuous jets. This may therefore explain why no lobes or hot spots have yet be seen in these sources.

3. Relativistic HD and Emission Simulations

Computation of the synchrotron emission from the hydrodynamical models requires several assumptions. Firstly, the particle and energy density of the nonthermal emitting particles is considered to be a constant fraction of the thermal gas. Secondly, the hydrodynamical internal energy density has to be distributed among the emitting particles, which is usually done by assuming a power law energy spectrum (Gómez *et al.*, 1995). Finally, it is still largely unknown what may be the role played by the magnetic field in the jet dynamics of AGNs and microquasars. There are some evidence pointing towards a small contribution of the magnetic field in the dynamics (Hughes, 2000), although only future observations and magnetohydrodynamical simulations (van Putten, 1993; Komissarov, 1999; Koide, Shibata and Kudoh, 1999) could answer this question. So far, and mainly due to the fact that the emission computations have been performed for purely hydrodynamical models, the magnetic field has been assumed to be dynamical negligible. In this case, adhoc magnetic field structures are considered, with an energy density proportional to the fluid pressure.

The presence of emitting gas at velocities close to that of the speed of light enhance the importance of the relativistic effects in the final emission structure of the simulated maps. Besides light aberration (Doppler boosting), time delays is the most important effect determining the final emission structure (no superluminal motions can be obtained from these simulations without considering the time delays between different jet regions). Once the absorption and emission coefficients are determined (accounting for the retarded times), the transfer equations for the synchrotron radiation are solved along the line of sight, allowing the computation of simulated intensity maps.

3.1. Relativistic shocks

Shock-in-jet models (Marscher and Gear, 1985; Hughes, Aller and Aller, 1985) have been proven to provide a general explanation for the emission variability observed in components of relativistic jets. Numerical relativistic HD and emission simulations provide a new powerful tool to improve upon these previous analytical models. With these new numerical techniques it is now possible to study with great detail the generation, structure, and evolution of strong shocks, and analyze its importance in the overall dynamical evolution and emission of jets through comparison with recent high resolution VLBI observations.

In the models presented by Gómez *et al.* (1997), Komissarov and Falle (1997), and Mioduszewski, Hughes and Duncan (1997), superluminal components are ob-

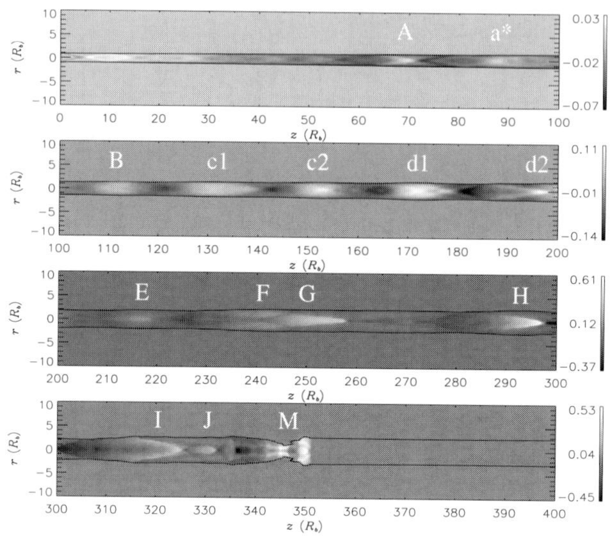

Figure 2. Relative variation with respect to the undisturbed steady jet of the Lorentz factor (logarithmic scale) for a jet in which a perturbation in Lorentz factor was introduced at the jet inlet. The panels show the appearance of multiple conical shocks (labeled 'A' to 'J') trailing the main shock ('M') produced by the perturbation.

tained from perturbations in steady relativistic jets. In Gómez *et al.* (1997) the introduction of a short living square-wave perturbation in the flow Lorentz factor (from an initial value of 4 to 10) is observed to quickly evolve into a traveling shock followed by a rarefaction of slower flow. Because of the enhanced pressure in the shocked plasma, these shocks appear in the simulated intensity maps as regions of increased emission moving downstream at superluminal apparent velocities.

The evolution of the shocked plasma is determined by its interaction with the external medium as well as the underlying jet, which was neglected in the previous simplified analytical models. This results in the appearance of an internal changing structure in the shocked plasma. Time delays stretches the emitting region in the observer's frame, amplifying this inhomogeneities and leading to the appearance of discrete knots within the superluminal component associated with the shocked plasma (Gómez *et al.*, 1997). Furthermore, the passage of the shock wave triggers a local pinch instability that leads to the formation of a series of conical trailing shocks (Agudo *et al.*, 2001a,b) that follow the main shock (Figure 2). These trailing shocks can be easily distinguished because they appear in the simulated maps as components being released on the wake of primary superluminal component (associated with the leading shock), instead of being ejected from the core of the jet. Those trailing components appearing closer to the core show small apparent motions and a very slow secular decrease in brightness, from which they

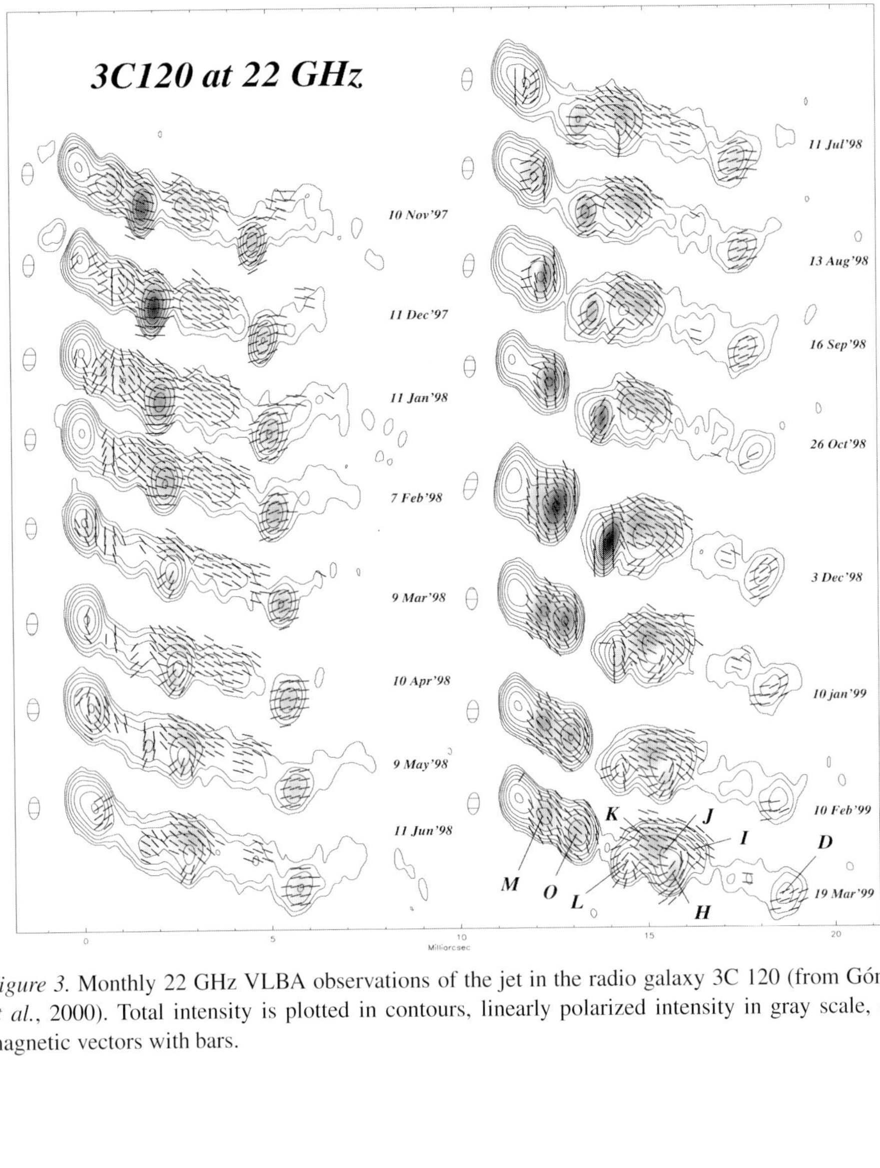

Figure 3. Monthly 22 GHz VLBA observations of the jet in the radio galaxy 3C 120 (from Gómez *et al.*, 2000). Total intensity is plotted in contours, linearly polarized intensity in gray scale, and magnetic vectors with bars.

could be identified as stationary components. Those appearing farther downstream are weaker and can reach superluminal apparent motions. Their oblique nature should also result in different polarization properties from that of the main planar leading shock. The existence of these trailing components indicates that not all observed components necessarily represent a major perturbation at the jet inlet; rather, multiple emission components can be generated by a single disturbance in the jet.

4. Comparison with Observations

High resolution millimeter VLBI observations have confirmed some of the above mentioned results. In the radio galaxy 3C 120 Gómez *et al.* (1998, 2000) found several components being associated with the ejection of a single major superluminal component. Furthermore, the passage of this new superluminal component ('O' in Figure 3) triggered the appearance of a stationary feature ('M' in Figure 3) that presents also enhanced linearly polarized emission. This behavior is in agreement with the numerical simulations of the trailing shocks by Agudo *et al.* (2001a).

VLBI observations of the superluminal microquasars GRO J1655-40 (Hjellming and Rupen, 1995; Tingay *et al.*, 1995) and GRS 1915+105 (Fender *et al.*, 1999; Dhawan, Mirabel and Rodríguez, 2000) show a more complex jet structure than usually observed in jets of AGNs. The jets in GRO J1655-40 present alternate brightening and fading that cannot be interpreted in terms of relativistic beaming alone (Hjellming and Rupen, 1995). These observations may be revealing some of the complex hydrodynamical process that are expected to manifest in high resolution images, such as interactions of the jet with the external medium (the jets in GRO J1655-40 are believed to be precessing with a period of three days). Similar complex components evolution have also been observed in 3C 120, interpreted as the interaction of the jet with a cloud of ionized gas located at about 8 pc from the center of the galaxy (Gómez *et al.*, 2000).

4.1. Magnetic fields

Polarimetric observations of the jets in AGNs and microquasars is of special relevance. Determination of the magnetic field structure and strength is needed to obtain a better understanding of the superluminal components nature (shocks or blobs), as well as to test the models for jet formation and collimation.

The low fluxes exhibited by the jets in microquasars make polarimetric VLBI observations of these objects a very difficult task. One of the few observations corresponds to MERLIN observations of GRS 1915+105 (Fender *et al.*, 1999). Linearly polarized emission was detected in the leading superluminal component of the approaching jet, revealing a changing polarization angle with time, accompanied by a depolarization. This is interpreted as evidence for a randomization of the magnetic field in the component. These observations contrast with the predictions of the shock-in-jet models, suggesting that components in GRS 1915+105 may instead consist of discrete plasmons.

Aloy *et al.* (2000) have studied the radio emission properties of 3D relativistic hydrodynamical jets. The interaction between the relativistic jet and the ambient medium results in a jet stratification where a fast spine is surrounded by a slow high energy shear layer. By introducing an ad-hoc helical magnetic field in the sheath, the authors found that the emission shows a cross section asymmetry, which is a function of the viewing angle and flow velocity. This may explain the otherwise puzzling asymmetric polarization structure observed in the blazar 1055+018 (Attridge, Roberts and Wardle, 1999).

5. Conclusions

Numerical relativistic (magneto)hydrodynamical and emission simulations have proven to be a powerful method to understand the physics of jets in AGNs and microquasars through a direct comparison with observations. Some perspectives for the future in numerical simulations would be: i) the exploration of different equations of state that would help to determine the jet composition (pairs or baryonic); ii) inclusion of radiative losses and particle acceleration (i.e., Jones, Ryu, and Engel, 1999) as well as implementation of high energy emission mechanisms (inverse Compton) that would help to obtain a better understanding of the emission at high energies (X and γ rays); iii) study of the jet-external medium interactions, that seems to be more common than previously thought (i.e., Gómez *et al.*, 2000); and iv) computation of the emission from relativistic magnetohydrodynamical models that would be of great help to understand the magnetic field influence in the jet dynamics, helping to explain recent and upcoming polarimetric VLBI observations.

Acknowledgement

This research is supported in part by the Spanish DGICT under grant PB97-1164.

References

Agudo, I., Gómez, J.L., Martí, J.M., Ibáñez, J.M., Marscher, A.P., Alberdi, A., Aloy, J.M. and Hardee, P.E.: 2001a, *Astrophys. J.*, **549**, L183.

Agudo, I., Gómez, J.L., Martí, J.M., Ibáñez, J.M., Marscher, A.P., Alberdi, A., Aloy, J.M. and Hardee, P.E.: 2001b, these Proceedings.

Aloy, M.A., Ibáñez, J.M., Martí, J.M. and Müller, E.: 1999a, *Astrophys. J. Suppl.* **122**, 151.

Aloy, M.A., Ibáñez, J.M., Martí, J.M., Gómez, J.L. and Müller, E.: 1999b, *Astrophys. J.* **523**, L125.

Aloy, M.A., Gómez, J.L., Ibáñez, J.M., Martí, J.M. and Müller, E.: 2000, *Astrophys. J.* **528**, L85.

Attridge, J.M., Roberts, D.H. and Wardle, J.F.C.: 1999, *Astrophys. J.* **518**, L87.

Dhawan, V., Mirabel, I.F. and Rodríguez, L.F.: 2000, *Astrophys. J.* **543**, 373.

Duncan, G.C. and Hughes, P.A.: 1994, *Astrophys. J.* **436**, L119.

Fender, R.P., Garrington, S.T., McKay, D.J., Muxlow, T.W.B., Pooley, G.G., Spencer, R.E., Stirling, A.M. and Waltman, E.B.: 1999, *Mon. Not. R. Astron. Soc.* **304**, 865.

Ferrari, A.: 1998, *Annu. Rev. Astron. Astrophys.* **36**, 539.

Gómez, J.L., Martí, J.M., Marscher, A.P., Ibáñez, J.M. and Alberdi, A.: 1997, *Astrophys. J.* **482**, L33.

Gómez, J.L., Martí, J.M., Marscher, A.P., Ibáñez, J.M. and Marcaide, J.M.: 1995, *Astrophys. J.* **449**, L19.

Gómez, J.L., Marscher, A.P., Alberdi, A., Jorstad, S.G. and García-Miró, C.: 2000, *Science* **289**, 2317.

Gómez, J.L., Marscher, A.P., Alberdi, A., Martí, J.M. and Ibáñez, J.M.: 1998, *Astrophys. J.* **499**, 221.

Hardee, P.E.: 2000, *Astrophys. J.* **533**, 176.

Hjellming, R.M. and Rupen, M.P.: 1995, *Nature* **375**, 464.

Hughes, P.A., Aller, H.D. and Aller, M.F.: 1985, *Astrophys. J.* **298**, 301.

Hughes, P.A.: 2000, in: R.A. Laing and K.M. Blundell (eds.), *Particles and Fields in Radio Galaxies, ASP Conf. Ser.* **201**, in press (astro-ph/0011127).

Jones, T.W., Ryu, D. and Engel, A.: 1999, *Astrophys. J.* **512**, 105.

Koide, S., Shibata, K. and Kudoh, T.: 1999, *Astrophys. J.* **522**, 727.

Komissarov, S.S. and Falle, S.A.E.G.: 1997, *Mon. Not. R. Astron. Soc.* **288**, 833.

Komissarov, S.S. and Falle, S.A.E.G.: 1998, *Mon. Not. R. Astron. Soc.* **297**, 1087.

Komissarov, S.S.: 1999, *Mon. Not. R. Astron. Soc.* **308**, 1069.

Marscher, A.P. and Gear, W.K.: 1985, *Astrophys. J.* **298**, 114.

Martí, J.M.: 2000, in: B. Rocca-Volmerange and H. Sol (eds.), *AGNs in their Cosmical Environment*, EDP Series in Astronomy and Astrophysics (EDP Sciences), in press.

Martí, J.M. and Müller, E.: 1999, *Living Reviews in Relativity*, http://www.livingreviews.org/Articles.

Martí, J.M., Müller, E., Font, J.A. and Ibáñez, J.M.: 1995, *Astrophys. J.* **448**, L105.

Martí, J.M., Müller, E., Font, J.A., Ibáñez, J.M. and Marquina, A.: 1997, *Astrophys. J.* **479**, 151.

Martí, J.M., Müller, E. and Ibáñez, J.M.: 1994, *Astron. Astrophys.* **281**, L9.

Meier, D.: 2001, these Proceedings.

Mioduszewski, A.J., Hughes, P.A. and Duncan, G.C.: 1997, *Astrophys. J.* **476**, 649.

Mirabel, I.F. and Rodríguez, L.F.: 1999, *Annu. Rev. Astron. Astrophys.* **37**, 409.

Norman, M.L.: 1996, in: P.E. Hardee, A.H. Bridle and J.A. Zensus (eds.), *Energy Transport in Radio Galaxies and Quasars, ASP Conf. Ser.* **100**, p. 405.

Rodríguez, L.F. and Mirabel, I.F.: 1999, *Astron. Astrophys.* **340**, L47.

Rosen, A., Hughes, P.A., Duncan, G.C. and Hardee, P.E.: 1999, *Astrophys. J.* **516**, 729.

Tingay, S.J., *et al.*: 1995, *Nature* **374**, 141.

van Putten, M.H.P.M.: 1993, *Astrophys. J.* **408**, L21.

JETS IN XRB LOW/HARD STATES: MODELS OF XTE J1118+480 SUGGEST A STRONG SIGNATURE FROM RADIO THROUGH X-RAY *

S. MARKOFF** and H. FALCKE
Max-Planck-Institut für Radioastronomie, Bonn, Germany
R. FENDER
Astronomical Institute 'Anton Pannekoek', University of Amsterdam, Netherlands

Summary

Observations reveal the presence of powerful jets in the Low/Hard states of several black hole candidate X-ray binaries, so it is important to consider what signatures these outflows will have in the broadband Low/Hard spectrum. Correlations between the radio and X-rays, as well as in at least one case the optical, suggest that emission from a jet, via either synchrotron or inverse Compton (IC) processes, may play a role. The recently discovered X-ray transient XTE J1118+480 (Remillard *et al.*, 2000), a black hole candidate in the Low/Hard X-ray state, provides an excellent opportunity for multi-wavelength modeling, since it has been observed not only in the radio through X-rays (see Fender *et al.*, 2000, and references therein), but is also at high enough Galactic latitude to allow the first ever EUV detections of an X-ray transient (Hynes *et al.*, 2000, hereafter H00).

We consider a simplistic scenario of symmetric, magnetized relativistic plasma outflows. The jet plasma encounters a shock, which accelerates the particle distribution into a power-law, creating a jump in the spectrum where the shock-accelerated leptons produce an optically thin power-law extending to the X-ray regime. The details of this calculation, as well as a more evolved model which addresses the presence of an accretion disk component, can be found in Falcke and Markoff (2000), Markoff, Falcke and Fender (2001), respectively, and references therein. We include a representative figure, which includes only synchrotron emission from the jet. Surprisingly, we can account for the radio-IR and the X-ray data reasonably well without invoking an IC component or the presence of significant disk emission.

As an interesting note, the hard X-rays have been seen lagging the soft in several XRBs (e.g., Ford *et al.*, 2000). A multi-component model such as that presented here, or including disk emission (see Markoff *et al.*, 2001) is favorable.

* We are grateful to the (late) Bob Hjellming for sharing his data with us.
** Supported by a Humboldt Research Fellowship

Astrophysics and Space Science is the original source of publication of this article. It is recommended that this article is cited as: *Astrophysics and Space Science* **276** (suppl.): 289–290, 2001.

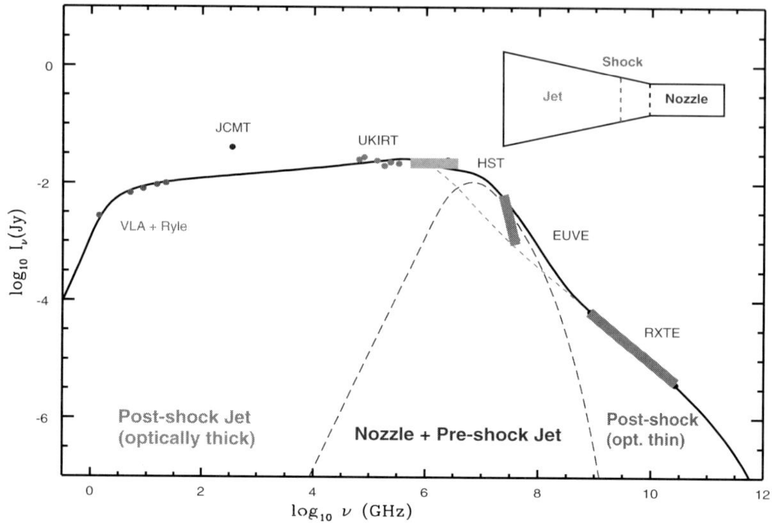

Figure 1. Fit for power-law spectral index 2.6, $M_{\rm bh} = 5M_\odot$, $L = 0.04L_{\rm Edd}$, $z_{\rm sh} = 16r_s$, and inclination angle 40°. Data points from H00 except radio (Hjellming *et al.*, in prep.), and JCMT (Fender *et al.*, 2000). We cannot account for the entire optical range from the jet alone, but the detection of strong H_α lines (Garcia *et al.*, 2000) could be indicative of disk emission at these frequencies. We use the maximum absorption case presented in H00 for the EUVE data. Included is a schematic explaining the different emission regions. The JCMT point obviously is well outside the range of this model, which could be accounted for either by the addition of another shock further out, resulting in enhanced emission, or by thermal emission from a torus.

References

Falcke, H. and Markoff, S.: 2000, *Astron. Astrophys.*, **362**, 113.

Fender, R.P., Hjellming, R.M., Tilanus, R.J., Pooley, G.G., Deane, J.R., Ogley, R.N. and Spencer, R.E.: 2000, *Mon. Not. R. Astron. Soc.*, submitted.

Ford, E.C., van der Klis, M., Méndez, M., Wijnands, R., Homan, J., Jonker, P.G. and van Paradijs, J.: 2000, *Astrophys. J.* **537**, 368.

Garcia, M., Brown, W., Pahre, M., McClintock, J., Callanan, P. and Garnavich, P.: 2000, IAU Circ. 7392.

Hynes, R.I., Mauche, C.W., Haswell, C.A., Shrader, C.R., Cui, W. and Chaty, S.: 2000, *Astrophys. J. Lett.* **539**, L37.

Markoff, S., Falcke, H. and Fender, R.: 2001, *Astron. Astrophys. Lett.*, in press.

Remillard, R., Morgen, E., Smith, D. and Smith, E.: 2000, IAU Circ. 7389.

THE DISC – JET CONNECTION IN GRS 1915+105

M. KLEIN-WOLT[1], R.P. FENDER[1], G.G. POOLEY[2], T. BELLONI[3], S. MIGLIARI[3,4], E.H. MORGAN[5] and M. VAN DER KLIS[1]

[1] *API Amsterdam*

[2] *MRAO Cambridge*

[3] *OAB Merate*

[4] *Univ. Milano*

[5] *MIT Cambridge*

Abstract. We present an analysis of simultaneous X-ray and radio observations of the black hole candidate GRS 1915+105.

1. Introduction

The black hole candidate GRS 1915+105 is known for its extremely complex X-ray behaviour, which is not observed in any other X-ray source. Belloni *et al.* (2000) succesfully described this behaviour as transitions between three basic spectral states: the soft states A and B, and the hard state C. GRS 1915+105 was discovered in 1994 to be the first galactic source to show superluminal motions of radio-emitting ejecta (Mirabel and Rodriguez, 1994). Besides these big radio flares, the source also shows smaller radio oscillation events with periods around 20–40 minutes (Pooley and Fender, 1997).

2. Comparing the Simultaneous Radio and the X-Ray Observations

We have cross-correlated simultaneous (15 GHz) Ryle Telescope (RT) radio and Rossi X-Ray Timing Explorer (RXTE; PCA, 2–60 KeV) X-ray observations of GRS 1915+105 made between June 12, 1996 and December 13, 1999, to find a total of 101 overlapping observations (about 9170 minutes of data). In Figure 1 we show one representative observation. The X-ray light curve (lower panel) is broken up into states A, B, C. From spectral fits (see Migliari, Vignarca and Belloni, this volume) it can be concluded that during the spectrally hard state-C dips the inner part of the accretion disc is either missing or just unobservable. The radio emission, at the same time, shows an oscillating flaring type of behaviour for which the period is found to be between 20 and 40 minutes. We also found observations which do not have state C intervals and consist of state A and B alone. For these observations the inner radius of the accretion disc does not change much and importantly, the

Astrophysics and Space Science is the original source of publication of this article. It is recommended that this article is cited as: *Astrophysics and Space Science* **276** (suppl.): 291–292, 2001.

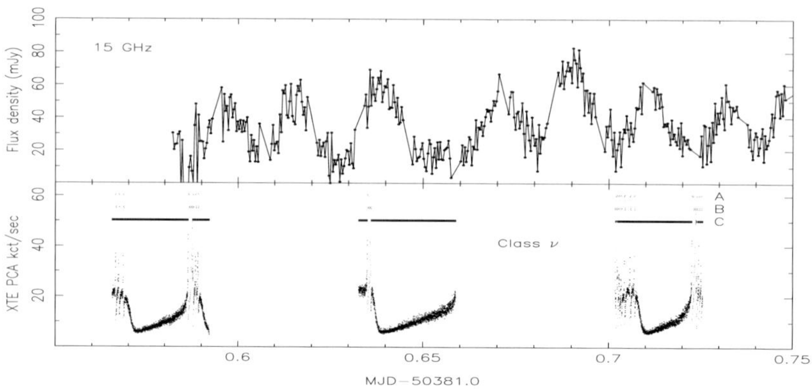

Figure 1. Overlapping Radio/X-ray observation showing the hard X-ray dips (state C), and the radio oscillation events. The rms noise is 3 mJy per point in the radio.

radio emission is strongly suppressed: the flux density is typically less than a few mJy, despite up to 50% variability in the X-ray flux.

We find that radio oscillations require the presence of long ($\geq$ 100 s) state C intervals; for observations consisting of state A and B only the radio emission is suppressed (typically less than a few mJy). With short state C intervals less than 100 s in length only very weak radio emission ($\sim$ 5 mJy) is observed. The long state-C intervals correspond to *hard* dips in the X-ray light curve which seem to have a one-to-one relation with the radio events: each hard dip produces a new radio 'flare'.

During long 'plateau' states, when the source is believed to be in state C for up to weeks at a time, a (large) quasi-continuous jet is formed with an almost flat synchrotron spectrum extending to at least the near-infrared. We suggest that both for these large jets and the radio oscillation events shown in Figure 1 the same mechanism could be responsible for the production of the radio emission: during a state C interval the corona and the jet are coupled (or are the same structure) and an out flow takes place. As the material moves downstream in the jet, it becomes optically thin at progressively lower frequencies. This causes the observed delay between the radio and the X-ray emission. The plateau states are just longer and more stable versions of the state C intervals: continuous ejection causes a superposition of optically thick and thin parts to be observed as spectrally flat.

References

Belloni, T., Klein-Wolt, M., *et al.*: 2000, *Astron. Astrophys.* **355**, 271.
Mirabel, I.F. and Rodríguez, L.F.: 1994, *Nature* **371**, 46.
Pooley, G.G. and Fender, R.P.: 1997, *Mon. Not. R. Astron. Soc.* **292**, 925.

HYDRODYNAMICAL AND EMISSION SIMULATIONS OF RELATIVISTIC JETS: STABILITY AND GENERATION OF SUPERLUMINAL AND STATIONARY COMPONENTS

IVÁN AGUDO and JOSÉ-LUIS GÓMEZ
Instituto de Astrofísica de Andalucía, CSIC, Granada, Spain

JOSÉ-MARÍA MARTÍ and JOSÉ-MARÍA IBÁÑEZ
Departamento de Astronomía y Astrofísica, Universidad de Valencia, Valencia, Spain

ALAN P. MARSCHER
Institute for Astrophysical Research, Boston University, Boston, U.S.A.

ANTONIO ALBERDI
Instituto de Astrofísica de Andalucía, CSIC, Granada, Spain

MIGUEL-ÁNGEL ALOY
Max-Planck-Institut für Astrophysik, Garching, Germany

PHILIP E. HARDEE
Department of Physics and Astronomy, University of Alabama, Tuscaloosa, U.S.A.

Abstract. We present 2D hydrodynamical and emission simulations of the jet stability after the introduction of strong perturbations on a relativistic jet. These simulations show that the interaction of a *single* strong perturbation with the underlying jet results in the formation of multiple conical shocks with very specific observational properties.

1. The Jet Model

Stationary and superluminal components are commonly observed in jets of AGNs and microquasars, usually associated with stationary and moving shocks. Here we describe a general mechanism for the generation of superluminal and stationary components in relativistic jets. We have used a 2D hydrodynamical code (see Martí *et al.*, 1997 and references therein) to compute the time evolution of a relativistic jet after the introduction of a strong square-wave perturbation at the jet inlet. Our steady jet model consists of a supersonic relativistically ideal gas in pressure equilibrium with a denser external medium and with a Lorentz factor of 4 at the jet inlet. The jet has an opening angle of 0.3° induced by a decreasing pressure gradient in the external medium. The square-wave perturbation in the steady jet model consists in an increase of the inlet Lorentz factor and pressure during a short time interval (for more details see Agudo *et al.*, 2000). The hydrodynamical results are used to compute the synchrotron radio emission (see Gómez *et al.*, 1997 and references

Astrophysics and Space Science is the original source of publication of this article. It is recommended that this article is cited as: *Astrophysics and Space Science* **276** (suppl.): 293–294, 2001.

therein), at a viewing angle of 10°, considering all the relativistic effects, including the time delays.

2. Results

Our simulations show that the passage of a strong perturbation through a jet triggers a local pinch instability (Hardee, 2000 and references therein) in the surface of the jet, leading to the formation of a series of conical *trailing shocks*. The analysis of the emission simulations in time predicts very specific observational properties for these features (Agudo *et al.*, 2000). The *trailing components*, as we call the counterpart of the trailing shocks in the emission simulations, appear upstream of the superluminal component associated with the perturbation induced at the jet inlet. Therefore, trailing components can be easily *observationally* distinguished from those generated at the jet inlet (which could be associated with the VLBI core): they would emanate not from the core, but rather would seem to appear spontaneously downstream, created in the wake of strong perturbations. Those components appearing close to the core are very stable both in position and flux densities and therefore may be associated with stationary components, in contrast to components appearing further downstream that show larger apparent motions (episodically superluminal, but always with lower velocities than the main superluminal component) as well as weaker flux densities. They are related to conical shocks, and hence should exhibit different polarization properties from those of the main component. VLBI observations can provide dynamic ranges of the order of 200:1 or larger. All components seen in our simulated radio maps are above this limit, and therefore are expected to be detected by actual VLBI arrays. Our simulations support the idea that jets can be steadier than previously thought, because multiple components in radio maps can be produced by a single ejection event.

References

Agudo, I., Gómez, J.L., Martí, J.M., *et al.*: 2001, *Astrophys. J. Lett.*, 1/**549**, L183–L186.
Gómez, J.L., Martí, J.M., Marscher, P.A., *et al.*: 1997, *Astrophys. J.* **482**, L33–L36.
Hardee, P.E.: 2000, *Astrophys. J.* **533**, 176–193.
Martí, J.M., Müller, E., Font, J.A., *et al.*: 1997, *Astrophys. J.* **479**, 151–163.

MAGNETO-ROTATIONAL EJECTION

N.V. ARDELJAN
Department of Computational Mathematics and Cybernetics, Moscow State University, Vorobjevy Gory, Moscow B-234 119899, Russia
E-mail: ardel@redsun.cs.msu.su

G.S. BISNOVATYI-KOGAN and S.G. MOISEENKO
Space Research Institute, Profsoyuznaya str. 84/32, Moscow 117810, Russia
E-mail: gkogan@mx.iki.rssi.ru

We have performed 2D simulations of the magneto-rotational ejection mechanism, suggested by Bisnovatyi-Kogan (1970). Amplification of the toroidal component of the magnetic field due to differential rotation leads to the transformation of a part of the rotational energy to kinetic energy of ejection. The first numerical simulations devoted to the magneto-rotational mechanism were made by Le Blanc and Wilson (1970) and investigated analytically in the paper by Meier *et al.* (1976).

One of the most important parameters for this problem is the ratio of magnetic energy to gravitational energy in the star: $\alpha = E_{\mathrm{mag}}/|E_{\mathrm{grav}}|$. We have performed simulations for initial values of $\alpha = 10^{-2}, 10^{-4}, 10^{-6}$. The initial magnetic field for the problem should satisfy the following requirements: 1. It should be divergence-free; 2. It should be balanced (i.e. the force produced by the magnetic field should be in balance with other forces in the star); 3. It should be stable. Details of the definition of the initial magnetic field are described in the paper by Ardeljan *et al.* (2000).

For the simulations we used a specially-developed Lagrangian 2D numerical implicit scheme on a triangular grid with mesh reconstruction (see Ardeljan and Kosmachevskii, 1995 and references therein).

The initial magnetic field chosen for our simulations has quadrupole-like symmetry (i.e. its 'z'-component is equal to zero at the equatorial plane). This is why we obtain radial ejection (Figure 1), instead of a jet. Application of an initial magnetic field with dipole-like symmetry should produce a jet, as in Le Blanc and Wilson (1970).

The toroidal magnetic field grows with time and produces an MHD shock, which moves to the boundary of the star.

Part of the matter of the envelope of the star (about 7% of the mass of the star) has radial kinetic energy larger than its potential energy and can be ejected. This ejected matter carries about 3.3% of the total energy of the star. Figure 2 presents the time evolution of the ejected mass and energy.

Astrophysics and Space Science is the original source of publication of this article. It is recommended that this article is cited as: *Astrophysics and Space Science* **276** (suppl.): 295–296, 2001.

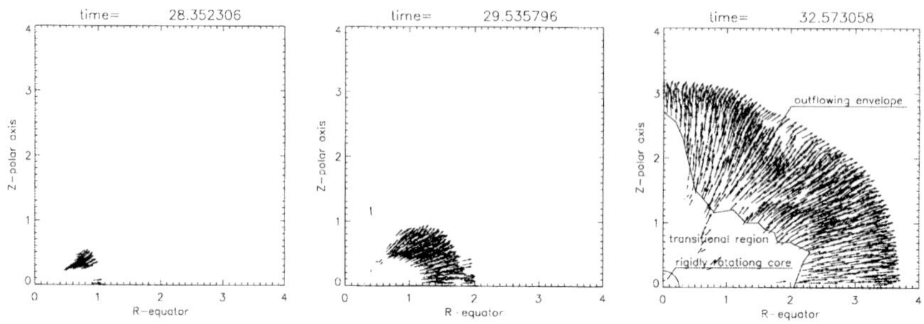

Figure 1. Ejected part of the envelope of the star at $t = 28.352$ (*left*), $t = 29.536$ (*center*) and at $t = 32.573$ (*right*).

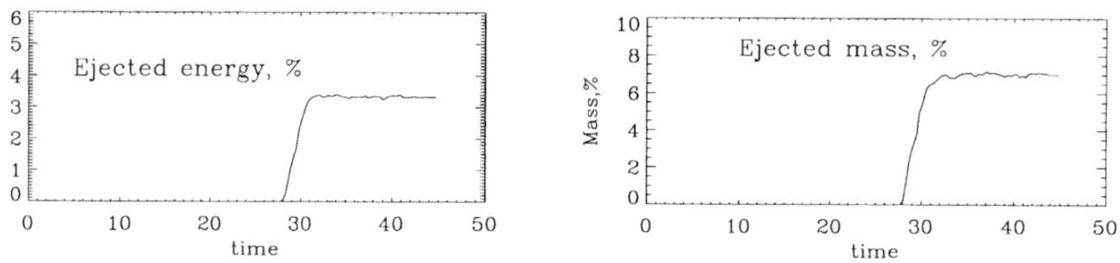

Figure 2. Left: Time evolution (in % to the total initial energy of the cloud) of the energy of the ejected matter of the envelope of the star. *Right:* Time evolution (in %) of the ejected mass of the envelope of the star.

Acknowledgements

G.S.B.-K. and S.G.M. thank Dr A. Castro-Tirado for hospitality, RFBR for partial support by grants 98-01-00352, 99-02-18180 and ESA-INTAS grant No. 120.

References

Ardeljan, N.V., Bisnovatyi-Kogan, G.S. and Moiseenko, S.G.: 2000, *Astron. Astrophys.* **355**, 1181.
Ardeljan, N.V. and Kosmachevskii, K.V.: 1995, *Comp. Math. Mod.* **6**, 209.
Bisnovatyi-Kogan, G.S.: 1970, *Astron. Zh.* **47**, 813.
Le Blanc, L.M. and Wilson, J.R.: 1970, *Astrophys. J.* **161**, 541.
Meier, D.L., Epstein, R.I., Arnett, W.D. and Schramm, D.N.: 1976, *Astrophys. J.* **204**, 869.

STATIONARY RELATIVISTIC MAGNETIC JETS FROM BLACK HOLES

CHRISTIAN FENDT and ELISABETTA MEMOLA
Astrophysikalisches Institut, An der Sternwarte 16, 14482 Potsdam, Germany

Abstract. We investigate the jet formation region of relativistic magnetic jets using stationary axisymmetric MHD models. We numerically calculate the field structure of the collimating jet and the flow acceleration. Special or general relativistic effects are considered. Stationarity allows for a *global* treatment of the jet physics. Our results are applicable to highly relativistic jets of AGN and microquasars.

1. Stationary Relativistic MHD

The stationary and axisymmetric ideal MHD equations can be reduced to the Grad-Shafranov equation (GSE) and the wind equation, describing the force-balance across and along the flux surfaces $\Psi \sim \int \vec{B}_P \cdot d\vec{A}$. The GSE is a non linear partial differential equation of second order for Ψ with the source term depending on the total poloidal current $I(\Psi)$ and the iso-rotation parameter $\Omega_{\rm F}(\Psi)$ (cf. Okamoto, 1992; Fendt, 1997; Fendt and Memola, 2001). At the light surface of the magnetosphere the GSE becomes singular. The stationary polytropic wind equation can be converted in a polynomial equation for the poloidal velocity $u_{\rm p} \equiv \gamma v_{\rm p}/c$ (Camenzind, 1986; Takahashi *et al.*, 1990; Fendt and Camenzind, 1996; Fendt and Greiner, 2001).

2. Two-Dimensional Structure of the Magnetic Jet

Our example solutions of the special relativistic GSE with constant $\Omega_{\rm F}$ were calculated with increasing strength of the poloidal electric current demonstrating its collimating effect. We show meridional cuts across the axisymmetric magnetic flux surfaces (Figure 1, *left*). The outer boundary is the light cylinder imposing the regularity condition.

Global magnetic jets calculated from the GSE in Kerr metric were presented in Fendt (1997). In general, the asymptotic jet radius is several light cylinder radii. To consider differentially rotating field line foot points is an essential step since jets are anchored in differentially rotating disks (Fendt and Memola, 2001). As a consequence, the shape of the light surface is not known a priori and must be calculated

Astrophysics and Space Science is the original source of publication of this article. It is recommended that this article is cited as: *Astrophysics and Space Science* **276** (suppl.): 297–299, 2001.

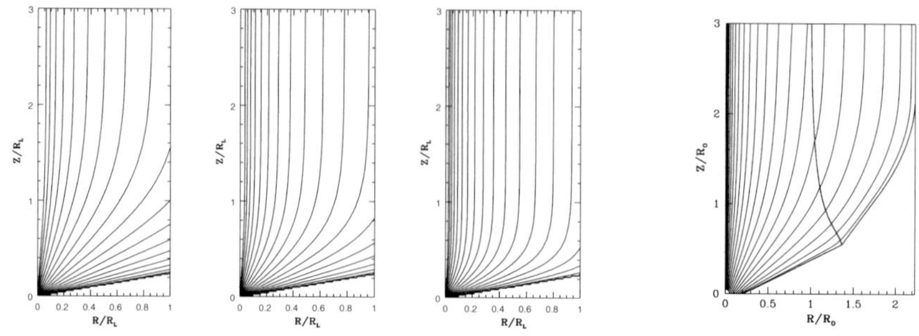

Figure 1. Jet magnetic flux distribution $\Psi(R, Z)$. Radius in units of light cylinder radii R_L or asymptotic light surface radii R_0. *Left*: influence of electric current (increasing by factors of 1, 2, 5 from left). *Right*: differentially rotating foot points.

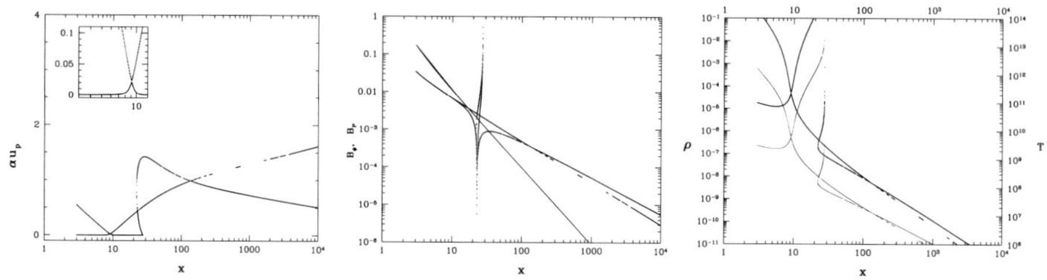

Figure 2. The two branches of the magnetic wind solution in Kerr metric ($a = 0.8$). Poloidal 4-velocity u_P and the corresponding branches of magnetic field B_p, B_ϕ, temperature T and density ρ. The jet solution is the branch with increasing velocity. Intersections indicate the 3 magnetosonic points. Radius x in units of the horizon.

iteratively together with the GSE solution. The global field is then determined by a smooth transition of the field across the light surface (Figure 1, *right*). Differential rotation, $\Omega_F(\Psi)$, leads to stronger collimation.

3. Magnetic Jet Acceleration in Kerr Metric

Solutions of the wind equation in Kerr metric prove that a relativistic jet speed $> 0.9\,c$ can be achieved for microquasars (Figure 2) (Fendt and Greiner, 2001). The field strength at the jet origin is 10^9 G. At large radii the field is toroidally dominated. A highly magnetized jet flow becomes highly relativistic, even if its origin is *not* close to the horizon ($x \sim 10$). Temperature and density decrease logarithmically (from 10^{10} K to 10^6 at a distance of ~ 1000). X- and γ-emission should be detected from the inner jet.

References

Okamoto, I.: 1992, *Mon. Not. R. Astron. Soc.* **254**, 192.
Camenzind, M.: 1986, *Astron. Astrophys.* **162**, 32.
Takahashi, M., Nitta, S., Tatematsu, Y. and Tomimatsu, A.: 1990, *Astrophys. J.* **363**, 206.
Fendt, Ch.: 1997, *Astron. Astrophys.* **323**, 999.
Fendt, Ch. and Camenzind, M.: 1996, *Astron. Astrophys.* **313**, 591.
Fendt, Ch. and Greiner, J.: 2001, *Astron. Astrophys.*, **369**, 308.
Fendt, Ch. and Memola, E.: 2001, *Astron. Astrophys.*, **365**, 631.

POSITRON-ELECTRON ANNIHILATION LINE OF RELATIVISTIC JETS IN AGNS AND QUASARS

JIAN-MIN WANG[1,*], PH. DUROUCHOUX[2] and TI-PEI LI[1]
[1] *Laboratory of Cosmic Ray and High Energy Astrophysics, IHEP, Beijing 100039, China*
[2] *CE Saclay, DSM, DAPNIA, Service d'Astrophysique, 91191 Gif sur Yvette Cedex, France*

Abstract. We determined there are two possible regions producing positron-electron annihilation line (PEAL) from the relativistic jet: close to and far away from the black hole in AGNs and quasars. The PEAL profile has been calculated for several cases. It is expected that future observations with *INTEGRAL* may discover these lines in order to test the paradigm of active galactic nuclei and quasars.

1. Introduction

It has been generally realized that there is a copious $e^{\pm}$ plasma in active galactic nuclei and quasars. Recently Wang (2000) argued that the interaction between the very high energy (VHE) photons and the synchrotron photons reflected by clouds in broad line region may produce a large electron-positron cloud. Such a cloud is moving with relativistic velocity. Therefore the observed PEAL is simply Doppler shifted towards to the blue for small viewing angle. The second source of producing PEAL is that $e^{\pm}$ blob may be ejected from the nucleus with highly relativistic speed (Li and Liang, 1996; Beloborodov, 1999; Spruit, 2000; Wang *et al.*, 2000a). PEAL from such a moving blob close to a rotating black hole may be seriously distorted by gravitational effects. Wang *et al.* (2000b) calculate the PEAL profiles from such a blob. In general the profiles are characterized by two peaks.

2. Profile of Positron-Electron Annihilation Line

2.1. Inner Region of Jet

PEAL profile should be affected by both special and general relativistic effects. The profiles from approaching jet are different from that from receding jet. The detail profiles received by an observer at infinity with different viewing angles can be found in Wang *et al.* (2000b).

* Send offprint requests to: Jian-Min Wang *wangjm@astrosv1.ihep.ac.cn*

Astrophysics and Space Science is the original source of publication of this article. It is recommended that this article is cited as: *Astrophysics and Space Science* **276** (suppl.): 301–302, 2001.

2.2. Outer Region of Jet

Although the deficiency of very high energy emission from γ-loud blazars may be due to the absorption of cosmological background radiation, Wang (2000) argues that there is an *intrinsic* mechanism responsible for the absorption of VHE photons by the reflected synchrotron (RSy) photons. We argue that the energy density of the RSy photons is higher than that of the local synchrotron photons. The VHE emission is thus constrained by RSy photons. The opacity of the pair production due to photon interaction can be expressed by the observable quantities

$$\tau_{\gamma\gamma}(\epsilon_{\rm obs}) = 2.0\, l_{\rm rsc}^{-1} \mathcal{D}_{10}^{4} \left(\frac{\epsilon_{\rm obs}}{\rm TeV}\right)^{0.7} \left(\frac{\Delta t_{\rm obs}}{\rm day}\right) \left(\frac{\nu_{\rm s}}{4.0\times 10^{3.7}}\right)^{3.7} \left(\frac{\nu_{\rm rsc}}{10^{25}{\rm Hz}}\right)^{-2} \quad (1)$$

where the Doppler factor $\mathcal{D}_{10} = \mathcal{D}/10$, $\Delta t_{\rm obs}$ is the timescale of the variation of γ-ray, $\nu_{\rm s}$ and $\nu_{\rm rsc}$ are the peak frequency at low and high energy bands, respectively. This formula is available for electron spectral index $\alpha = 2.4$. It should be addressed that here is the parameter $l_{\rm rsc}$, representing the ratio of the peak flux at high energy (γ) band to that at low energy band. The larger $l_{\rm rsc}$ the larger pair opacity is, namely, the larger is the $e^{\pm}$ cloud in quasars.

If the deficiency of VHE from γ-ray loud blazars is indeed due to this *intrinsic* mechanism, it is expected that future observation should discover the PEAL at $5\mathcal{D}_{10}$ MeV. Its profile is rather simple since only Doppler effect works in such an outer region.

3. Conclusions

We should choose the high $l_{\rm rsc}$ blazars as the observable objects in future observations. The complex profiles of an emission line depends on the viewing angle. It is expected that future observations with *INTEGRAL* may discover the complex PEALs.

References

Beloborodov, A.: 1999, *Mon. Not. R. Astron. Soc.* **305**, 181.
Li, H. and Liang, E.: 1996, *Astrophys. J.* **458**, 514.
Spruit, H.C.: 2000, astro-ph/0003043.
Wang, J.-M.: 2000, *Astrophys. J.* **538**, 181.
Wang, J.-M., Yuan, Y.-F., Wu, M. and Kusunose, M.: 2000a, *Astrophys. J.* **541**, L41.
Wang, J.-M., Zhou, Y.-Y., Yuan, Y.-F. and Cao, X.: 2000b, *Astrophys. J.* **544**, in press (astro-ph/0006268).

Chapter VI
PROSPECTS FOR THE FUTURE

FUTURE X-RAY TIMING MISSIONS

DIDIER BARRET
Centre d'Etude Spatiale des Rayonnements, France

MICHIEL VAN DER KLIS
University of Amsterdam, Netherlands

GERRY K. SKINNER
University of Birmingham, England

RÜDIGER STAUBERT
University of Tübingen, Germany

LUIGI STELLA
Astronomical Observatory of Roma, Italy

Abstract. Thanks to the Rossi X-ray Timing Explorer (RXTE), it is now widely recognized that fast X-ray timing can be used to probe strong gravity fields around collapsed objects and constrain the equation of state of dense matter in neutron stars. We first discuss some of the outstanding issues which could be solved with an X-ray timing mission building on the great successes of RXTE and providing an order of magnitude better sensitivity. Then we briefly describe the 'Experiment for X-ray timing and Relativistic Astrophysics' (EXTRA) recently proposed to the European Space Agency as a follow-up to RXTE and the related US mission 'Relativistic Astrophysics Explorer' (RAE).

1. Introduction

In the last few years the predicted sub-millisecond plasma orbital periods and millisecond spin periods of accreting low-magnetic-field neutron stars (NS) have finally been detected by RXTE. The same satellite also found 0.1 kHz Quasi-Periodic Oscillations (QPOs) in black hole candidates (BH) and, the first (and so far only) accreting millisecond pulsar (see van der Klis, 2000 for a review). RXTE has opened up a completely new field of investigation and it is now widely recognized that fast timing provides one of the most powerful tools *(a)* to test General Relativity (GR) in strong gravity fields, *(b)* to constrain the fundamental properties of collapsed stars (EOS of NS, mass and spin of BH).

The RXTE discoveries were made possible by its large X-ray detector area (0.67 m^2), sub-μs time resolution and high data rates. The RXTE All Sky Monitor (ASM) and the flexibility of operation and rapid response also played a major role in the success of the mission (Bradt *et al.*, 1993).

For X-ray timing, the signal to noise ratio at which non-coherent variability such as a QPO is detected scales as the source count rate (i.e. the detector area) in

Astrophysics and Space Science is the original source of publication of this article. It is recommended that this article is cited as: *Astrophysics and Space Science* **276** (suppl.): 305–312, 2001.

the strong source limit (for a photon counting experiment). Based on this, here we describe how an instrument with *ten times* the detector area of RXTE could help to solve some of the outstanding issues in the astrophysics of collapsed stars.

2. The Science of a Future X-Ray Timing Mission

High-frequency QPOs have been seen in both BHs and NSs. In BHs, the QPO models proposed invoke GR effects in the inner accretion disk and depend strongly on the BH spin, making these QPOs effective probes of spacetime near the event horizon (see e.g. McClintock, 1998). In NSs, three types of QPOs are commonly observed (see Figure 1, ν_{LF}, 15–60 Hz, ν_1, 200–800 Hz and ν_2). Evidence is building that there exists a close relationship between the QPO properties of NS and BH X-ray binaries. In particular a remarkable correlation between the centroid frequency of QPOs (or peaked noise components) has been found which extends over nearly 3 decades in frequency and encompasses both NS and BH systems (Psaltis, Belloni and van der Klis, 1999). In the relativistic precession model (Stella and Vietri, 1998), ν_{LF}, ν_1, ν_2 observed across a wide range of objects are identified with three fundamental frequencies characterizing the motion of matter in the strong field of a point mass as predicted by GR. The low-frequency QPO at ν_{LF} is thought to be due to *nodal precession*, dominated by the inertial-frame dragging predicted by GR in the vicinity of a fast rotating collapsed object. The lower frequency kHz QPO at ν_1 is identified with relativistic *periastron precession*, an early crucial test of GR. Unlike the cases of Mercury and the relativistic binary pulsar PSR1913+16, for which weak field expansions apply, the periastron precession close to a collapsed star is dictated by strong field effects, requiring a full general relativistic treatment. Finally ν_2 is the orbital ('Keplerian') frequency; its value alone restricting the allowed range of mass and radius of the NS (Miller *et al.*, 1998). Other models rely on a beat-frequency interpretation for both ν_1 and ν_{LF} (Alpar and Shaham, 1985; Miller *et al.*, 1998).

With an order of magnitude better sensitivity, QPOs from a few Hz to 1600 Hz should be detected from many objects with a high enough significance to use the data for crucial tests. Regardless of the physical origin of the QPOs at ν_{LF} and ν_1, the increased sensitivity and range will have dramatic benefits. At higher frequencies, either strong signatures of the innermost stable circular orbit (ISCO), which sets an upper bound on ν_2, will be discovered from several sources (evidence has been found for one source so far: 4U1820-30, Zhang *et al.*, 1998), or the frequencies themselves will allow the elimination of several candidate equations of state of dense matter (Miller *et al.*, 1998). In the *relativistic precession* interpretation, there are several predictions that could be tested. First, the epicyclic frequency $\Delta\nu = \nu_2 - \nu_1$ steeply falls to zero as ν_2 increases and the orbital radius approaches the ISCO. The behaviour of the epicyclic frequency in the vicinity of BH and NS is dominated by strong-field effects and drastically different from any Newtonian

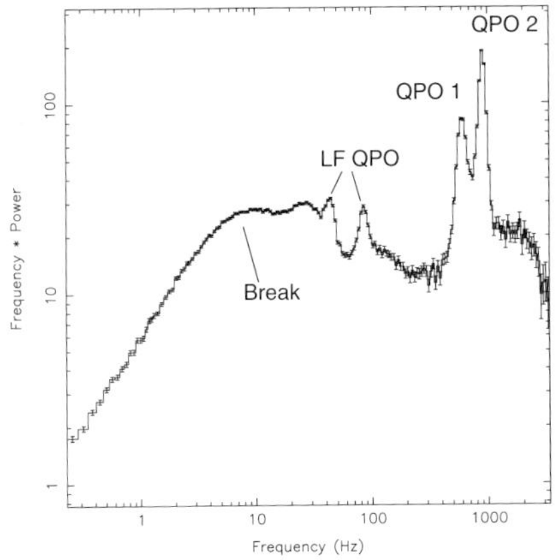

Figure 1. Power spectrum of Sco X-1 obtained with RXTE showing the kHz QPOs at ν_1 and ν_2 and the LF QPO (plus its harmonic). With EXTRA, about twenty kHz QPO sources will be measured at signal-to-noise similar to this (the fainter sources have stronger kHz QPOs). For Sco X-1 the signal-to-noise will improve tenfold.

or post-Newtonian approximation. Hence it provides a powerful test of the strong field properties of the metric (see Stella and Vietri, 1999). According to the model $\Delta\nu$ should also decrease for low values of ν_2. Second, ν_{LF} should scale as ν_2^2 over a wide range of frequencies (until 'classical' terms due to stellar oblateness become important). Observing such scaling would provide an unprecedented test of the $1/r^3$ radial dependence of ν_{LF} predicted in the Lense-Thirring interpretation (Stella and Vietri, 1999). We note that in the hydrodynamical model of the disk inner boundary developed by Psaltis and Norman (2000), the test particle frequencies (the same as in the relativistic precession model plus a few other additional frequencies, see also Psaltis, 2000) are selected by the response of the inner disk, when this is subject to a wide-band input noise.

In the above models, the QPO frequencies depend sensitively on the mass (M) and angular momentum (J) of the compact star, as well as on the radius at which the QPOs are produced. M and J could be independently estimated from waveform measurements at ν_2, thus overdeterminating the problem so that the underlying theories can be tested in critical ways. The increased sensitivity will enable QPOs to be detected within their coherence times. The cycle waveform, which it will be possible to reconstruct, depends on the Doppler shifts associated with the local velocity of the radiating matter in the emitting blob or spot, as well as on curved-spacetime light propagation effects. If the frequency ν_2 of the orbit is known, QPO waveform fitting yields the mass M (and Kerr spin parameter) of the compact object.

Nearly coherent oscillations at $\sim$ 300 Hz or $\sim$ 600 Hz have been observed during type I X-ray bursts from about 10 NS so far (see Strohmayer *et al.*, 1998 for a review). These oscillations are probably caused by rotational modulation of a hot spot on the stellar surface. The emission of the hot spot is affected by gravitational

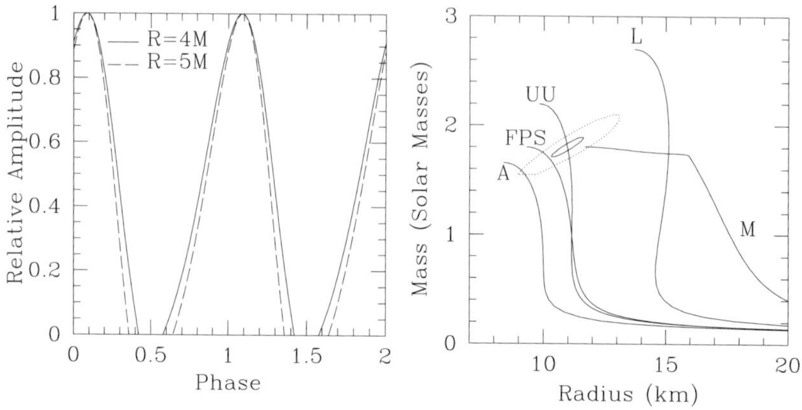

Figure 2. left) Simulated cycle waveforms for burst oscillations on a 1.8 $M_{\odot}$ neutron star with a spin frequency of 364 Hz of radius $R = 4GM/c^2$ (solid line) and radius of $R = 5GM/c^2$ (dashed line). The more compact star (with smaller R/M) has a broader waveform, because it has stronger gravitational light deflection. The less compact star has larger waveform asymmetry because its surface rotation velocity is higher. Therefore, mass and radius can be determined independently by fitting the waveform. *right)* Constraints on mass and radius from waveform fitting. The contours show the 1 σ confidence regions from RXTE (dotted) and from an instrument 10 times bigger than RXTE (drawn). The mass-radius relations labeled as in Miller *et al.* (1998) are also shown. (Courtesy of Cole Miller).

light deflection and Doppler shifts (e.g. Miller and Lamb, 1998). With the next generation timing mission, the oscillation will be detected within one cycle.

The composition and properties of the NS cores have been the subject of considerable speculation, and remain a major issue in modern physics: at the highest densities, matter could be composed of pion or kaon condensates, hyperons, quark matter, or strange matter (e.g. Heiselberg and Hjorth-Jensen, 2000). By fitting the waveform it will be possible to constrain simultaneously the mass and radius of the star (Figure 2), and hence determine the equation of state of its high density core.

Beside the few examples described above, with its increased sensitivity, a future timing mission will also address a broad range of other astronomical issues (accreting millisecond pulsars, micro-quasars, X-ray pulsars, CVs, Novae, Soft gamma-ray repeaters, Anomalous X-ray pulsars, …). For instance, there is only *one* accreting millisecond pulsar known so far: SAXJ1808-3658 (Wijnands and Van der Klis, 1998; Chakrabarti and Morgan, 1998). Its properties suggest that all NS systems should show pulsations at some level. In most models, pulse amplitudes cannot be suppressed below $\sim 0.1\%$ (rms) without conflicting with spectroscopic or QPO evidence. With an instrument 10 times larger than RXTE, the sensitivity to persistent millisecond pulsations will be well below this level (pulsations at the level of 0.01% rms would be detected in 10000 seconds in Sco X-1). Detection of such pulsations in objects also showing kHz QPOs and burst oscillations would immediately confirm or reject several models for these phenomena involving the NS spin (e.g. Miller *et al.*, 1998). In addition, it has been suggested that such objects could be among the brightest gravitational radiation sources in the sky, emitting a periodic gravitational wave signal at the star's spin frequency (Bildsten, 1998). Measuring the spin period very accurately would be therefore of great importance for periodicity searches with gravitational wave antennas (e.g. Brady *et al.*, 1997).

Similarly for micro-quasars, the link between the very fast disk transitions observed in X-rays and the acceleration process could be studied on very short time scales, allowing the non steady state disk properties and their link to the formation of relativistic jets to be explored (Belloni *et al.*, 1997; Fender *et al.*, 1999). This would be of direct relevance to understanding the properties of AGNs, where presumably similar jet formation mechanisms operate on a much larger scale.

3. Proposed Advanced Timing Missions: EXTRA and RAE

The 'Experiment for X-ray Timing and Relativistic Astrophysics' (EXTRA) was proposed to ESA in January 2000 as a F2/F3 flexi-mission. This mission was not at that stage selected, but studies continue. In addition the 'Relativistic Astrophysics Explorer' (RAE), a closely related concept, is being discussed in the US (Kaaret *et al.*, 2000). The objectives and requirements of the two missions are essentially identical and the two will be discussed together here.

Both EXTRA and RAE have a main instrument consisting of a very large X-ray detector array, together with an all-sky monitor to provide source monitoring and alerts to transients.

The main detector array has a geometric area of about 10 m^2, giving about ~ 6 m^2 effective area at 6–10 keV. An area of this size can be accommodated within the fairings available with a low-cost launcher (Starsem Soyuz-ST or Delta II) without the need for deployable mechanisms (Figure 3). The arrays are highly modular and in each case Si PIN diodes are being considered for the main detector. Si PIN diodes can now be manufactured with high yield at acceptable cost and provide a superior alternative to the proportional counters used on RXTE. The absence of absorption edges in the critical 4–6 keV range is important for spectral studies. Furthermore, they are rugged and require neither high voltages nor special handling conditions. Similar Si PIN diodes have been proposed for the first layer of the ASTRO-E Hard X-ray Detector (Sugizaki *et al.*, 1997).

By operating the detectors at reduced temperature, needing only passive cooling through the front face of the detector array, the leakage current can be reduced to a low level even for an area of ~ 1 cm^2 per element. The associated capacitance is reduced by using a thick detector (1–2 mm), also needed for good high energy response. In this way good energy resolution can be obtained with an off-chip preamplifier, allowing the silicon and processing to be optimised for X-ray detection.

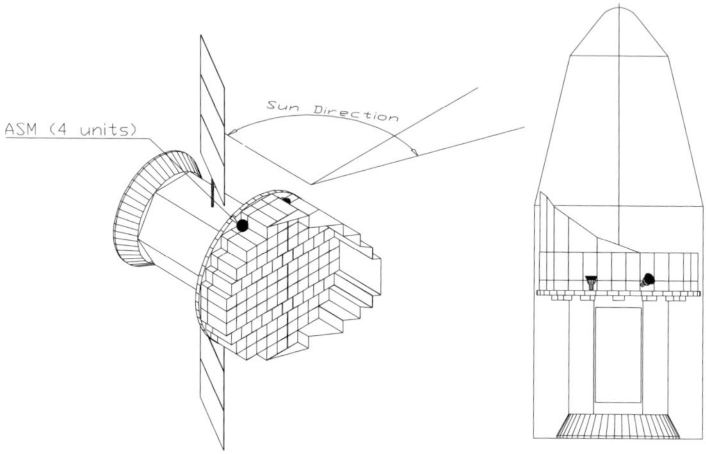

Figure 3. Left: The EXTRA science payload with the Matra Marconi Space LEOSTAR platform. One of the four ASM modules (WATCH design) is shown. *Right:* EXTRA in the STARSEM-Soyuz ST Ariane 4 type fairing.

Resolution of 600 eV or better can be expected – a factor of two better than the RXTE PCA.

For both missions Silicon Drift Detectors (SDDs) are being considered as a possible alternative to PIN diodes, potentially offering even better energy resolution and perhaps larger area for a single detector, so reducing the total number of electronic channels. SDDs are, however, at a less advanced stage of development.

The number of electronic channels will be large ($\sim$ 52416 for EXTRA) and the resulting electronic complexity requires careful optimisation and effective use of a modular design concept. Similar numbers of channels are, however, being dealt with for the IBIS instrument on INTEGRAL and on AGILE.

The detector background is reduced because charged particles passing through the relatively dense detectors will normally deposit an energy larger than the upper limit of the operating range. Simulations show that particle interactions in the collimator and surrounding material often produce showers of secondaries, so the background can be further reduced by operating nearby pixels in anti-coincidence with each other. Nevertheless, because of the unprecedented area the total background rate may approach 2500 Counts s^{-1} between 1 and 50 keV.

The missions are intended to study bright sources and the data rates will often be high, requiring careful optimisation of the on board electronics and data systems. With the proposed detector area, the Crab will produce about 400000 cts/s. A low, near equatorial, orbit is to be preferred in order that the background be low and stable and also because of considerations of data recovery. The objective is to telemeter information about the time and energy of every event, leading to considerable volumes of data. Nevertheless this turns out to be feasible with only minimal constraints. For example, the EXTRA studies led to the proposal to use the Matra-Marconi Space LEOSTAR bus, which provides 160 Gbit solid state memory. By reorienting to direct a X-band antenna toward the ground station during passes, the full memory can be dumped during a single pass, using a 240 Mbits s^{-1} link. Observations of the brightest sources ($\sim$ 10 Crab) would thus be restricted to orbits followed by ground station passes.

4. Conclusions

The need for a follow-up to RXTE has been clearly identified; fast timing studies are now widely recognized as a powerful tool to study the spacetime around collapsed stars, and to derive constraints on the fundamental properties of neutron stars and black holes. The studies carried out for EXTRA have demonstrated that a mission using solid-state detector technologies providing an order of magnitude improvement over RXTE can be done within the limits of a modest size mission. It has also shown that there are existing platforms with resources matching the specific needs of such a mission. More importantly, the large number of active astronomers that gathered around the proposals demonstrates that there is a very large community supportive of such a mission.

Acknowledgements

It is a real pleasure to thank all the people who have contributed to the EXTRA proposal (J.L. Atteia, T. Belloni, E. Bravo, A. Brunton, S. Campana, D. Chakrabarti, F. Cotin, D. Dal Fiume, C. Done, E.C. Ford, R.P. Fender, C. Hellier, W. Hermsen, G.L. Israel, P. Jean, P. Kaaret, E. Kuulkers, C. Labanti, S. Larsson, N. Lund, J.E. McClintock, S. Mereghetti, C.M. Miller, C. Motch, J.F. Olive, M. Orlandini, S. Paltani, A. Santangelo, R.A. Sunyaev, A. Zdziarski, P. Zycki), and those who have given their scientific or engineering support to EXTRA (A. Bazzano, L. Burderi, M. Böer, H. Bradt, S. Brandt, C. Castelli, A. Castro-Tirado, M. Cochi, T. Courvoisier, M. Cropper, A.M. Cruise, D. De Martino, C. Eyles, M. Gilfanov, J.M. Hameury, G. Henry, E. Kendziorra, W. Kluzniak, L. Kuiper, J.P. Lasota, F.K. Lamb, G. La Rosa, J. Poutanen, R. Rothschild, R. Svensson, M. Tavani, G. Vedrenne, N.E. White, R. Wijnands, J. Wilms, W. Zhang).

References

Alpar, A. and Shaham, J.: 1985, *Nature* **316**, 239.
Belloni, T., *et al.*: 1997, *Astrophys. J.* **479**, L145.
Bildsten, L.: 1998, *Astrophys. J.* **501**, L89.

Bradt, H., Rothschild, R. and Swank, J.H.: 1993, *Astron. Astrophys. Suppl.* **97**, 355.
Brady, P.R., *et al.*: 1997, *Phys. Rev.* **D57**, 2101.
Chakrabarti, D. and Morgan, E.H.: 1998, *Nature* **394**, 346.
Fender, R., *et al.*: 1999, *Mon. Not. R. Astron. Soc.* **304**, 865.
Heiselberg, H. and Horth-Jensen, M.: 2000, *Phys. Rep.*, **328**, 237.
Kaaret, P., *et al.*: 2000, *Astrophys. Lett. Comm.*, in press.
Van der Klis, M.: 2000, *Annu. Rev. Astron. Astrophys.*, **38**, 717.
McClintock, J.E.: 1998, *AIP Conf. Proc.* **431**, 290.
Miller, M.C., Lamb, F.K. and Psaltis, D.: 1998, *Astrophys. J.* **508**, 791.
Miller, M.C. and Lamb, F.K.: 1998, *Astrophys. J.* **499**, L39.
Psaltis, D., Belloni, T. and van der Klis, M.: 1999, *Astrophys. J.* **520**, 262.
Psaltis, D. and Norman, C.: 2000, *Astrophys. J.*, in press (astro-ph/0001391).
Psaltis, D.: 2000, *Astrophys. J. Lett.*, submitted (astro-ph/0010316).
Stella, L. and Vietri, M.: 1998, *Astrophys. J.* **492**, L59.
Stella, L. and Vietri, M.: 1999, *Phys. Rev. Lett.* **82**, 17.
Strohmayer, T.: 1998, *The Active X-ray Sky* **69**, 129.
Sugizaki, M., *et al.*: 1997, *Proc. SPIE* **3115**, 244.
Wijnands, R. and Van der Klis, M.: 1998, *Nature* **394**, 344.
Zhang, W., *et al.*: 1998, *Astrophys. J.* **500**, L71.

THE BALLERINA EXPERIMENT ON THE RØMER MISSION *

SOREN BRANDT
Danish Space Research Institute

Abstract. The Rømer mission has recently been approved as the next mission within the Danish Small Satellite Program. The scientific payload will consist of two separate experiments, the MONS and the Ballerina payloads. The primary objective of Ballerina is to provide accurate, real-time positions relayed to ground for $\approx$ 70 Gamma Ray Bursts (GRBs) per year, and to study the temporal and spectral evolution of the early GRB X-ray afterglow. As an additional goal, Ballerina will detect and study bright X-ray transients, in particular X-ray novae and micro-quasar systems. Rømer is currently scheduled for launch in late 2003.

1. Introduction

The Rømer mission has been selected for the next mission within the Danish Small Satellite Program (DSSP). Rømer is currently in the systems definition phase, and is expected to be launched in late 2003. Rømer is the result of merging two mission proposals with similar requirements for the spacecraft capabilities.

The MONS experiment will measure stellar oscillations in nearby solar-like stars, using a 40 cm telescope and a photometric detector, to probe the stellar interior for determination of its composition, age, mixing and internal rotation (MONS). MONS will typically require observation of the same target for $\approx$ 30 days. These observations will be interrupted by the autonomous re-orientations of the spacecraft as response to the detection of a gamma ray burst. The Ballerina experiment will study gamma ray bursts, X-ray transients, and monitor the brighter, persistent X-ray sources.

The merged mission promises a 90% science return for each of the two experiments onboard compared to individual missions. Some of the Rømer parameter are summarized in Table I.

2. The Ballerina Experiment

A crucial turning point in GRB research occurred in 1997 with the discovery of X-ray, optical, and radio afterglows of GRBs (Costa *et al.*, 1997). The observations

* On behalf of the Ballerina Collaboration

Astrophysics and Space Science is the original source of publication of this article. It is recommended that this article is cited as: *Astrophysics and Space Science* **276** (suppl.): 313–316, 2001.

TABLE I

The Rømer spacecraft

Spacecraft dimensions	$\approx 60 \times 60 \times 80$ cm^3
Mass	120 kg
Attitude stability	$> 0.5'$
Attitude slew rate	90 degrees in ≈ 20 sec
Orbit	12 hour period, 63 degree inclination
Observation efficiency (Ballerina)	60%, ≈ 7 hours per orbit
Launcher	Russian SOYUZ/FREGAT (secondary payload)

confirmed that GRBs are indeed located at very large distances with a typical red-shift of $z \approx 1-2$ and higher. There are two main avenues in the study of GRBs: The first is the quest for the understanding of the event itself. What is the progenitor, and how does the 'central engine' work? Are GRBs caused by the collapse of a massive star and the formation of a black hole (hyper-nova model), or are GRBs the result of the merging of two neutron stars and the formation of a black hole? The second line of research, and the cause for GRB research now entering center-stage of astrophysics, is the use of the bright afterglow to study the structure and evolution of the high red shift Universe.

The concept of the Ballerina experiment is to combine an all-sky monitor, providing rapid GRB alerts (5-30 seconds) with good, but limited accuracy ($\approx 1°$) with a rapid moving X-ray telescope and a small optical monitor to study the earliest phase of the afterglow, and giving high precision ($< 30''$) positions within a few minutes. In addition, real-time telemetry will ensure rapid dissemination of positions for follow-up observations. The system will also allow uplink of positions for new targets of opportunity from other sources.

2.1. The Ballerina Science Objectives

The primary objective of Ballerina is the study of gamma ray bursts, through rapid reaction on-board, and by enabling multiwavelength campaigns of observations of the afterglow. Ballerina will be able to study ≈ 70 GRBs per year.

An important secondary objective of the Ballerina experiment is to detect bright X-ray transients, in particular X-ray novae and micro-quasar systems. These systems are prime laboratories for the study of stellar mass black holes. The Ballerina all-sky monitor will also monitor the brighter persistent X-ray sources for time variability and state changes. In addition to the capabilities in its own right, Ballerina on Rømer will be an important complement to current and upcoming missions such as Chandra, XMM, Integral, and Swift.

TABLE II
The Ballerina all-sky monitor

Number of cameras	4 rotation modulation collimators (RMC)
Detector type	NaI/YAP scintillator mosaic
Field of view (per camera)	65 degrees radius
Energy range	6–120 keV
Source position accuracy	< 1 degree
GRB detection efficiency	160/year of efficient operation
X-ray transient detection sensitivity	50 mCrab in one day

2.2. The all-sky monitor

The all-sky monitor on the Rømer mission is based on the WATCH wide field X-ray monitor. WATCH has a proven record of detection of GRBs Sazonov *et al.*, 1998) and X-ray transients from the GRANAT and EURECA missions, like the discovery of the first micro-quasar, GRS 1915+105 (Castro-Tirado *et al.*, 1994), and the black hole candidate Nova Muscae 1991 (Brandt *et al.*, 1992).

Four WATCH X-ray cameras will cover the full sky, and determine the position of GRBs and X-ray transients with and accuracy better than 1 degree. The WATCH rotation modulation collimators will also double as momentum wheels for attitude stabilization, and performance of fast slews in response to GRBs. The capabilities of the WATCH all-sky monitor is summarized in Table II.

2.3. The X-ray telescope

The purpose of the Ballerina narrow-field X-ray telescope (see Table III) is to provide accurate positions of the GRBs after the spacecraft has made a fast slew to the position determined by the all-sky monitor. After the fine position is determined and relayed to the ground, the objective is to study the temporal and spectral evolution of the X-ray afterglow. This study will last for the remainder of the current orbit, typically 4 hours. In the following orbit the control of the spacecraft will be given back to the MONS stellar oscillation experiment. As the objective is the study of the relatively bright early GRB afterglow, a coded mask telescope has been selected over focusing X-ray optics. This choice ensures full utilization of the spectral capabilities of the X-ray CCD, which has a flight heritage derived from the XMM program.

2.4. The optical monitor

The optical monitor on Rømer is a dual use instrument for the MONS and Ballerina payloads. This monitor has a 5 cm aperture and a 40 cm focal length. This will

TABLE III

The Ballerina coded mask X-ray telescope

'Focal' length	60 cm
Field of view (at zero response)	3×3 degrees
Mask area	6×6 cm^2
Mask pixel size	0.5×0.5 mm^2
Detector type	Pn-CCD
Detector area	6×6 cm^2
Quantum efficiency	$> 90\%$ (0.35–10 keV)
Source position accuracy	$< 30''$

enable the possible detection of the optical emission of GRBs in the transition between the gamma ray burst and the afterglow phase. So far this has only been observed in one case, in GRB 990123 (Akerlof *et al.*, 1999). The sensitivity of the optical monitor will be $m_V \approx 15$ for a 10 second integration time.

3. Conclusion

The Ballerina experiment on the Rømer mission promises exciting returns in the fields of gamma ray bursts, and the study of black hole X-ray transients. It is interesting to speculate that a common denominator, and a key to progress in these two fields, may be the understanding of highly relativistic jets.

References

Akerlof, C.W., *et al.*: 1999, *Nature* **398**, 400 .
Brandt, S., *et al.*: 1992, *Astron. Astrophys.* **254**, L39.
Castro-Tirado, A.J., *et al.*: 1994, *Astrophys. J. Suppl.* **92**, 469.
Costa, E., *et al.*: 1997, *Nature* **387**, 876 .
Sazonov, Y., et al.: 1998, *Astron. Astrophys. Suppl.* **129**, 1.
http://astro.ifa.au.dk/MONS/english.
http://www.dsri.dk (see Danish Small Satellite Program).

MICROQUASARS: OPEN QUESTIONS AND FUTURE PERSPECTIVES

I.F. MIRABEL
CEA/DSM/SAp. CE-Saclay, France & IAFE/CONICET, Argentina

Abstract. The discovery and subsequent study of microquasars lead to major progress in our understanding of: 1) the nature of relativistic jets seen elsewhere in the universe, and 2) the connection between the accretion onto compact objects and the formation of collimated jets. A detailed account of the major progress accomplished until present was published in *Annual Review of Astronomy & Astrophysics* (Mirabel and Rodríguez, 1999). Here I review the questions that remain unanswered, as well as the future perspectives that this new field of research is opening.

1. Major Progress Accomplished

In relativistic sources located in the Milky Way, two-sided moving jets can be observed, and therefore can be overcome by several of the ambiguities that had dominated the physical interpretation of extraglactic jets, where so far only the motions of one-side of the jets could be followed. From the observation of two-sided moving jets in microquasars, the system of equations can be solved. For first time, an upper limit of the distance to a microquasar was derived from the proper motions using special relativity constraints (see Mirabel and Rodríguez, 1999).

Major progress has also been made in the understanding of the accretion/ejection phenomenology. From multiwavelength observations, the connection between accretion flow instabilities – observed in the X-rays – with the ejection of relativistic plasma – observed at radio, infrared, and possibly optical and X-ray wavelengths – has been established on a firm basis. After the softer component of the X-ray emitting plasma disappears, the inner accretion disk is rapidly re-established and the plasma that produces the hard X-ray component, is blown away (Mirabel *et al.*, 1998) in the form of collimated relativistic jets (Dhawan, Mirabel and Rodríguez, 2000).

The X-ray source GRS 1915+105 (Castro-Tirado *et al.*, 1994) has become a main target to study accreting black holes of stellar mass. However, the discovery and study of other microquasars is important not only to enhance the statistical sample, but also to make progress on the different aspects of the diverse phenomenology that each new object of this class has revealed (Mirabel and Rodríguez, 1999).

Astrophysics and Space Science is the original source of publication of this article. It is recommended that this article is cited as: *Astrophysics and Space Science* **276** (suppl.): 319–327, 2001.

2. Open Questions

2.1. WHAT IS A MICROQUASAR?

By this term we usually design stellar-mass black holes that mimic, on a smaller scale, many of the phenomena seen in quasars. The question is whether under this concept we should also include neutron star systems with jets. Although the physical meaning of the quasar-microquasar analogy as proposed by Mirabel and Rodríguez (1998) is valid for black holes, because in many cases the nature of the compact object is unknown, we may currently include in this subclass of stellar sources, neutron star binaries that produce relativistic jets.

2.2. ARE ALL ACCRETING BLACK HOLES OF STELLAR MASS MICROQUASARS?

All microquasars are accreting compact objects of stellar mass, but is the reverse true? Do all X-ray black hole binaries produce jets? From the theoretical point of view jets are needed to liberate angular momentum. On the other hand, observations show that X-ray emitting black holes, in addition to possible synchrotron emission associated to sporadic outbursts, always exhibit flat-spectrum compact counterparts of synchrotron emission. It has been proposed (Fender, 2000) and demonstrated by high resolution images (Dhawan, Mirabel and Rodríguez, 2000), that these flat spectrum compact radio counterparts are thick collimated jets of AU size scales. Furthermore, from multiwavelength observations at radio, infrared, and X-ray wavelengths of the large-scale jets in SS433/W50, we know that in some circumstances jets exist without being seen. Therefore, the answer to this question is that probably all accreting black holes of stellar mass are microquasars.

2.3. ARE THE LORENTZ FACTORS OF THE BULK MOTIONS IN THE JETS FROM QUASARS DIFFERENT FROM THOSE IN MICROQUASARS?

E.g. is the quasar-microquasar analogy valid? The microquasar jets observed so far have bulk speeds that are statistically smaller than in quasars. For instance, in quasars jets have bulk speeds of up to $\sim$ 20c, but no microquasar with such jets has so far been found. Is this due to some fundamental difference that would invalidate the physical analogy between quasars and microquasars, or is it rather due to selection effects related to the Doppler favoritism needed in order to measure the proper motion of the ejecta in distant quasars? Although from the observational point of view this remains an open question, theoretically it is not clear why the bulk motion in jets from supermassive black holes would be different from that in the jets from stellar mass black holes.

2.4. Could the Terminal Velocity of the Jet be a Diagnostic for the Nature of the Collapsed Object (NS versus BH)?

In all areas of astrophysics there have been increasing evidences that accretion is always related to the production of collimated jets. The observed outflow speeds seem to be of the order of the escape velocity from the surface of the accreting object. Objects as diverse as very young stars, nuclei of planetary nebulae, and accreting white dwarfs have jets with non-relativistic velocities ($\sim 100-10000$ km s^{-1}), whereas neutron stars and black holes produce jets with relativistic speeds V $\geq 0.1c$ (Livio, 1999; Mirabel and Rodríguez, 1999). If the present trends between the escape velocities from the accreting objects and the velocity of the outflows were confirmed as a strong correlation, it would imply that gravity is important and dominates over MHD mechanisms (Meier *et al.*, 1997). However, it is not clear whether we can *discriminate between neutron stars and black holes by knowing the terminal velocity of the collimated outflows*. For instance, Sco X-1 which probably contains neutron star, has lobes moving with an average velocity of 0.45c (Fomalont, Geldzahler and Bradshaw, 2000). Since only a handful of relativistic jet sources have been discovered so far in the Galaxy (Mirabel and Rodríguez, 1999), we need the discovery and study more sources to obtain a statistically significant number to answer this question.

2.5. Why are QPOs in Microquasars and not in AGNs?

The scale of time of the instabilities in the accretion disk of black holes is proportional to the mass of the black hole. QPOs with periods of ~ 1 sec in black holes of 10 $M_\odot$ would correspond to QPOs of ~ 1 day to ~ 3 years in black holes with masses of 10^6 and 10^9 $M_\odot$ respectively. QPOs of 1 sec in microquasars correspond to typical fluctuations of $\sim 5\%$ of the X-ray flux. At X-rays the companion binary stars are much dimmer than the accretion disk. On the other hand, the accretion disk in AGNs is cooler (Mirabel and Rodríguez, 1999), and QPOs should be observed in the UV and optical, wavelengths at which stars mostly radiate, and interstellar absorption is important. Since AGNs may be blurred by nuclear starbursts, it may be difficult to detect fluctuations of $\sim 5\%$ from AGNs that are embedded in starburst nuclei.

One then may ask about the analogous of the large scale oscillations of the flux ($\geq 50\%$) with periods of ~ 30 min regularly observed in GRS 1915+105 (Greiner, Morgan and Remillard, 1996). In black holes with masses in the range of 10^6 to 10^9 $M_\odot$ analogous QPOs would correspond to periods in the range of 5 to 500 years, which are difficult to monitor in human time-scales (Sams, Eckart and Sunyaev, 1996).

2.6. To image the jet close to the black hole, do microquasars offer an advantage with respect to AGNs?

No. The Schwarzschild radii (r_s) of BHs are proportional to their mass and in these units nearby AGNs can be imaged closer to the BH than microquasars. For instance, M87 (in the Virgo cluster at 15 Mpc and containing a BH of 3 10^9 $M_\odot$) has been imaged with a resolution of $\sim$ 50 (r_s) (Junor, Biretta and Livio, 1999). A given angular size projected on the supermassive BH of M87 is in terms of r_s 10^4 smaller than the same distance projected on a BH of 10 $M_\odot$ at a distance of 2 kpc from the Sun. Therefore, relative to supermassive BHs in the Local Universe, the study of stellar mass BHs presents an advantage because of the time scales, but not as far as the length dimension is concerned.

2.7. What is the connection between accretion and ejection?

In GRS 1915+105 large amounts of X-ray emitting plasma ($\sim$ 10^6 $L_\odot$) disappear in less than a few seconds and soon after synchrotron jets are formed (Fender and Pooley, 1998; Eikenberry *et al.*, 1998; Mirabel *et al.*, 1998). Recent analysis shows that the X-ray flux that suddenly disappears only corresponds to the softer component (2–20 keV). The formation of the jets starts later, at the time of a spike that consists of a sudden increase in the flux of the soft component simultaneous with a decrease of the flux of the harder component (20–60 keV). In the context of current models these observations imply that the inner accretion disk first disappears (Belloni *et al.*, 1997), and when the inner accretion disk is re-established, a shock is produced which triggers the blow up of plasma that was emitting hard X-rays in the form of collimated jets at relativistic speeds (Dhawan, Mirabel and Rodríguez, 1999). In GRS 1915+105, an analogous correlation between the sudden disappearance of the 20–100 keV measured by BATSE and major ejection events seen in the radio with the VLA had been observed (see Mirabel and Rodríguez, 1999 for references and the discussion of this issue).

2.8. What fraction of the inflow goes into the jets?

The observations described above indicate that the matter and energy emitting in the soft X-rays that suddenly disappears is perhaps advected into the BH, since it is not immediately translated neither in hard X-rays nor in synchrotron jets. Because of the uncertanties in estimating the mass of the X-ray emitting plasma it is difficult to estimate the fraction of the inflow that goes into the jets. On the contrary, assuming equipartition, the amount of mass and energy of the jets can be calculated (Fender and Pooley, 1998; Mirabel *et al.*, 1998).

Now there are increasing evidences that the power of the jets may be a large fraction of the accretion power. For instance, in its very high state, GRS 1915+105 may have a short-term jet power of $\sim$ 10^{39} erg s^{-1} (Mirabel and Rodríguez, 1999; Fender, 2000), which is a large fraction of the observed accretion power. This is consistent with the model for a rapidly rotating BH with a high accretion rate by Meier (2000).

2.9. WHAT IS THE MECHANISM THAT LAUNCHES THE JETS?

It is MHD power, which was nicely reviewed by Meier (2001) and may play an important role. However, it is still unclear the relative importance of the rotating thin accretion disk mechanism by Blandford and Payne (1982) and that from the frame-dragged accreting matter inside the ergosphere of a rotating BH (Blandford and Znajek, 1977). The X-ray spectral properties of microquasars with powerful jets (e.g. GRS 1915+105, GRO J1655-40) indicate that these are rotating BHs with spins near maximum (Zhang, Cui and Chen, 1997), which is consistent with the Blandford and Znajek mechanism. On the other hand, the Blandford and Payne mechanism is consistent with the following two recent observations: 1) the large opening angle of $\sim 60^\circ$ of the jet in M87 (Junor, Biretta and Livio, 1999) which seems to be the two dimensional image of a jet with a magnetic polar field angle $\geq 30^\circ$. The M87 jet strongly collimates at 30–100 (r_s) from the BH, collimation continuing out to $\sim 1000\ r_s$. 2) The semicontinuous emanation of the jets in the infrared and radio during time intervals of the order of minutes, as observed in GRS 1915+105 (Eikenberry *et al.*, 1998; Fender and Pooley, 1998; Mirabel *et al.*, 1998).

Despite several numerical simulations there are still no definitive tests to discriminate the relative role of these two MHD mechanisms.

2.10. WHAT ARE THE QPOS OF MAXIMUM FIX FREQUENCY?

Because QPOs with maximum fix frequency have been observed many times in some microquasars (e.g. 67 Hz in GRS 1915+105; Morgan, Remillard and Greiner, 1997), it is believed that they are related to fundamental properties of the BHs, such as its mass and spin. The problem is that there are more than 3 alternative explanations in terms of General Relativity (see Mirabel and Rodríguez, 1999). Theories should provide observational tests that would discriminate among these models.

2.11. ARE THE JETS DISCRETE PLASMONS OR SEMI-CONTINUOUS FLOWS WITH INTERNAL SHOCKS?

In the internal shock models (Bodo and Ghisellini, 1995; Kaiser, Sunyaev and Spruit, 2000) the plasma velocity is smaller than the pattern velocity, and the jets are lit up by shock fronts travelling along the jets during outbursts, as proposed for gamma-ray bursts (Mészáros and Rees, 1997). The shocks accelerate relativistic particles which emit synchrotron radiation. Although the integrated energy in both models is the same, the internal shock model relaxes the requirements on the power of the central engine because much of the energy underlying the outbursts may be stored in the continuous jet. So far there has been no observational test to discriminate between these two physical interpretations.

Despite the large differences in Lorentz factors, the jets in microquasars and GRB afterglows show analogous phenomenologies. It is interesting that the internal shock model for GRBs originally developed for GRBs by Mészáros and Rees (1997) was applied later to microquasars by Kaiser, Sunyaev and Spruit (2000). Conversely, the plasmon model originally proposed to interpret the microquasar jets has been proposed in the 'cannonball' model of GRBs by Dar and De Rújula (2000).

2.12. Have extragalactic microquasars been identified?

Ultraluminous X-ray compact sources were identified in several nearby spiral galaxies (Colbert and Mushotzky, 1999; Makishima *et al.*, 2000) as well as in dwarf galaxies (Mirioni and Pakull, 2000). These enigmatic sources have luminosities in the range of $L_X = 10^{38-40}$ erg s^{-1}, and Colbert and Mushotzky (1999) proposed that they are BHs of 10^{2-4} $M_\odot$. Makishima *et al.* (2000) propose that some of these superluminous sources are BHs as the microquasars GRS 1915+105 and GRO J1655-40, with masses below $\sim$ 100 $M_\odot$ and high disk temperatures because rapid rotation gets the disk closer to the BH, hence hotter. Unfortunately, the VLBA array does not have the sensitivity to image in its plateau state extragalactic 'super-microquasars' as GRS 1915+105.

2.13. Why superluminal microquasars have unbroken power law photon spectra?

Microquasars with superluminal jets have power-law gamma-ray states with an index of 2.5–3 (Grove, 1999). Contrary to Cygnus X-1 and 1E 1740.7-2942, the microquasars GRO J1655-40 and GRS 1915+105 have unbroken power laws up to almost 1 MeV with no indication of Comptonization. A similar photon spectrum has been observed by SAX in XTE 1118+48 (Frontera *et al.*, 2001). Since we don't know the origin of the electrons that produce the gamma-ray photons it is still unclear why some BHs have broken, whereas others have unbroken power law photon spectra. The sensitivity of INTEGRAL in this energy domain will certainly provide new perspectives on this issue.

2.14. Will microblazars be found?

In most microquasars where θ (the angle between the line of sight and the axis of ejection) has been determined, large values are found ($\theta \geq 70^\circ$). Namely, except some exceptions (as in the case of Sco X-1 where $\theta = 45^\circ$; Fomalont, Geldzahler and Bradshaw, 2000), the axis of ejection in most cases is close to the plane of the sky. This is consistent with the statistical expectation since the probability of finding a source with a given θ is proportional to sin θ. We then expect to find as many objects in the $60° \leq \theta \leq 90°$ range as in the $0° \leq \theta \leq 60°$ range. However, this argument suggests that we should eventually detect objects with a small θ. For objects with $\theta \leq 10°$ we expect the timescales to be shortened by $0.5\gamma^2$ and the flux densities to be boosted by $8\gamma^3$ with respect to the values in the rest frame of the condensation. For instance, for motions with $v = 0.98c$ ($\gamma = 5$), the timescale will shorten by a factor of ~ 50, the flux densities will be boosted by a factor of $\sim 10^3$, and the photon spectrum of the source will be very hard. Then, for a galactic source with relativistic jets and small θ we expect fast and intense variations in the observed flux. Microblazars may be hard to detect in practice, because both the low probability of small θ and the fast flux decline.

2.15. Could Microquasars be Unidentified Sources of Gamma-Rays Detected by EGRET?

There is indication that the synchrotron component of microquasar jets could reach the X-ray domain (Markoff, Falcke and Fender, 2000). This would require the ejection of plasma containing electrons with Lorentz factors $\geq 10^7$, as in blazars. In this context, the recent observation by Paredes *et al.* (2000) of persistent radio jets from LS 5039, which is located in the error box of an EGRET source raises the possibility that microquasars could be persistent sources of gamma-rays.

2.16. Could Microquasars be Sources of $\geq 10^{20}$ eV Cosmic Rays?

Due to opacity by cosmic infrared photons, the sources of $\sim 10^{20}$ eV cosmic rays must be within 50 Mpc from the Galaxy. AGNs have been proposed as possible sources of such high energy particles. As shown above, the bulk and intrinsic Lorentz factors of the electrons in microquasars are comparable to those in AGNs. Since they are much closer and numerous than the later, the very high energy cosmic rays could be produced by shocks in microquasar jets.

2.17. Are Gamma-Ray-Bursts Extreme Microquasars?

Gamma-ray bursts are at cosmological distances and ultra-relativistic bulk motion and beaming appear as essential ingredients to solve the enormous energy requirements (Castro-Tirado *et al.*, 1999). Beaming reduces the energy release by a factor $f = \Delta\Omega/4\pi$, where $\Delta\Omega$ is the solid angle of the beamed emission. Additionally, the photon energies can be boosted to higher values. BHs formed by core collapse producing outflows with bulk Lorentz factors > 100 have been proposed as GRB sources (Mészáros and Rees, 1997). Recent studies of GRB afterglows suggest that they are highly collimated jets since breaks and a steepening of the light curve from a power law to $t^{-1.2}$, ultimately approaching $t^{-2.5}$ have been observed (Castro-Tirado *et al.*, 1999).

Interestingly, the power laws describing the light curves of the ejecta in microquasars show similar breaks and steepening of the radio flux density (Rodríguez and Mirabel, 1999). In microquasars, these breaks and steepenings have been interpreted (Hjellming and Johnston, 1988) as a transition from slow intrinsic expansion to free expansion in two dimensions. Linear polarizations of about 2% were recently measured in the optical afterglows, providing strong evidence that the afterglow radiation from gamma-ray bursts is, at least in part, produced by synchrotron processes. Linear polarizations in the range of 2-10% have also been measured in microquasars at radio (Rodríguez *et al.*, 1995; Hannikainen *et al.*, 2000), and optical (Scaltriti *et al.*, 1997) wavelengths.

In this context, microquasars in our own Galaxy seem to be less extreme local analogs of the super-relativistic jets associated to the more distant GRBs, which are jets with Lorentz factors 2 orders of magnitude larger. Therefore, the physical mechanism that launches the jets in GRBs are likely different to microquasars. Furthermore, GRBs do not repeat and seem to be related to catastrophic events, and have much larger super-Eddington luminosities. Therefore, the scaling laws in terms of the black hole mass that are valid in the analogy between microquasars and quasars do not seem to apply in the case of GRBs.

Acknowledgement

This work was partially supported by Consejo Nacional de Investigaciones Científicas y Técnicas of Argentina.

References

Belloni, T., Méndez, M., King, A.R., *et al.*: 1997, *Astrophys. J.* **479**, L145.
Blandford, R.D. and Payne, D.G.: 1982, *Mon. Not. R. Astron. Soc.* **199**, 883.
Blandford, R.D. and Znajek, R.L.: 1977, *Mon. Not. R. Astron. Soc.* **179**, 433.
Bodo, G. and Ghisellini, G.: 1995, *Astrophys. J.* **441**, L69.
Castro-Tirado, A.J., *et al.*: 1994, *Astrophys. J. Suppl.* **92**, 469.
Castro-Tirado, A.J., *et al.*: 1999, *Science* **283**, 2069.
Colbert, E.J.M. and Mushotzky, R.F.: 1999, *Astrophys. J.* **519**, 89.
Dar, A. and De Rújula, A.: 2000, submitted (astro-ph/0008474).
Dhawan, V., Mirabel, I.F. and Rodríguez, L.F.: 2000, *Astrophys. J.* **543**, 373.
Eikenberry, S.S., Matthews, K., Morgan, E.H., *et al.*: 1998, *Astrophys. J.* **494**, L61.
Fender, R.P.: 2000, *Mon. Not. R. Astron. Soc.*, in press (astro-ph/0008447).
Fender, R.P. and Pooley, G.G.: 1998, *Mon. Not. R. Astron. Soc.* **300**, 573.
Fomalont, E.B., Geldzahler, B.J. and Bradshaw, C.F.: 2000, *Astrophys. J. Lett.*, submitted.
Frontera, L., *et al.*: 2000, in: *Fourth INTEGRAL Workshop*, ESA-SP, in press.
Greiner, J., Morgan, E.H. and Remillard, R.A.: 1996, *Astrophys. J.* **473**, L107.
Grove, J.E.: 1999, *ASP Conf. Ser.* **161**, 54.
Hannikainen, D.C., *et al.*: 2000, *Astrophys. J.* **540**, 521.
Hjellming, R.M. and Johnston, K.J.: 1988, *Astrophys. J.* **328**, 600.
Junor, W., Birettta, J.A. and Livio, M.: 1999, *Nature* **401**, 891.
Kaiser, C.R., Sunyaev, R. and Spruit, H.C.: 2000, *Astron. Astrophys.* **356**, 975.

Livio, M.: 1999, *Phys. Rep.* **311**, 225.
Makishima, K., *et al.*: 2000, *Astrophys. J.* **535**, 632.
Markoff, S., Falcke, H. and Fender, R.: 2000, *Astrophys. J. Lett.*, submitted (astro-ph/0010560).
Meier, D.L., Edgington, S., Godon, P., Payne, D. and Lind, K.: 1997, *Nature* **388**, 350.
Meier, D.L.: 2001, these Proceedings.
Mészáros, P. and Rees, M.J.: 1997, *Astrophys. J.* **482**, L29.
Mirabel, I.F. and Rodríguez, L.F.: 1998, *Nature* **392**, 673.
Mirabel, I.F. and Rodríguez, L.F.: 1999, *Annu. Rev. Astron. Astrophys.* **37**, 409.
Mirabel, I.F., *et al.*: 1998, *Astron. Astrophys.* **330**, L9.
Mirioni, L. and Pakul, M.: 2000, private communication.
Morgan, E.H., Remillard, R.A. and Greiner, J.: 1997, *Astrophys. J.* **482**, 993.
Paredes, J.M., Martí, J., Ribó, M. and Massi, M.: 2000, *Science* **288**, 2340.
Rodríguez, L.F., *et al.*: 1995, *Astrophys. J. Suppl.* **101**, 173.
Rodríguez, L.F. and Mirabel, I.F.: 1999, *Astron. Astrophys.* **340**, L47.
Sams, B.J., Eckart, A. and Sunyaev, R.: 1996, *Nature* **382**, 47.
Scaltriti, F., *et al.*: 1997, *Astron. Astrophys.* **325**, L29.
Zhang, N.S., Cui, W. and Chen, W.: 1997, *Astrophys. J.* **482**, L155.

AUTHOR INDEX

SUBJECT INDEX

Note: Page numbers in **bold** *indicate article titles.*

OBJECT INDEX

Note: Page numbers in **bold** *indicate article titles.*